Lecture Notes in Mathematics 2065

Editors:
J.-M. Morel, Cachan
B. Teissier, Paris

For further volumes:
http://www.springer.com/series/304

FONDAZIONE CIME ROBERTO CONTI

CENTRO INTERNAZIONALE MATEMATICO ESTIVO
INTERNATIONAL MATHEMATICAL SUMMER CENTER

Fondazione C.I.M.E., Firenze

C.I.M.E. stands for *Centro Internazionale Matematico Estivo*, that is, International Mathematical Summer Centre. Conceived in the early fifties, it was born in 1954 in Florence, Italy, and welcomed by the world mathematical community: it continues successfully, year for year, to this day.

Many mathematicians from all over the world have been involved in a way or another in C.I.M.E.'s activities over the years. The main purpose and mode of functioning of the Centre may be summarised as follows: every year, during the summer, sessions on different themes from pure and applied mathematics are offered by application to mathematicians from all countries. A Session is generally based on three or four main courses given by specialists of international renown, plus a certain number of seminars, and is held in an attractive rural location in Italy.

The aim of a C.I.M.E. session is to bring to the attention of younger researchers the origins, development, and perspectives of some very active branch of mathematical research. The topics of the courses are generally of international resonance. The full immersion atmosphere of the courses and the daily exchange among participants are thus an initiation to international collaboration in mathematical research.

C.I.M.E. Director
Pietro ZECCA
Dipartimento di Energetica "S. Stecco"
Università di Firenze
Via S. Marta, 3
50139 Florence
Italy
e-mail: zecca@unifi.it

C.I.M.E. Secretary
Elvira MASCOLO
Dipartimento di Matematica "U. Dini"
Università di Firenze
viale G.B. Morgagni 67/A
50134 Florence
Italy
e-mail: mascolo@math.unifi.it

For more information see CIME's homepage: http://www.cime.unifi.it

CIME activity is carried out with the collaboration and financial support of:

- INdAM (Istituto Nazionale di Alta Matematica)

- MIUR (Ministero dell'Universita' e della Ricerca)

Anna Capietto • Peter Kloeden • Jean Mawhin
Sylvia Novo • Rafael Ortega

Stability and Bifurcation Theory for Non-Autonomous Differential Equations

Cetraro, Italy 2011

Editors:
Russell Johnson
Maria Patrizia Pera

 Springer

Anna Capietto
Università di Torino
Dipartimento di Matematica
Torino, Italy

Peter Kloeden
Universität Frankfurt
Institut für Mathematik
Frankfurt, Germany

Jean Mawhin
Université Catholique de Louvain
Institut de Recherche
en Mathématique et Physique
Louvain-la-Neuve, Belgium

Sylvia Novo
E. de Ingenierías Industriales
Departamento de Matemática Aplicada
Universidad de Valladolid
Valladolid, Spain

Rafael Ortega
Universidad de Granada
Departamento de Matemática Aplicada
Granada, Spain

ISBN 978-3-642-32905-0   ISBN 978-3-642-32906-7 (eBook)
DOI 10.1007/978-3-642-32906-7
Springer Heidelberg New York Dordrecht London

Lecture Notes in Mathematics ISSN print edition: 0075-8434
                ISSN electronic edition: 1617-9692

Library of Congress Control Number: 2012951625

Mathematics Subject Classification (2010): 34B15, 37B55, 34C25, 37E40, 37G35, 34K12

Printed on acid-free paper

Springer is part of Springer Science+Business Media (www.springer.com)

Preface

The CIME session "Stability and Bifurcation Theory for Non-Autonomous Differential Equations" was held at Cetraro Italy, from 19 to 25 June 2011. This volume contains the notes of the five lecture courses which were held on that occasion.

One of our goals in organizing the session was to foster a comparison between the "topological" and "dynamical" approaches to the study of nonautonomous differential equations. Another goal was to facilitate the interaction between specialists versed in the one approach or the other. We are amply convinced that those goals were fulfilled.

In these course notes, the reader will find a systematic introduction to many of the themes and methods which make up the modern theory of nonautonomous differential and dynamical systems. Topics pertaining to differential equations in finite and infinite dimensions receive sustained attention. Also, discrete equations and systems have an important place in the notes. This is natural both because a differential equation is often studied via an appropriate discretization process and because nonautonomous discrete systems are of fundamental importance in their own right.

Here is a partial list of the various themes which were taken up in the course of the lectures: bounded orbits and stability in non-periodic monotone twist maps; properties of the minimal subsets of nonautonomous monotone differential-delay systems; resonance phenomena in nonautonomous ordinary differential equations; existence and properties of pullback attractors in skew-product dynamical systems; and the use of the Maslov index in bifurcation problems regarding nonautonomous Hamiltonian systems. Of course an impressive range of other topics was considered as well, and an ample quantity of specific problems was discussed.

The methods introduced by the speakers in the theoretical developments and in the treatment of specific problems may be divided into two classes. First, the use of "classical" techniques drawn from the topological degree theory, the calculus of variations, the search for upper/lower solutions, and others of a similar vein. Second, the use of "dynamical" constructs such as processes and skew-product flows, minimal sets and omega-limit sets, pullback attractors, invariant measures, and the like.

Here is a brief description of each of the courses which made up the session.

- Anna Capietto (Torino) considered a broad class of boundary value problems posed for nonautonomous nonlinear Hamiltonian systems. She stated and proved bifurcation results of "Rabinowitz" type for these problems. She showed how the Maslov index can be used as an effective tool in deriving such results.
- Peter Kloeden (Frankfurt) gave an introduction to the language and concepts of nonautonomous discrete dynamical systems. He discussed the theory of pullback attractors and went on to mention some results from nonautonomous bifurcation theory. He also took up some questions in the area of random dynamical systems.
- Jean Mawhin (Louvain-la-Neuve) discussed a number of illustrative nonlinear nonautonomous resonance problems. He effectively used a mix of methods drawn from the Leray–Schauder theory, the calculus of variations, and the technique of upper and lower solutions. He presented many results regarding existence and multiplicity of periodic solutions of certain paradigmatic periodically forced ODEs.
- Sylvia Novo (Valladolid) considered a significant class of nonautonomous functional differential equations having monotonicity properties. She studied the existence and the stability properties of minimal sets, together with the existence and structural properties of global attractors. She gave several applications, e.g., to the theory of neural networks and to that of compartmental systems.
- Rafael Ortega (Granada) first discussed the existence of bounded orbits and invariant curves for exact symplectic twist maps on the cylinder and especially on the plane. The results on invariant curves have stability statements as corollaries. He made use of a variational principle of Mather type. He then analyzed certain impact problems, especially the so-called ping-pong model.

The session was attended by about 50 scientists of "topological" and "dynamical" extractions. Their good-natured and active participation in the courses and their individual discussions helped to create a positive atmosphere which certainly facilitated the exchange of scientific ideas. We believe that the interaction between specialists in the topological and in the dynamical approaches to nonautonomous differential equations was greatly enriched by this CIME session.

Firenze, Italy   Russell Johnson
19 Dec 2011   Maria Patrizia Pera

Contents

Non-autonomous Functional Differential Equations and Applications

The Maslov Index and Global Bifurcation for Nonlinear Boundary Value Problems*

Alberto Boscaggin, Anna Capietto, and Walter Dambrosio

Abstract We first describe the notion of Maslov index and relate it to the concepts of moment of verticality and of phase-angles. We then illustrate the role of the Maslov index in the development of global bifurcation results for nonlinear first order differential systems in $\mathbb{R}^{2N}$ and for nonlinear planar Dirac-type systems.

1 Introduction and Classical Results

This lecture note describes a topological method for the study of boundary value problems associated to systems of nonlinear ordinary differential equations. The abstract setting is provided by bifurcation theory and a crucial role is played by a "generalized phase plane analysis".

In what follows we first introduce the reader to our work by describing the celebrated Rabinowitz global bifurcation theorem [43], together with one of its classical applications to the Dirichlet boundary value problem associated to an ordinary differential equation (cf. (1)). In doing this, we give suggestions on the difficulties which arise when dealing, as it is our aim, with boundary value problems more general than (1).

*Lectures given by the second author at the C.I.M.E. course "Stability and Bifurcation for non-autonomous differential equations", Cetraro, Italy, June 20–June 25, 2011.

A. Boscaggin
SISSA, International School for Advanced Studies, Via Bonomea 265, 34136 Trieste, Italy
e-mail: boscaggi@sissa.it

A. Capietto (✉) · W. Dambrosio
Dipartimento di Matematica - Università di Torino, Via Carlo Alberto 10, 10123 Torino, Italy
e-mail: anna.capietto@unito.it; walter.dambrosio@unito.it

A. Capietto et al., *Stability and Bifurcation Theory for Non-Autonomous Differential Equations*, Lecture Notes in Mathematics 2065, DOI 10.1007/978-3-642-32906-7_1,
© Springer-Verlag Berlin Heidelberg 2013

Consider thus the Dirichlet problem

$$\begin{cases} -u'' + q(t)u = f(t, u, u', \lambda), \\ \\ u(0) = 0 = u(\pi) \end{cases} \tag{1}$$

where $q \in C([0, \pi])$, $q(t) \geq 0$ for all $t \in [0, \pi]$ and $f \in C([0, \pi] \times \mathbb{R}^3)$.

In order to enter an abstract bifurcation setting, we need to assume that f is a perturbation of a linear term, i.e. we assume that f can be written as

$$f(t, \xi, \eta, \lambda) = \lambda a(t)\xi + h(t, \xi, \eta, \lambda), \quad \forall\, (t, \xi, \eta, \lambda) \in [0, \pi] \times \mathbb{R}^3,$$

where $a \in C([0, \pi])$ satisfies $a(t) \geq a_0 > 0$, for every $t \in [0, \pi]$, and the function h fulfills the condition

$$h(t, \xi, \eta, \lambda) = o(\sqrt{\xi^2 + \eta^2}), \ (\xi, \eta) \to (0, 0),$$

uniformly in $t \in [0, \pi]$ and λ in bounded subsets of $\mathbb{R}$.

Rabinowitz's approach to (1) is based on the transformation of the given Dirichlet problem into an abstract equation in a Banach space. More precisely, one considers an equation of the form

$$u = \lambda L u + H(\lambda, u), \quad \lambda \in (a, b), \ u \in X, \tag{2}$$

where X is a Banach space equipped with a norm denoted by $||\cdot||_X$, L is a compact linear operator and H is a completely continuous nonlinear operator s.t. $H(\lambda, u) = o(||u||_X)$, $u \to 0$, uniformly in bounded λ-intervals. Let us point out that if $\lambda \in \mathbb{R}$ in (1) then $(a, b) = \mathbb{R}$ in (2).

In [43] it is proved a bifurcation result for (2) which describes the global behaviour of branches of nontrivial solutions emanating from the trivial ones. In order to state this result, we denote by Σ the closure of the set

$$\{(\lambda, u) \in (a, b) \times X : u \neq 0\}.$$

Then we can write

Theorem 1.1. *(Rabinowitz [43]). Let L be a compact linear operator and let H be a completely continuous nonlinear operator s.t. $H(\lambda, u) = o(||u||_X)$, $u \to 0$ uniformly in bounded λ-intervals. Then, if $\bar{\mu}$ is an eigenvalue of L of odd multiplicity it follows that Σ contains a continuum C such that $(\bar{\mu}^{-1}, 0) \in C$ and either*

(A1) *There exists $(\lambda_n, u_n) \in C$ such that*

$$|\lambda_n| + ||u_n||_X \to +\infty \quad \text{or } \lambda_n \to a \quad \text{or} \quad \lambda_n \to b$$

or

(A2) *There exists $(\hat{\mu}^{-1}, 0) \in C$ such that $\hat{\mu}$ is an eigenvalue of L and $\hat{\mu} \neq \bar{\mu}$.*

Let us observe that when $(a, b) = \mathbb{R}$ then condition (A1) means that the continuum C is unbounded in $\mathbb{R} \times X$.

It is well-known that problem (1) is equivalent to an equation of the form (2), with $X = C_0^1([0, \pi])$ and $(a, b) = \mathbb{R}$. The application of Theorem 1.1 to (1) requires the knowledge of the spectral theory of the linear operator $u \mapsto -u'' + qu$ defined in X; this is a very classical topic in the theory of ODEs. From the fact that $a(t) \geq a_0 > 0$ for all t it follows that the BVP

$$\begin{cases} -u'' + q(t)u = \lambda a(t)u, \ t \in [0, \pi], \\[2mm] u(0) = 0 = u(\pi) \end{cases} \tag{3}$$

has a sequence of simple eigenvalues

$$0 < \lambda_1 < \lambda_2 < \cdots < \lambda_k < \ldots$$

such that $\lim_{k \to +\infty} \lambda_k = +\infty$.

It is important to recall the elementary fact that the eigenfunction φ_k associated to the eigenvalue λ_k of (3) has $(k - 1)$ simple zeros in $(0, \pi)$. Moreover, if one considers the first order planar system associated to the equation in (3) and denotes by γ the orbit of $z(t) = (u(t), u'(t)) = (x(t), y(t))$ in the phase plane, then it is defined the rotation number

$$\mathrm{rot}_z = \frac{1}{2\pi} \int_0^\pi \frac{x(t)y'(t) - x'(t)y(t)}{x(t)^2 + y(t)^2} \, dt. \tag{4}$$

Moreover, if we think of the parametrization $x(t) = \rho(t) \sin \theta(t)$, $y(t) = \rho(t) \cos \theta(t)$ then for the angular coordinate $\theta(\cdot)$ we have

$$\theta(t) = k(t)\pi + \alpha(t),$$

being $k(t) \in \mathbb{N}$ and $\alpha(t) \in (0, \pi]$. Thus, it is easily seen that the number of zeros in $(0, \pi)$ of a solution u to (3), rot_z and $k(\pi)$ are related one to the other (cf. Remark 4.8). The understanding of this relation in the case when the phase plane is substituted by a $2N$-dimensional phase space is one of the main goal of these lectures.

The linear theory guarantees the applicability of Theorem 1.1; thus, for every $k \in \mathbb{N}$ there exists a continuum $C_k \subset \mathbb{R} \times C_0^1([0, \pi])$ of solutions of (1) bifurcating from $(\lambda_k, 0)$ and such that

(A1)′ C_k is unbounded in $\mathbb{R} \times C_0^1([0, \pi])$

     or

(A2)′   There exists $j \in \mathbb{N}$, $j \neq k$, such that $(\lambda_j, 0) \in C_k$.

In [43] the author then shows that alternative (A2)′ cannot hold; in order to prove this, the crucial point is the study of the nodal properties of the nontrivial solutions to (1). Indeed, it is possible to prove that nontrivial solutions to (1) have a finite number of zeros in $(0, \pi)$ and that this number is constant on every bifurcating branch.

Thus, it is obtained the following global bifurcation result for (1).

Theorem 1.2. *(Rabinowitz [43]). For every $k \in \mathbb{N}$, $(\lambda_k, 0)$ is a bifurcation point for (1). Moreover, the bifurcating branch $C_k \subset \mathbb{R} \times C_0^1([0, \pi], \mathbb{R})$ is unbounded in $\mathbb{R} \times C_0^1([0, \pi], \mathbb{R})$ and for every $(\lambda, u) \in C_k$, with $u \neq 0$, u has $(k - 1)$ simple zeros in $(0, \pi)$.*

We point out that the argument leading to the exclusion of alternative (A2)′ can be considered from a slightly different point of view by observing that it is based on a continuity property of the number of zeros of solutions to (1). Indeed, let us define $\Phi : \Sigma \to \mathbb{N}$ by setting

$$\Phi(\lambda, u) = \begin{cases} \text{card}(u^{-1}(0) \cap (0, \pi)) & \text{if } u \neq 0 \\ (k - 1) & \text{if } (\lambda, u) = (\lambda_k, 0), \end{cases}$$

for every $(\lambda, u) \in \Sigma$. Then the preservation of the number of zeros on the bifurcating branches C_k follows from the continuity of Φ on C_k. This remark can be formulated in the setting of the abstract equation (2); more precisely, it is possible to prove [10, Prop. 2.1] that alternative (A2) in Theorem 1.1 does not hold if we can define a continuous integer-valued functional $\Phi : \Sigma \to \mathbb{Z}$. We are now able to summarize the approach introduced in this section; in order to obtain the existence of solutions to (1) it is possible to use a bifurcation method. To this aim, some steps have been made:

(I) The obtention of an abstract global bifurcation result, which also describes the behaviour of the bifurcating branches (with, possibly, some alternatives);

(II) The study of the spectral theory of the linear operator in (1), in order to apply the abstract theorem;

(III) The use of a suitable topological invariant (e.g. the functional Φ), in order to exclude some of the alternatives in the abstract result.

The boundary value problem (1) is a particular case of a BVP of the form

$$\begin{cases} Jz' = S(t, z, \lambda)z, \ z = (x, y) \in \mathbb{R}^{2N}, \\ x_{|\partial I} = 0, \end{cases} \tag{5}$$

where $I \subset \mathbb{R}$, J is the standard symplectic matrix (cf. (7)) and $S : I \times \mathbb{R}^{2N} \times (a, b) \to \mathbb{R}^{2N}$ is a symmetric matrix. In case I is unbounded, e.g. $I = [1, +\infty)$, the boundary condition is intended as $\lim_{t \to +\infty} x(t) = 0$. This lecture note focuses on the question whether by the above described procedure it is possible to obtain a global bifurcation result for problems of the form (5). At least two main difficulties are clear. On the one hand, in case I is unbounded Rabinowitz global bifurcation result cannot be applied. On the other hand, we have to deal with some notion which substitutes the number of zeros in the description of the "nodal properties" of the solutions.

Section 2 is devoted to the description of a geometric index (called the Maslov index) according to the paper by Robbin–Salamon [44] (see also [36]). More precisely, it is introduced the Maslov index first for a pair (L, V) of paths of lagrangian subspaces in $\mathbb{R}^{2N}$ (cf. Definition 2.9) and then for pairs of the form $(\psi V, V)$, where V is the lagrangian subspace $\{0\} \times \mathbb{R}^N$ and ψ is a C^1-path of symplectic matrices (cf. Definition 2.12).

In Sect. 3 we focus on a system of $2N$ first order equations of the form

$$Jz' = B(t)z, \ t \in [0, \pi], \tag{6}$$

where $B(t)$ is a $2N \times 2N$ symmetric matrix. Following [2], we introduce the notion of "number of moments of verticality" (cf. Definition 3.2) and relate it with the Maslov index of the (symplectic) path $(\psi V, V)$, where ψ is now the fundamental matrix solution associated to (6).

In Sect. 4 we explain the reason why we call "generalized phase-plane analysis" the topological method we are describing. The main references for this section are the book of Atkinson [3] and the lecture note by Greenberg [26]. Indeed (cf. Remark 4.8), when we deal with system (6), N "angles" $\theta_j : [0, \pi] \to \mathbb{R}$ can be defined which play the role of angular coordinates of a solution $z(t)$ of (6) in the phase space $\mathbb{R}^{2N}$. In analogy with the elementary situation when the Prüfer angular coordinate of $z(t)$ in the phase plane $\mathbb{R}^2$ describes the number of rotations of the orbits around the origin, in Theorem 4.4 we describe the connection between the phase angles and the number of moments of verticality.

Section 5 contains a brief survey of various notions of "index" associated to a differential system that are available in the literature.

The remaining sections (which are taken from [10] and [11], to whom we refer for details) are devoted to the study of nonlinear boundary value problems of the form (5).

More precisely, in Sect. 6 we consider a first order problem of the form (5) with $N \geq 2$ and $I = [0, \pi]$ (cf. (38)). The crucial steps for the application of Theorem 1.1 are the study of the spectral properties of the linear operator associated to the differential system (cf. Proposition 6.1) and the fact that a (suitably defined) Maslov index of the solutions of the nonlinear problem is preserved along the bifurcating branches. The main result of Sect. 6 is Theorem 6.4. It is important to remark that in [43] an application of Theorem 1.1 provides the existence of multiple solutions to (1), for fixed λ, when f has a superlinear growth at infinity. To do this, a priori estimates on the solutions $(\lambda, u) \in C_k$ (i.e. on solutions with a fixed number of zeros) are developed. This procedure, which has been widely applied in the literature, can be found in [10], for $N = 1$, as well (see also Remark 6.5).

In Sect. 7 we focus on a Dirac-type system of the form (5) with $N = 1$ and $I = [1, +\infty)$ (cf. (42)). Due to the unboundedness of I, Rabinowitz Theorem 1.1

is not applicable. Moreover, no complete spectral theory for the associated linear operator (44) is available. By the use of an abstract bifurcation result due to Stuart [49] and a (suitably defined) rotation index (cf. Definition 7.4) the linear aspects of the problem are developed in Proposition 7.3 and Proposition 7.6. The main result of Sect. 7 is the global bifurcation result contained in Theorem 7.7.

We end this section with some remarks on the possible future developments of the approach described.

First, we observe that the results on Dirac-type systems given in Sect. 7 can be extended to the (physically relevant) case when the linear differential operator τ in (44) is singular. Indeed, in the forthcoming paper [12] it is considered the case when $I = (0, +\infty)$ (cf. (42)). In [12] the focus is on the nonlinear term Q, recalling that the systems (of ODEs) of the form (42) arise from the decomposition of the (partial differential) Dirac operator in the so-called partial wave subspaces [50]. More precisely, it is shown that those nonlinearities which leave the partial wave subspaces invariant fit into the framework of the set of functions $\mathscr{Q}$ (cf. (46), (47)) in which bifurcation occurs.

Another interesting topic (cf. Remark 6.5) is the investigation of the existence of multiple solutions to systems of the form (5) for fixed λ. To this end, one has to perform (by the use of an additional assumption of "superlinear" kind on the nonlinearity) a priori estimates on the solutions belonging to a fixed branch. A serious difficulty is in the fact that no explicit expression of the form (4) is available in the more general context of Sects. 6 and 7.

A major interest may also be devoted to the generalization of the results in Sect. 7 to the case when in (42) the unknown z belongs to $\mathbb{R}^{2N}$. To our knowledge, in this general context a satisfactory linear theory is available only when $t \in [0, \pi]$ (cf. [9]). Results in this direction will be obtained by an accurate investigation of the index in Definition 7.4 in relation with the abstract notion of Maslov index and with the Levinson theorem (cf. also [30, 34]).

Finally, we point out that it is interesting to attack systems in which the nonlinearity is periodic or "asymptotically periodic". To this end, one should focus (instead of Rabinowitz theorem) on bifurcation results obtained in the framework of the Fredholm index. We have in mind, in particular, the papers by Rabier–Stuart [42] and by Secchi–Stuart [47], which deal with an elliptic equation and with an hamiltonian system, respectively.

In what follows we shall use the following notation.

xy denotes the scalar product of $x, y \in \mathbb{R}^n$.

A^t, A^* denote the transpose and the transpose conjugate of a matrix A.

$\| \cdot \|$ denotes both the norm of a matrix and the norm of a finite dimensional vector.

M_S^n is the set of $n \times n$ symmetric matrices.

$\mathscr{L}(X)$ is the space of linear operators from a given Banach space X to itself. For $G \in \mathscr{L}(X)$, $\mathrm{Gr}G$ denotes the graph of G.

2   The Maslov Index

In this section we introduce the Maslov index, which is the topological invariant that we will use in the study of the nonlinear eigenvalue problem for first order systems of differential equations of Sect. 6. We follow the approach of [44], in which the authors define the Maslov index for paths of lagrangian subspaces of $\mathbb{R}^{2N}$; in view of our application, we will also describe the particular situation when the lagrangian subspaces are the images of the fundamental matrix of a system of first order linear differential equations in $\mathbb{R}^{2N}$.

Before giving the definition of the Maslov index, we recall some standard notions of symplectic geometry (cf. [4, 38]).

Consider a linear space V of finite dimension and a bilinear, antisymmetric and nondegenerate form $\omega : V \times V \rightarrow \mathbb{R}$. The space (V, ω) is called a symplectic vector space. The classical example is $(\mathbb{R}^N \times \mathbb{R}^N, \omega_0)$, where, setting $z = (x, y)$, $z' = (x', y')$, it is $\omega_0(z, z') = xy' - x'y$. Let

$$
J = \begin{pmatrix} 0 & -\mathrm{Id} \\ \mathrm{Id} & 0 \end{pmatrix} \tag{7}
$$

be the standard symplectic matrix. Note that $\omega_0(z, z') = (Jz)z'$. Every symplectic space (V, ω) has even dimension; moreover, there exists a symplectic isomorphism $\phi : (V, \omega) \rightarrow (\mathbb{R}^{2N}, \omega_0)$, i.e. a map such that $\omega_0(\phi(z), \phi(z')) = \omega(z, z')$ for all $z, z' \in V$. In what follows we shall write ω instead of ω_0.

Definition 2.1. A subspace L of $\mathbb{R}^{2N}$ is called lagrangian if $\dim L = N$ and $\omega(z, z') = 0$ for all $z, z' \in L$.

The subspaces $L_1 = \{0\} \times \mathbb{R}^N$ and $L_2 = \mathbb{R}^N \times \{0\}$ are lagrangian subspaces. If $N = 1$, every line through the origin is a lagrangian subspace.

It is useful to know various ways to describe a lagrangian subspace. First, we can use a linear injective map $Z : \mathbb{R}^N \rightarrow \mathbb{R}^{2N}$ such that $L = \mathrm{Im} Z$, together with an additional condition which can be expressed by writing

$$
Z = \begin{pmatrix} X \\ Y \end{pmatrix},
$$

being X, Y two $N \times N$ matrices s.t. Z has rank N. Precisely,

Proposition 2.2. *The subspace L is lagrangian if and only if $Y^t X = X^t Y$.*

Proof. For $z \in L$ there exists a unique $u \in \mathbb{R}^N$ s.t. $z = (Xu, Yu)$. Then we have

$$
\omega(z, z') = \omega((Xu, Yu), (Xu', Yu')) = Xu\,Yu' - Yu\,Xu' = (Y^t X - X^t Y)u\,u'.
$$

$\square$

The map Z is called a lagrangian frame.

Another useful expression for lagrangian subspaces is obtained as follows. Consider $G \in \mathscr{L}(\mathbb{R}^N)$ and

$$L = \operatorname{Gr} G = \{(x, Gx), x \in \mathbb{R}^N\}.$$

Then we have

Proposition 2.3. *The subspace L is lagrangian if and only if the matrix G is symmetric.*

Proof. Consider the frame

$$\begin{pmatrix} \operatorname{Id} \\ G \end{pmatrix}$$

and observe that $G^t \operatorname{Id} - \operatorname{Id}^t G = 0$.   □

Note that $(\operatorname{Gr} G) \cap (\{0\} \times \mathbb{R}^N) = \{0\}$. Moreover, it is useful to observe that if L is a lagrangian subspace s.t.

$$L \cap (\{0\} \times \mathbb{R}^N) = \{0\} \tag{8}$$

and if $Z = \begin{pmatrix} X \\ Y \end{pmatrix}$ is a frame for L then X is invertible and $L = \operatorname{Gr}(Y\, X^{-1})$ (note that the lagrangian subspace $\{0\} \times \mathbb{R}^N$ does not satisfy (8)). Indeed, (8) implies that $\operatorname{Ker} X = \{0\}$ and hence X^{-1} is defined. Moreover,

$$\begin{aligned} L &= \{(Xu, Yu), u \in \mathbb{R}^n\} = \{(XX^{-1}v, YX^{-1}v), v \in \mathbb{R}^n\} \\ &= \{(v, YX^{-1}v), v \in \mathbb{R}^n\} = \operatorname{Gr}(Y\, X^{-1}). \end{aligned}$$

Theorem 2.4. *The space $\Lambda(2N)$ of lagrangian subspaces of $\mathbb{R}^{2N}$ is a C^∞-manifold of dimension $N(N + 1)/2$.*

The space $\Lambda(2N)$ is called the lagrangian grassmannian.

We are now ready to give the definition of the Maslov index for paths of lagrangian subspaces. Consider $L \in C^1([0, \pi], \Lambda(2N))$, a curve of lagrangian subspaces. The Maslov index will be defined by means of a quadratic form which we now describe.

Consider $\bar{t} \in [0, \pi]$ and $JL(\bar{t})$; it is easy to see that this lagrangian subspace is a complement of $L(\bar{t})$. Take $z \in L(\bar{t})$ and t in a neighbourhood of $\bar{t}$. Denote by $w(t) = (w_1(t), w_2(t))$ the (unique) vector in $JL(\bar{t})$ such that $z + w(t) \in L(t)$.

Then we can define the quadratic form $Q(L, \bar{t}) : L(\bar{t}) \to \mathbb{R}$ by

$$Q(L, \bar{t})(z) = \frac{d}{dt}\omega(z, w(t))_{|t=\bar{t}}, \quad \forall z \in L(\bar{t}).$$

In the particular case $N = 1$ and $L(\bar{t}) = \mathbb{R} \times \{0\}$, then $z = (x, 0)$ and $w_2(t) = a(t)x$, for some real function a. As a consequence, $Q(L, \bar{t})(z) = a'(\bar{t})x^2$.

We now give an useful expression for Q.

Proposition 2.5. *Let $Z(t)$ be a frame for $L(t)$. Then*

$$Q(L, \bar{t})(z) = X(\bar{t})u\, Y'(\bar{t})u - Y(\bar{t})u\, X'(\bar{t})u, \qquad (9)$$

being u s.t. $z = Z(\bar{t})u \in L(\bar{t})$.

Proof. Without loss of generality, assume $L(\bar{t}) = \mathbb{R}^N \times \{0\}$. We know that, being $z = (X(\bar{t})u, Y(\bar{t})u)$ and $w(t) = (0, w_2(t))$, the condition $z + w(t) \in L(t)$ can be written as

$$Y(\bar{t})u + w_2(t) = Y(t)\, X(t)^{-1}\, X(\bar{t})u$$

(note that in a neighbourhood of $\bar{t}$ it is guaranteed that $L(t) \cap (\{0\} \times \mathbb{R}^N) = \{0\}$). Observe that

$$\omega(z, w(t)) = X(\bar{t})u\, w_2(t)$$

and

$$Q(L, \bar{t})(z) = X(\bar{t})u\, w_2'(t).$$

Moreover,

$$w_2'(t) = Y'(t)\, X(t)^{-1}\, X(\bar{t})u - Y(t)\, X(t)^{-1}\, X'(t)\, X(t)^{-1}\, X(\bar{t})u.$$

Then (9) follows using the fact that $Y^t X = X^t Y$. $\square$

In general, the Maslov index is defined for a pair (L_1, L_2) of curves of lagrangian subspaces. We shall consider the particular case $L_2 \equiv V = \{0\} \times \mathbb{R}^N$. We first give the following:

Definition 2.6. Given $L_1 \in C^1([0, \pi], \Lambda(2N))$, a point $\bar{t} \in [0, \pi]$ is a crossing for (L_1, V) if $L_1(\bar{t}) \cap V \neq \{0\}$.

Then we set

$$\Gamma(L_1, V, \bar{t}) := Q(L_1, \bar{t})_{|L_1(\bar{t}) \cap V}.$$

Note that if $\bar{t}$ is not a crossing for (L_1, V) then $\Gamma(L_1, V, \bar{t}) \equiv 0$.

Definition 2.7. A crossing $\bar{t}$ is regular if the quadratic form $\Gamma(L_1, V, \bar{t})$ is nondegenerate.

Remark 2.8. Using the fact that the eigenvalues of a smooth family of symmetric matrices are continuous functions of t, we know that regular crossings are isolated. Thus, there are at most a finite number of regular crossings in $[0, \pi]$.

We shall now define the Maslov index for a pair (L_1, V) having only regular crossings. To do this, we use the notion of *signature* of a quadratic form, which is the difference between the number of positive eigenvalues and the number of negative eigenvalues.

Definition 2.9. The Maslov index of the pair (L_1, V), with only regular crossings in $[0, \pi]$, is defined as follows

$$\mu(L_1, V) = \frac{1}{2}\text{sign}\Gamma(L_1, V, 0) + \sum_{0 < t < \pi} \text{sign}\Gamma(L_1, V, t) + \frac{1}{2}\text{sign}\Gamma(L_1, V, \pi), \quad (10)$$

where the summation is taken over all crossings.

Let us note that from the above definition it follows that $\mu(L_1, V) \in \mathbb{Z}/2$.

Definition 2.9 is easily understandable in case $N = 1$. Indeed, the fact that $\bar{t}$ is a crossing for (L_1, V) means that $L_1(\bar{t}) = \{0\} \times \mathbb{R}$; as a consequence, for every t in a neighbourhood of $\bar{t}$ the lagrangian subspace $L_1(t)$ is the line $x = b(t)y$, for some $b(t) \in \mathbb{R}$. Moreover, for $z = (0, y)$, $\Gamma(L_1, V, \bar{t})(z) = -b'(\bar{t})y^2$ and the signature of this quadratic form is determined by the sign of $b'(\bar{t})$. The Maslov index is thus an algebraic count of those t for which $L_1(t) = \{0\} \times \mathbb{R}$.

Remark 2.10. The restriction on the regularity of the crossings in the above definition can be eliminated. Indeed, in [44] it is proved that every lagrangian path is homotopic with fixed endpoints to a lagrangian path having only regular crossings and that two homotopic (with fixed endpoints) lagrangian paths with regular crossings have the same Maslov index.

These facts allow to define the Maslov index of any pair (L_1, V) as the Maslov index of a lagrangian path with regular crossings homotopic with fixed endpoints to L_1.

We now consider the particular case in which $L_1(t)$ arises from a symplectic matrix. To this end, recall that the space of symplectic matrices is

$$Sp(2N) = \{A \in \mathcal{L}(\mathbb{R}^{2N}) : A^t J A = J\}.$$

Consider again $V = \{0\} \times \mathbb{R}^N$ and $\psi \in C^1([0, \pi], Sp(2N))$.

Proposition 2.11. *For every* $t \in [0, \pi]$, *the subspace* $\psi(t)V$ *is a lagrangian subspace.*

Proof. Let us fix $t \in [0, \pi]$ and write

$$\psi(t) = \begin{pmatrix} X_0(t) & X(t) \\ Y_0(t) & Y(t) \end{pmatrix}, \quad (11)$$

where $X_0(t), Y_0(t), X(t), Y(t)$ are $N \times N$ block matrices; from a straightforward computation we deduce that

$$Z(t) = \begin{pmatrix} X(t) \\ Y(t) \end{pmatrix} \quad (12)$$

is a lagrangian frame for $\psi(t)V$. Then it is easy to check that

$$\psi(t)^t J \psi(t) = J \implies Y(t)^t X(t) = X(t)^t Y(t).$$

$\square$

We can now give the following:

Definition 2.12. The Maslov index of the symplectic path ψ is defined as follows

$$\mu(\psi) = \mu(\psi V, V).$$

For the computation of $\mu(\psi)$, it is useful to observe that if $\bar{t}$ is a crossing and $z \in \psi(\bar{t})V \cap V$ then $z = (X(\bar{t})u, Y(\bar{t})u)$ with $X(\bar{t})u = 0$. Then, using (9),

$$\Gamma(\psi V, V, \bar{t})(z) = Q(\psi V, \bar{t})_{|\psi(\bar{t})V \cap V}(z) = -Y(\bar{t})u \, X'(\bar{t})u.$$

Note also that, according to (11),

$$\dim(\psi(t)V \cap V) = \dim(\mathrm{Ker} X(t)), \ \forall\, t \in [0, \pi]. \tag{13}$$

An important example of path of symplectic matrices arises from a linear system of first order differential equations in $\mathbb{R}^{2N}$ of the form

$$J \, z' = B(t)\, z, \tag{14}$$

where $B \in C([0, \pi], M_S^{2N})$. Indeed, denoting by ψ the fundamental matrix associated to (14), we have the following straightforward result:

Proposition 2.13. *For every $t \in [0, \pi]$, the matrix $\psi(t)$ is symplectic.*

Example 2.14. Let us now compute the Maslov index in an elementary situation in the case $N = 1$. Consider the planar system (14) with

$$B(t) = \begin{pmatrix} t+1 & 0 \\ 0 & t+1 \end{pmatrix}, \quad \forall\, t \in [0, \pi].$$

As above, let ψ be the fundamental matrix of (14) and $V = \{0\} \times \mathbb{R}$; a simple computation shows that a frame for $\psi(t)V$ is given by

$$\begin{pmatrix} \sin\left(\dfrac{t^2 + 2t}{2}\right) \\ \cos\left(\dfrac{t^2 + 2t}{2}\right) \end{pmatrix}, \quad \forall\, t \in [0, \pi].$$

Hence, according to Definition 2.6, $\bar{t} \in [0, \pi]$ is a crossing for $(\psi V, V)$ if $\psi(\bar{t})V \cap V \neq \{0\}$, i.e. $\sin\left(\dfrac{\bar{t}^2 + 2\bar{t}}{2}\right) = 0$; we then deduce that the only crossings in $[0, \pi]$ are $t_0 = 0, t_1 = -1 + \sqrt{1 + 2\pi}, t_2 = -1 + \sqrt{1 + 4\pi}$.

Let us now observe that all the crossings are regular, being

$$\Gamma(\psi V, V, t_i)(z) = -Y(t_i) X'(t_i) y^2 \neq 0, \quad \forall z = (0, y) \in V, \ y \neq 0, \quad i = 0, 1, 2.$$

Using (9), it follows that

$$\text{sign} \Gamma(\tilde{\psi} V, V, t_i) = -1, \quad i = 0, 1, 2.$$

Hence, we conclude that

$$\mu(\psi) = \frac{1}{2}(-1) + (-1) + (-1) + 0 = -\frac{5}{2}. \tag{15}$$

3 The Number of Moments of Verticality

The subject of this section is an index (defined in [2]) associated to a system of linear first order differential equations. We also illustrate the relation between this index (the so-called "number of moments of verticality") and the Maslov index of the symplectic path associated to the fundamental matrix of the system and discussed in the previous section.

Consider the system of $2N$ first order equations

$$J \, z' = B(t) \, z, \quad z = (x, y) \in \mathbb{R}^{2N}, \tag{16}$$

where $B \in C([0, \pi], M_S^{2N})$. Assume that $B(t)$ is positive definite for all $t \in [0, \pi]$.

Definition 3.1. We say that $\bar{t} \in (0, \pi]$ is a moment of verticality for (16) if the boundary value problem

$$\begin{cases} J \, z' = B(t) \, z \\ x(0) = 0 = x(\bar{t}) \end{cases} \tag{17}$$

has a nontrivial solution. The number $m(\bar{t})$ of linearly independent solutions of (17) is called the multiplicity of $\bar{t}$.

Then we can give the following:

Definition 3.2. The number of moments of verticality of B is

$$j(B) = \sum_{0 < t < \pi} m(t),$$

where the summation is taken over all moments of verticality.

The fact that $j(B)$ is a finite sum is a consequence of the positive definiteness of $B(t)$, for every $t \in [0, \pi]$. This will be clarified in the proof of Theorem 3.4 (see also Remark 3.5).

In what follows we shall give a formula which relates $j(B)$ and the Maslov index of the symplectic path generated by the fundamental matrix ψ of (16).

Proposition 3.3. *A point $\bar{t} \in (0, \pi]$ is a crossing for $(\psi V, V)$ if and only if $\bar{t}$ is a moment of verticality for* (16). *Moreover, if we denote*

$$\psi(t) = \begin{pmatrix} X_0(t) & X(t) \\ Y_0(t) & Y(t) \end{pmatrix},$$

then $m(\bar{t}) = \dim(\mathrm{Ker}X(\bar{t}))$.

Proof. Let us first recall that

$$Z(t) = \begin{pmatrix} X(t) \\ Y(t) \end{pmatrix}$$

is a frame for $\psi(t)V$, for every $t \in [0, \pi]$. Now, it follows from the definition that $\bar{t} \in (0, \pi]$ is a crossing of $(\psi V, V)$ if and only if $\psi(\bar{t})V \cap V \neq \{0\}$, i.e. if and only if there exists $u \in \mathbb{R}^N \setminus \{0\}$ such that $(X(\bar{t})u, Y(\bar{t})u) \in V$. As a consequence, $\bar{t}$ is a crossing for $(\psi V, V)$ if and only if there exists $u \in (\mathrm{Ker}X(\bar{t})) \setminus \{0\}$.

On the other hand, $\bar{t} \in (0, \pi]$ is a moment of verticality for (16) if and only if (17) has a nontrivial solution; using the fundamental matrix ψ it is easy to see that the solutions $z = (x, y)$ of (16) satisfying the condition $x(0) = 0$ can be written as

$$z(t) = \begin{pmatrix} X(t) \\ Y(t) \end{pmatrix} u, \tag{18}$$

for some $u \in \mathbb{R}^N$. Therefore, $\bar{t} \in (0, \pi]$ is a moment of verticality for (16) if and only if $X(\bar{t})u = 0$ for some $u \in \mathbb{R}^N \setminus \{0\}$; hence, the conditions for $\bar{t}$ being a crossing or a moment of verticality coincide.

Finally, from (18) we plainly obtain $m(\bar{t}) = \dim(\mathrm{Ker}X(\bar{t}))$. $\qquad\square$

The result stated below establishes a relation between the Maslov index and the number of moments of verticality (cf. [7, 19]).

Theorem 3.4. *The following relation holds true:*

$$\mu(\psi) = -\left[\frac{N}{2} + j(B) + \frac{m(\pi)}{2}\right]. \tag{19}$$

Proof. Remember that $z \in \psi(t)V \cap V$ can be represented by $z = (X(t)u, Y(t)u)$ with $X(t)u = 0$, i.e. $z = (0, Y(t)u)$. In order to compute $\mu(\psi)$, let us note that, according to (9), we have

$$\Gamma(\psi(t)V, V, t)(z) = -Y(t)u\, X'(t)u. \tag{20}$$

Thus we need to compute $X'(t)$. To this end, we use the fact that the fundamental matrix ψ is a (matrix) solution of (16); hence, we have

$$
\psi'(t) = J^{-1} B(t) \psi(t) = \begin{pmatrix} 0 & \mathrm{Id} \\ -\mathrm{Id} & 0 \end{pmatrix} \begin{pmatrix} B_{11}(t) & B_{12}(t) \\ B_{21}(t) & B_{22} \end{pmatrix} \begin{pmatrix} X_0(t) & X(t) \\ Y_0(t) & Y(t) \end{pmatrix},
$$

which implies

$$
\begin{pmatrix} X_0'(t) & X'(t) \\ Y_0'(t) & Y'(t) \end{pmatrix} = \begin{pmatrix} \dots & B_{21}(t)X(t) + B_{22}(t)Y(t) \\ \dots & \dots \end{pmatrix}.
$$

Thus, (20) reduces to

$$
\Gamma(\psi(t)V, V, t)(z) = -Y(t)u \left(B_{21}(t)X(t) + B_{22}(t)Y(t) \right) u
$$

$$
= -B_{22}(t)^t\, Y(t)u\, Y(t)u = -B_{22}(t)\, Y(t)u\, Y(t)u,
$$

(21)

where we have used the fact that $B(t)$ is symmetric. On the other hand, an easy computation shows that

$$
B(t)z\, z = -B_{22}(t)\, Y(t)u\, Y(t)u,
$$

for every $z = (0, Y(t)u) \in \psi(t)V \cap V$; as a consequence, from (21) we obtain

$$
\Gamma(\psi(t)V, V, t)(z) = -B(t)z\, z,
$$

(22)

for every $z \in \psi(t)V \cap V$. From this relation and the fact that $B(t)$ is positive definite, for every $t \in [0, \pi]$, we first deduce that $\Gamma(\psi(t)V, V, t)$ is non degenerate at every crossing in $[0, \pi]$, i.e. the crossings are regular; moreover, at every crossing $\Gamma(\psi(t)V, V, t)$ is negative definite and, according to (13) and Proposition 3.3, we have

$$
\mathrm{sign}\Gamma(\psi V, V, t) = -\dim(\psi(t)V \cap V) = -\dim(\mathrm{Ker}X(t)) = -m(t).
$$

For the completion of the proof, recalling (10), it is sufficient to consider $t = 0$ and prove that $\mathrm{sign}\Gamma(\psi(0)V, V, 0) = -N$. This is a consequence of the fact that $\psi(0) = \mathrm{Id}$: indeed, from this condition we deduce that $\psi(0)V \cap V = V \neq \{0\}$ and then

$$
\mathrm{sign}\Gamma(\psi(0)V, V, 0) = -\dim(\mathrm{Ker}X(0)) = -N.
$$

$\square$

Remark 3.5. Let us observe that the proof of Theorem 3.4 shows that that the sum in the definition of $j(B)$ is finite under the weaker assumption that $B_{22}(t)$ is positive definite, for every $t \in [0, \pi]$ (cf. relation (21)). Indeed, even in this case, every crossing of $(\psi V, V)$ is regular. As already observed, we can then deduce that the crossings are isolated and in a finite number in $[0, \pi]$; according to Proposition 3.3, we conclude that there exist only finitely many moments of verticality in $[0, \pi]$.

This observation is crucial when dealing with first order systems arising from second order equations; indeed, let us consider

$$X'' + A(t) X = 0, \tag{23}$$

with $A \in C([0, \pi], M_S^N)$. It is well-known that (23) can be written in the form (16) with

$$B(t) = \begin{pmatrix} A(t) & 0 \\ 0 & \mathrm{Id} \end{pmatrix}, \quad \forall\, t \in [0, \pi]; \tag{24}$$

therefore, in order to define $j(B)$ no assumption on the definiteness of A is needed. Let us also observe that in the case when A is constant it is possible to show that

$$j(B) = \sum_{i:\lambda_i > 0} \left(\left\lceil \sqrt{\lambda_i} \right\rceil - 1 \right),$$

where $\lambda_1, \ldots, \lambda_N$ are the eigenvalues of A and, for every $r > 0$, we have set $\lceil r \rceil = n$ if $n - 1 < r \leq n$ for some $n \in \mathbb{N}$, $n \geq 1$ (cf. [39], also in a more general setting). A formula for the computation of $j(B)$ in the framework of linear Hamiltonian systems can be found in [41].

Example 3.6. (Continuation of Example 2.14) Consider again (16) with

$$B(t) = \begin{pmatrix} t+1 & 0 \\ 0 & t+1 \end{pmatrix}, \quad \forall\, t \in [0, \pi].$$

The solutions $z = (x, y)$ satisfying the condition $x(0) = 0$ are given by

$$z(t) = c \left(\sin\left(\frac{t^2 + 2t}{2} \right), \cos\left(\frac{t^2 + 2t}{2} \right) \right), \quad \forall\, t \in [0, \pi], \ c \in \mathbb{R}.$$

The moments of verticality in $(0, \pi]$ are then the zeros of $\sin\left(\dfrac{t^2 + 2t}{2} \right)$ in $(0, \pi]$, i.e. (cf. Example 2.14) the points $t_i, i = 1, 2$, whose multiplicity is obviously one. Therefore we have $j(B) = 2$.

Recalling (15), formula (19) reads then as $-\dfrac{5}{2} = -\left(\dfrac{1}{2} + 2 + 0 \right)$.

4 The Phase-Angles and the Number of Moments of Verticality

In this section we give a formula for the number of moments of verticality introduced in Definition 3.2 based on the notion of "phase-angles" introduced in [3, 26]. They are defined using the symplectic structure described in the previous sections; in the case of planar systems they are strictly related to the usual Prüfer angle coordinate.

As in the previous sections, we denote by

$$\psi(t) = \begin{pmatrix} X_0(t) & X(t) \\ Y_0(t) & Y(t) \end{pmatrix}, \tag{25}$$

for every $t \in [0, \pi]$, the fundamental matrix of (16) and we recall that

$$Z(t) = \begin{pmatrix} X(t) \\ Y(t) \end{pmatrix}$$

is a frame for $\psi(t)V$, for every $t \in [0, \pi]$. As a consequence, $Z(t)$ has rank N; hence the matrix

$$\begin{pmatrix} Y(t) & X(t) \\ -X(t) & Y(t) \end{pmatrix}$$

is invertible and thus the matrix $(Y(t) - iX(t))$ is invertible as well.

For every $t \in [0, \pi]$ we can then define the matrix

$$\Theta(t) = (Y(t) + iX(t))(Y(t) - iX(t))^{-1}. \tag{26}$$

The following fact will be useful in the sequel.

Lemma 4.1. *The matrix $\Theta(t)$ is unitary, for all $t \in [0, \pi]$, i.e.*

$$\Theta^*(t)\,\Theta(t) = \mathrm{Id}, \quad \forall\, t \in [0, \pi],$$

where $\Theta^*(t)$ denotes the adjoint of $\Theta(t)$.

The computation needed for the proof is based on the fact that $\psi(t)V$ is lagrangian, for every $t \in [0, \pi]$ (cf. Propositions 2.11 and 2.13).

Proposition 4.2. *(Atkinson [3, Chap. V.4 - Th. 10.8.1]). There exist N continuous functions $\theta_j : [0, \pi] \to \mathbb{R}$, $j = 1, \ldots, N$, such that:*

(i) $e^{2i\theta_j(t)}$ is an eigenvalue of $\Theta(t)$, for $j = 1, \ldots, N$;

(ii) $\theta_1(0) = 0 = \cdots = \theta_N(0);$
(iii) $\theta_1(t) \leq \theta_2(t) \leq \cdots \leq \theta_N(t) \leq \theta_1(t) + \pi.$
(iv) $\theta_j : [0, \pi] \to \mathbb{R}$ *is strictly increasing, for every* $j = 1, \ldots, N.$

The functions θ_j are called "phase angles". The existence of θ_j satisfying Proposition 4.2 is proved in [3] using the fact that the eigenvalues of a symmetric matrix are continuous functions of t.

Statements (i) and (ii) also follow from Kato Selection theorem (cf. Theorem 5.2 in [29]). Let us also observe that (iii) implies the uniqueness of the phase-angles.

The proof of Proposition 4.2 is based on the fact that eigenvalues of a family of differentiable hermitian matrices with positive definite derivative are increasing functions and on the relation

$$\Theta'(t) = i \, \Theta(t) \, \Omega(t), \quad \forall \, t \in [0, \pi],$$

with

$$\Omega(t) = 2 \left[\left(Y(t) - i X(t) \right)^{-1} \right]^t \begin{pmatrix} X(t) \\ Y(t) \end{pmatrix}^t B(t) \begin{pmatrix} X(t) \\ Y(t) \end{pmatrix} \left(Y(t) - i X(t) \right)^{-1}, \forall \, t \in [0, \pi].$$

We point out that a crucial role is also played by the fact that $\Omega(t)$ is positive definite, for every $t \in [0, \pi]$, which is a consequence of the assumption on the positive definiteness of $B(t)$.

The relation between the concepts of phase angle and of moment of verticality is explained in the following:

Theorem 4.3. *Let* $\bar{t} \in (0, \pi]$. *The following facts are equivalent:*

(a) $\bar{t}$ *is a moment of verticality for* (16) *of multiplicity* m;
(b) 1 *is an eigenvalue of* $\Theta(\bar{t})$ *of algebraic multiplicity* m;
(c) *There exist* $j_1, \ldots, j_m \in \{1, \ldots N\}$ *s.t.* $\theta_{j_k}(\bar{t}) = 0 (\mathrm{mod}\pi)$, $k = 1, \ldots m.$

Proof. The equivalence between *(b)* and *(c)* is straightforward. Let 1 be an eigenvalue of $\Theta(\bar{t})$ with algebraic multiplicity m. Being $\Theta(\bar{t})$ unitary, it follows that the geometric multiplicity of 1 coincides with its algebraic multiplicity; this means that $m = \dim(\mathrm{Ker}(\Theta(\bar{t}) - \mathrm{Id}))$. Note also that

$$\left(\Theta(\bar{t}) - \mathrm{Id} \right) \left(Y(\bar{t}) - i X(\bar{t}) \right) = \left[\left(Y(\bar{t}) + i X(\bar{t}) \right) \left(Y(\bar{t}) - i X(\bar{t}) \right)^{-1} - \mathrm{Id} \right] \left(Y(\bar{t}) - i X(\bar{t}) \right)$$

$$= Y(\bar{t}) + i X(\bar{t}) - Y(\bar{t}) + i X(\bar{t}) = 2 i X(\bar{t}).$$

Hence,

$$\left(\Theta(\bar{t}) - \mathrm{Id} \right) = 2 i X(\bar{t}) \left(Y(\bar{t}) - i X(\bar{t}) \right)^{-1}$$

and

$$\dim(\mathrm{Ker}(\Theta(\bar{t}) - \mathrm{Id})) = \dim(\mathrm{Ker}(X(\bar{t}))) = m.$$

This is sufficient to conclude that part (a) and (b) are equivalent.   □

We are now in position to state and prove a formula for the number of moments of verticality based on the phase-angles. To this aim, for every $t \in [0, \pi]$ and for every $j = 1, \ldots, N$, let us write

$$\theta_j(t) = k_j(t)\pi + \alpha_j(t), \qquad (27)$$

with $\alpha_j(t) \in (0, \pi]$ and $k_j(t) \in \mathbb{N}$.

Theorem 4.4. *The number of moments of verticality of B satisfies*

$$j(B) = k_1(\pi) + \cdots + k_N(\pi). \qquad (28)$$

Proof. Formula (28) is a consequence of the equivalence of parts (a) and (c) in Theorem 4.3; indeed, for every $j = 1, \ldots, N$, according to (27) and the fact that θ_j is increasing we infer that $k_j(\pi)$ is exactly the number of $t \in (0, \pi)$ such that $\theta_j(t)$ is a multiple of π. Hence, the sum $k_1(\pi) + \ldots + k_N(\pi)$ counts (with multiplicity) the number of $t \in (0, \pi)$ such that there exists $j \in \{1, \ldots, N\}$ for which $\theta_j(t)$ is a multiple of π, i.e. it coincides with the number of moments of verticality.   □

In the proof of formula (28) we used the fact that the phase-angles are increasing (which is a consequence of the positive definitiveness of $B(t)$, for every $t \in [0, \pi]$); a more careful analysis shows that for (28) it is sufficient that the phase-angles are increasing in a neighbourhood of the moments of verticality. Indeed we have

Proposition 4.5. *Assume that $B_{22}(t)$ is positive definite for all $t \in [0, \pi]$ and that there exist $t^* \in [0, \pi]$ and $j \in \{1, \ldots, N\}$ such that*

$$\theta_j(t^*) = 0 \mod \pi. \qquad (29)$$

Then θ_j is stricty increasing in t^*.

Proof. We use Theorem V.6.2 in [3]. Thus, we have to prove that for all $u \in \mathbb{R}^N \setminus \{0\}$ such that

$$\Theta(t^*)u = e^{2i\theta_j(t^*)}u = u, \qquad (30)$$

then $\Omega(t^*)u \; u > 0$.
By the definition of Ω (and omitting t^*), we get

$$\Omega u \, u = \left(2[(Y - iX)^{-1}]^* \begin{pmatrix} X \\ Y \end{pmatrix}^t B \begin{pmatrix} X \\ Y \end{pmatrix} (Y - iX)^{-1}\right)u \, u$$

$$= \left(2B \begin{pmatrix} X \\ Y \end{pmatrix} (Y - iX)^{-1}\right)u \left(\begin{pmatrix} X \\ Y \end{pmatrix} (Y - iX)^{-1}\right)u$$

$$= \left(2B \begin{pmatrix} X(Y - iX)^{-1} \\ Y(Y - iX)^{-1} \end{pmatrix}\right)u \begin{pmatrix} X(Y - iX)^{-1} \\ Y(Y - iX)^{-1} \end{pmatrix} u.$$

A simple computation (cf. the proof of Theorem 4.3) shows that

$$\Theta - \mathrm{Id} = 2i X (Y - iX)^{-1}, \qquad \Theta + \mathrm{Id} = 2Y(Y - iX)^{-1}$$

and, as a consequence,

$$\Omega u\, u = \left(2B \left(\frac{\frac{\Theta - \mathrm{Id}}{2i}}{\frac{\Theta + \mathrm{Id}}{2}} \right) \right) u \, \left(\frac{\frac{\Theta - \mathrm{Id}}{2i}}{\frac{\Theta + \mathrm{Id}}{2}} \right) u.$$

Using (30), it follows that $\Theta - I = 0$ and hence, setting $\tilde{u} = \frac{\Theta + I}{2} u \neq 0$, the positive definiteness of B_{22} implies that

$$\Omega u\, u = \left(2B \begin{pmatrix} 0 \\ \tilde{u} \end{pmatrix} \right) \begin{pmatrix} 0 \\ \tilde{u} \end{pmatrix}$$

$$= 2 \begin{pmatrix} B_{12}\tilde{u} \\ B_{22}\tilde{u} \end{pmatrix} \begin{pmatrix} 0 \\ \tilde{u} \end{pmatrix}$$

$$= \left(2 B_{22}\tilde{u} \right) \tilde{u} > 0.$$

$\square$

Remark 4.6. Note that the above Proposition follows from the less restrictive assumption that $B_{22}(t)$ is positive definite, for every crossing $t \in [0, \pi]$. Observe also that (on the lines of Remark 3.5), this same condition is indeed sufficient for the finiteness of $j(B)$. We also refer to Remark 4.9. We finally recall that the situation in which only the matrix $B_{22}(t)$ is definite positive is met in the study of second order systems of differential equations (cf. Remark 3.5). In this framework, the monotonicity of the phase angles in a neighbourhood of a crossing is treated (quoting a result of Conley [16]) in [26, Lemma 8.2].

It is also worth noticing that in [32] it is shown that the positive definiteness of $B_{22}(t)$ is a sufficient condition for the monotonicity of the eigenvalues of a matrix analogue to $\Theta(t)$; more precisely (cf. (26)), in [32] the author deals with the first row in (25) instead of the second column.

Example 4.7. (Conclusion of Examples 2.14 and 3.6) Consider again (16) with

$$B(t) = \begin{pmatrix} t+1 & 0 \\ 0 & t+1 \end{pmatrix}, \qquad \forall\, t \in [0, \pi].$$

As already observed, in this situation we have $X(t) = \sin\left(\frac{t^2 + 2t}{2} \right)$ and $Y(t) = \cos\left(\frac{t^2 + 2t}{2} \right)$, for every $t \in [0, \pi]$; hence, the matrix Θ defined in (26) is

$$\Theta(t) = \frac{\cos\left(\frac{t^2 + 2t}{2}\right) + i \sin\left(\frac{t^2 + 2t}{2}\right)}{\cos\left(\frac{t^2 + 2t}{2}\right) - i \sin\left(\frac{t^2 + 2t}{2}\right)} = \cos(t^2 + 2t) + i \sin(t^2 + 2t), \quad \forall\, t \in [0, \pi].$$

By writing

$$\Theta(t) = e^{i(t^2 + 2t)} = e^{2i\left(\frac{t^2 + 2t}{2}\right)}, \quad \forall\, t \in [0, \pi],$$

we deduce that the phase-angle θ_1 is given by

$$\theta_1(t) = \frac{t^2 + 2t}{2}, \quad \forall\, t \in [0, \pi].$$

In particular, for $t = \pi$ we have

$$\left[\frac{\theta_1(\pi)}{\pi}\right] = 2,$$

which implies that $k_1(\pi) = 2$. The right hand side of formula (28) is then 2, which is exactly $j(B)$, as computed in Example 3.6.

Remark 4.8. It is important to observe that in the case $N = 1$ the phase-angle $\theta = \theta_1$ given in Proposition 4.2 coincides with the usual Prüfer angular coordinate ϑ, which is defined by

$$\begin{cases} X(t) = \rho(t) \sin \vartheta(t) \\[2mm] Y(t) = \rho(t) \cos \vartheta(t), \end{cases} \tag{31}$$

where, as above, (X, Y) is the solution of (16) satisfying $(X(0), Y(0)) = (1, 0)$. Indeed, recalling that

$$\Theta(t) = \frac{Y(t) + iX(t)}{Y(t) - iX(t)} = \frac{Y^2(t) - X^2(t)}{X^2(t) + Y^2(t)} + 2i \frac{X(t)Y(t)}{X^2(t) + Y^2(t)} = e^{2i\theta(t)}, \quad \forall\, t \in [0, \pi],$$

a straightforward computation shows that $\theta = \vartheta$.

The consequence of this fact is that in the case $N = 1$ we can establish a relation between the number of moments of verticality (or the Maslov index) and the usual rotation number in the plane. Indeed, let us recall that for every nontrivial solution $z = (x, y)$ of (16) (with $N = 1$) we can define the rotation number rot_z by means of

$$\mathrm{rot}_z = \frac{1}{2\pi} \int_0^\pi \frac{x(t)y'(t) - x'(t)y(t)}{x(t)^2 + y(t)^2}\, dt;$$

moreover, when the condition $x(0) = 0$ is imposed, then rot_z does not depend on z. In [7, Chap. 2.6] it is proved that

$$2\text{rot}_z = -k_1(\pi) - \frac{\alpha_1(\pi)}{\pi} = -j(B) - \frac{\alpha_1(\pi)}{\pi};$$

therefore $j(B) \in \mathbb{N}$ is the integer satisfying the relation

$$-2\text{rot}_z - 1 \leq j(B) < -2\text{rot}_z.$$

Remark 4.9. In view of Remark 4.8 we are now able to show directly in the case $N = 1$ why the positive definiteness of $B(t)$ or $B_{22}(t)$, for every $t \in [0, \pi]$, implies the monotonicity of θ_1 in $[0, \pi]$ or at the moments of verticality, respectively (cf. statement (*iv*) in Proposition 4.2 and Proposition 4.5). The angle θ_1 coincides with the angular coordinate ϑ, which satisfies the well-known equation

$$\theta'(t) = B_{11}(t)\sin^2\theta(t) + 2B_{12}(t)\sin\theta(t)\cos\theta(t) + B_{22}(t)\cos^2\theta(t)$$

$$\tag{32}$$

$$= Q_{B(t)}(\sin\theta(t), \cos\theta(t)), \quad t \in [0, \pi],$$

where $Q_{B(t)}$ denotes the quadratic form associated to the symmetric matrix $B(t)$, for every $t \in [0, \pi]$. It is then immediate to see that the fact that $B(t)$ is positive definite for every $t \in [0, \pi]$ implies that θ is increasing in $[0, \pi]$.

On the other hand, from (32) we also deduce that

$$\theta'(\bar{t}) = B_{22}(\bar{t})\cos^2\theta(\bar{t}),$$

for every moment of verticality $\bar{t} \in [0, \pi]$; as a consequence, an assumption on the positive definiteness of $B_{22}(t)$, for every $t \in [0, \pi]$, implies that θ is increasing in a neighbourhood of any moment of verticality.

5   Some Related Notions

The index theory for linear Hamiltonian systems is an extremely wide topic. In this Section, we try to give a brief account on this subject, indicating some notions and questions related to the ones discussed before.

First of all, it is worth mentioning that, besides the approach in [44], there are other possibilities for the construction of the abstract Maslov index for paths of Lagrangian subspaces (see, for instance, [13, 19]). Each of them leads to a slightly different object, so that some attention is needed. See also [18].

Secondly, we notice that part of the discussion in Sects. 2–4 can be carried out in a more general (non-Hamiltonian) setting. In what follows we briefly discuss this possibility, following [9].

Consider a system of the form

$$Jz' + A(t)Jz = B(t)z, \quad z = (x, y) \in \mathbb{R}^{2N}, \tag{33}$$

where the matrix $B(t)$ is symmetric and $A(t)$ is "Hamiltonian-like", i.e. there exists $c(t) \in \mathbb{R}$ such that $(A(t)J)^t = A(t)J + c(t)J$. Notice that system (33) is not Hamiltonian, unless $c(t) \equiv 0$, which means that the matrix $A(t)$ is Hamiltonian, according to the standard definition. Moreover, no definiteness assumptions on $B(t)$ are imposed. Notice that the important Dirac-type systems we treat in Sect. 7

$$2q(t)Jz' + q'(t)Jz + P(t)z = S(t)z, \quad z = (x, y) \in \mathbb{R}^{2N},$$

being $q \in C^1([0, \pi])$ with $q(t) > 0$ and $P(t), S(t)$ $2N \times 2N$ symmetric matrices, are of the form (33) (cf. [5, 23, 54] and [6, 8, 51, 52]).

The first remark is that the fundamental matrix ψ is "symplectic-like", i.e. there exists $d(t) \neq 0$ such that

$$\psi(t)^t J \psi(t) = d(t)J, \quad \forall\, t. \tag{34}$$

This guarantees that Proposition 2.11 holds true and, as a consequence, that the Maslov index $\mu(\psi)$ can be defined as in Definition 2.12.

On the other hand (thinking of Sect. 4), one can show that (34) is sufficient to guarantee that the matrix $\Theta(t)$—defined as in Sect. 4 using $\psi(t)$—is unitary; this fact enables to define the phase-angles as the N continuous functions $\theta_1(t), \ldots, \theta_N(t)$ satisfying conditions $(i), (ii), (iii)$ of Proposition 4.2. It is important to remark that, due to the lack of definiteness assumptions on $B(t)$, the phase angles do not satisfy monotonicity properties (cf. (iv) of Proposition 4.2). Hence, even if the mere definition of a moment of verticality for (33) still makes sense, equality (28) is in general not available (it is not even possible to guarantee that the sum in Definition 3.2 is finite).

However, a relation between the Maslov index and the phase-angles still holds true in this more general setting. More precisely, with the same arguments as in Proposition 3.3 and Theorem 4.3, it is possible to see that

$$\dim(\psi(t)V \cap V) = \operatorname{card} M(t),$$

where $M(t) := \left\{ j \in \{1, \ldots, N\} \mid \theta_j(t) = 0 \bmod \pi \right\}$. Hence, $\bar{t}$ is a crossing for the pair $(\psi V, V)$ if and only if $M(\bar{t}) \neq \emptyset$. Moreover (see [14, Proposition 5] for details), if the crossing $\bar{t}$ is regular then the monotonicity of θ_j in a neighborhood of $\bar{t}$ is strictly related with the signature of $\Gamma(\psi V, V, \bar{t})$.

Having in mind Steps (II) and (III) in the Introduction, it turns out that a topological invariant defined through the phase-angles enables to develop a complete linear theory for (33). More precisely, in [9] it is proved the existence of a sequence of eigenvalues; moreover, the corresponding eigenfunctions are characterized by means of an index defined by

$$i(A, B) = \sum_{j=1}^{N} \left\lfloor \frac{\theta_j(\pi)}{\pi} \right\rfloor$$

where, for every $r \in \mathbb{R}$, we have set $\lfloor r \rfloor = m$ if $m \le r < m+1$ for $m \in \mathbb{Z}$ (see also [31] for a similar definition, in the Hamiltonian case). Notice that in the case $N = 1$ every linear first order system is of the form (33) and, as it is easily seen arguing as in Remark 4.8, it holds that $i(A, B) = \lfloor -2\mathrm{rot}_{z^*} \rfloor$, being $z^*(t) = (x^*(t), y^*(t))$ a nontrivial solution to (33) with $x^*(0) = 0$.

Another widely investigated topic is the index theory for linear Hamiltonian systems with periodic boundary conditions. A classical reference, in the context of stability problems, is the book by Yakubovich–Starzinski [55] (we refer, in particular to Sects. III.5.3 and III.5.4 of Volume 1). A more modern and very common tool in this framework is the *Conley-Zehnder index*, whose definition goes back to [17] (we refer to the books [1,33] for an exhaustive treatment of the subject). Here we limit ourselves to mention the fact that also this index can be expressed in terms of the abstract Maslov index for Lagrangian subspaces (see [44, Remark 5.4]); moreover, when the matrix $B(t)$ is positive definite (in case of convex Hamiltonian systems) a formula on the lines of (19) is still available (cf. [22,37]). We also recall that, in the case $N = 1$, a characterization of the Conley-Zehnder index in terms of the rotation number of the solutions in the plane has been given in [35].

We then touch on the relationship between the Maslov index and the Morse index. This question, which goes back to the Sturm oscillation theorem for second order differential equations and to the celebrated Morse index theorem in Riemannian geometry, is very deep and many efforts have been devoted to prove some general "Index Theorems", that is, results that relate Maslov-type and Morse type indices.

In the case of systems of second order equations like (23), it can be proved that the number of moments of verticality $j(B)$, being $B(t)$ as in (24), equals the Morse index of the origin as a critical point of the Lagrange functional

$$\varphi_A(X) = \frac{1}{2} \int_0^{\pi} \|X'(t)\|^2 \, dt - \int_0^{\pi} A(t)X(t)X(t) \, dt, \qquad \forall \, X \in H_0^1([0, \pi]).$$

Since formula (19) relates $j(B)$ to the Maslov index (cf. Remark 3.5), a relationship between the Maslov index and the Morse index is established. Such a result is essentially due to [19].

In the general case of a linear Hamiltonian system, a great problem appears since the natural variational formulation of $Jz' = B(t)z$ leads to a strongly indefinite quadratic functional (that is, the usual Morse index of the origin is always infinite). In this case, more sophisticated tools (dual variational techniques in the convex case, relative Morse indices, spectral flow) have been introduced by various authors. We refer, among others, to [15,25,40,45]. According to [1], the analogous question for periodic Hamiltonian system is even better understood.

Finally, we mention another notion of Maslov index which has been given (starting from the paper by Johnson–Moser [27]) by Johnson–Nerurkar in [28]. More recent improvements and developments can be found in the paper by Fabbri–Johnson–Nuñez [24] and references therein. In [28] it is considered a system of the form

$$Jz' = A(t)z + \lambda \Gamma(t)z, \ z \in \mathbb{R}^{2N}, \ t \in \mathbb{R}, \ \lambda \in \mathbb{R}, \tag{35}$$

where $A(t), \Gamma(t) \in M_S^2$, for every $t \in \mathbb{R}$. The fact that the system is defined on the real line, without any assumption of periodicity, implies that some difficulties in the definition of an index arise; the authors then define a topological invariant which can be considered as an average Maslov index.

The first step in the construction in [28] is to enter in the framework of topological dynamics [48] and to embed the original system (35) in a family of systems of the form

$$Jz' = a_y(t)z + \lambda \gamma(\tau_t(y))z, \ z \in \mathbb{R}^2, \ t \in \mathbb{R}, \ \lambda \in \mathbb{R}, \ y \in Y, \tag{36}$$

where Y is a compact translation invariant subset of $L_{\text{loc}}^p(\mathbb{R}, \mathscr{L}(\mathbb{R}^{2N})) \times C_b(\mathbb{R}, \mathscr{L}(\mathbb{R}^{2N}))$, $a_y(t)$ is symmetric, for every $t \in \mathbb{R}$ and $y \in Y$, $\gamma : Y \to \mathscr{L}(\mathbb{R}^{2N})$ is continuous, $\gamma(y)$ is symmetric and positive semi-definite, for every $y \in Y$, and τ_t is a flow on Y (by $C_b(\mathbb{R}, \mathscr{L}(\mathbb{R}^{2N}))$ we denote the set of matrices having bounded continuous entries).

This can be done in a standard way assuming for instance that the functions A and Γ in (35) are uniformly locally L^p-integrable, with $p \geq 1$, and uniformly bounded and continuous, respectively.

When (36) is considered, for every $\lambda \in \mathbb{R}$ and $y \in Y$ it is possible to define the Maslov index $\mu_{\lambda,y,t}(\psi)$, where ψ denotes, as usual, the fundamental matrix of the system (36); then, a suitable index is

$$\alpha(\lambda) = \lim_{T \to +\infty} \frac{\pi \mu_{\lambda,y,t}(\psi)}{t}, \tag{37}$$

whenever defined. For the existence, in a suitable sense, of the above limit it is necessary to consider system (36). More precisely, it is possible to show that, given an ergodic measure on Y, then there exists $Y_1 \subset Y$, whose complement has μ-measure zero, such that for every $y \in Y_1$ the limit in (37) exists and is independent on the choice of $y \in Y_1$.

We finally point out that in [28] the index defined in (37) is used in order to characterize the set of real numbers λ for which (36) has an exponential dichotomy.

6 Nonlinear First Order Systems in $\mathbb{R}^{2N}$

In this section we illustrate a result contained in [10] concerning a nonlinear eigenvalue problem associated to a first order system in $\mathbb{R}^{2N}$, $N \geq 2$. We adopt the bifurcation approach described in Sect. 2, using Theorem 1.1; as already observed, the crucial points are the study of the spectral properties of the linear operator given in the differential system (cf. Proposition 6.1) and the introduction of the topological invariant which will be preserved on the bifurcating branches.

In this case this invariant is the Maslov index previously described; a fundamental point is the study of its continuity (cf. Proposition 6.3).

Let us then consider the Dirichlet problem

$$
\begin{cases}
J z' = -\lambda S z + F(t, z, \lambda) z, \ t \in [0, \pi], \ z = (x, y) \in \mathbb{R}^{2N}, \ \lambda > 0, \\
x(0) = 0 = x(\pi)
\end{cases}
\tag{38}
$$

where $S \in M_S^{2N}$, $F(t, z, \lambda) \in M_S^{2N}$, for every $(t, z, \lambda) \in [0, \pi] \times \mathbb{R}^{2N} \times (0, +\infty)$, and $F(t, 0, \lambda) = 0$, for every $(t, \lambda) \in [0, \pi] \times (0, +\infty)$. As usual, we denote by Σ the closure of the set of nontrivial solutions to (38).

In order to study the linear problem

$$
\begin{cases}
J z' = -\lambda S z, \\
x(0) = 0 = x(\pi),
\end{cases}
\tag{39}
$$

we introduce some assumptions on S. First of all, let us denote by $\lambda_1, \ldots, \lambda_{2N}$ the (real) eigenvalues of $S \subset M_S^{2N}$ and let $D = \mathrm{diag}(\lambda_1, \ldots, \lambda_{2N})$; moreover, let P be an orthogonal matrix such that

$$
P^T S P = D.
$$

We denote by $\mathscr{S}$ the class of constant matrices $S \in M_S^{2N}$ such that:

1. For every $i \neq j, i, j = 1, \ldots, N$, we have

$$
\frac{\sqrt{\lambda_i \lambda_{N+i}}}{\sqrt{\lambda_j \lambda_{N+j}}} \notin \mathbb{Q}.
$$

2. The matrix P has the form

$$
P = \begin{pmatrix} P_{11} & P_{12} \\ -P_{12} & P_{11} \end{pmatrix}
\tag{40}
$$

and P_{11} is invertible.
3. The matrix $Q = -P_{12} P_{11}^{-1}$ is diagonal.

We remark that it would be possible to define a different class $\mathscr{S}$, suitable for the proof of our results, by requiring that P_{12} is invertible and by replacing the matrix Q with the matrix $Q' = -P_{11} P_{12}^{-1}$.

We recall that every symplectic orthogonal matrix P can we written in the form (40); hence, starting from a symplectic ortogonal matrix satisfying condition 3. it is possibile to construct matrices $S \in \mathscr{S}$. Indeed, let us for instance take $P_{11} \in M_S^N$ such that $P_{11}^2 = $ Id and set $P_{12} = P_{11}$. Then, for every diagonal matrix $D = \mathrm{diag}(\lambda_1, \ldots, \lambda_{2N})$ whose eigenvalues satisfy condition 1, the matrix $S = P \, D \, P^T$ belongs to the class $\mathscr{S}$.

When $S \in \mathscr{S}$ the behaviour of the linear problem (39) is completely known. Indeed, we have the following result:

Proposition 6.1. *Assume that $S \in \mathscr{S}$. Then the eigenvalues of the linear problem* (39) *are*

$$\lambda_{j,k} = \frac{k}{\sqrt{\lambda_j \lambda_{N+j}}}, \quad \forall \, j = 1, \ldots, N, \, k \in \mathbb{Z}.$$

Moreover, for every $k \neq 0$ and for every $j = 1, \ldots, N, \, \lambda_{j,k}$ is simple.

We observe that, according to the previous sections, it is possible to associate to every eigenvalue $\lambda_{j,k}$ the number of moments of verticality of the corresponding linear system. By rearranging the eigenvalues, we obtain the following result.

Theorem 6.2. *Assume that $S \in \mathscr{S}$. Then* (39) *has a double sequence of simple eigenvalues μ_k such that*

$$\mu_k \to \pm \infty, \quad as \quad k \to \pm \infty.$$

Moreover, for every $k \in \mathbb{Z}, k \neq 0$,

$$j(-\mu_k S) = |k| - 1.$$

Let us now focus on the notion of index of a solution to the initial nonlinear problem (38); consider a solution (λ, z) of (38) and let

$$S_{\lambda,z}(t) = -\lambda S + F(t, z(t), \lambda), \quad \forall \, t \in [0, \pi].$$

We then define

$$\Phi(\lambda, z) = j(-S_{\lambda,z}).$$

The continuity of Φ, which plays a crucial role in the bifurcation argument, is a consequence of the fact that the moments of verticality of (39) are simple when $S \in \mathscr{S}$. Indeed, it is possible to prove the following result:

Proposition 6.3. *[10, Prop. 2.2-Lemma 3.1] There exists $\xi : \mathbb{R} \setminus \{0\} \to \mathbb{R}$ such that if*

$$\|F(t, z, \lambda)\| \leq \xi(\lambda), \quad \forall \, (t, z, \lambda) \in [0, \pi] \times \mathbb{R}^{2N} \times \mathbb{R} \setminus \{0\} \tag{41}$$

then Φ is continuous in Σ.

In view of the results stated in this Section, the existence of bifurcating branches of nontrivial solutions of (38) can be proved:

Theorem 6.4. *Assume that $S \in \mathscr{S}$ and $F(t, 0, \lambda) = 0$, for every $t \in [0, \pi]$, $\lambda > 0$. Moreover, suppose that F satisfies (41). Then, for every $k \in \mathbb{N}$, $k \neq 0$, Σ contains a continuum C_k such that $(\mu_k, 0) \in C_k$, there exists $(\lambda_n, u_n) \in C_k$ such that*

$$\Phi(\lambda, u) = k - 1, \quad \text{for every } (\lambda, u) \in C_k$$

and

$$\lambda_n + ||u_n|| \to +\infty \quad \text{or } \lambda_n \to 0^+, \ n \to +\infty,$$

where $|| \cdot ||$ is the sup-norm in the space $C^0([0, \pi], \mathbb{R}^{2N})$

Remark 6.5. In the case $N = 1$ the conclusion of Theorem 6.4 still holds, assuming $\det S > 0$ and $F(t, 0, \lambda) = 0$, for every $t \in [0, \pi]$, $\lambda > 0$. Moreover, in this situation a multiplicity result (Theorem 3.10 in [10]) can be proved. More precisely, a suitable growth condition at infinity on F enables to give a priori bounds on the solutions with a fixed Maslov index.

7 Nonlinear Dirac-Type Systems in the Half-Line

In this section we present a result contained in [11] which deals with the existence of solutions to a nonlinear planar Dirac-type system in the half-line. The method of proof follows the same lines of the one used in Sect. 6; in this situation the study of the spectral theory of the linear Dirac operator is the main difficulty. Indeed, it is necessary to use some functional analysis results [53], together with the classical Levinson theorem [21].

Let us consider the planar system

$$Jz' + P(t)z = \lambda z + Q(t, z)z, \quad t \in [1, +\infty), \quad \lambda \in \mathbb{R}, \quad z = (u, v) \in \mathbb{R}^2, \quad (42)$$

where $P : [1, +\infty) \to M_S^2$ and $Q : [1, +\infty) \times \mathbb{R}^2 \to M_S^2$ are continuous functions.

When $Q \equiv 0$ and $t \in [a, b]$, for some bounded interval $[a, b] \subset \mathbb{R}$, then (42) is of the form (33); therefore, it is possible to use one of the topological invariants introduced in Sect. 5 to characterize the solutions. However, the major interest in the study of this kind of system is the case when the equations are defined in an unbounded interval; unfortunately, in this situation a complete index theory is not available.

In order to overcome this problem, we shall take advantage of the fact that (42) is a planar system and we consider the usual angular coordinate in order to define an index associated to nontrivial solutions.

First of all, let us observe that we are interested in nontrivial solutions $z = (u, v)$ of (42) belonging to the space

$$D_0 = \{z \in H^1(1, +\infty) : v(1) = 0\}; \tag{43}$$

the choice of this space is strictly related to the properties of the linear operator τ formally defined by

$$\tau z = J z' + P(t) z \tag{44}$$

(see the discussion following Theorem 7.1).

We denote by $\mathscr{P}$ the class of continuous maps $P : [1, +\infty) \longrightarrow M_S^2$ such that

$$\lim_{t \to +\infty} P(t) = \begin{pmatrix} \mu^- & 0 \\ & \mu^+ \end{pmatrix} := P_0, \tag{45}$$

for some $\mu^- < \mu^+$, and there exists $q \geq 1$ such that

$$\int_1^{+\infty} |R(t)|^q \, dt < +\infty,$$

where $R(t) = P(t) - P_0$, for every $t \geq 1$. Moreover, we denote by $\mathscr{Q}$ the set of continuous functions $Q : [1, +\infty) \times \mathbb{R}^2 \longrightarrow M_S^2$ satisfying the conditions

1. There exist $\alpha \in L^\infty(1, +\infty)$, $\eta_1, \eta_2, \eta_{12} : \mathbb{R}^2 \longrightarrow \mathbb{R}$, continuous and with $\eta_1(0) = \eta_2(0) = \eta_{12}(0) = 0$, and $p \geq 1$ for which

$$|Q_i(t, z)| \leq \alpha(t) \eta_i(z), \quad \forall t \geq 1, \quad z \in \mathbb{R}^2, \ i = 1, 2,$$
$$\tag{46}$$
$$|Q_{12}(t, z)| \leq \alpha(t) \eta_{12}(z), \quad \forall t \geq 1, \quad z \in \mathbb{R}^2;$$

2. For every compact $K \subset \mathbb{R}^2$ there exists $A_K > 0$ such that

$$\|Q(t, z) - Q(t, z')\| \leq A_K \|z - z'\|, \quad \forall t \geq 1, \quad z, z' \in K. \tag{47}$$

In order to use the approach described in Sect. 1, we first need an abstract bifurcation theorem suitable for the application to (42). We observe that in this situation there is a lack of compactness, due to the fact that the planar system is defined in the unbounded interval $[1, +\infty)$; as a consequence, Theorem 1.1 cannot be applied and a different result is needed. It turns out that a slight variant of a bifurcation theorem of Stuart [49] is suitable for our application. Let us briefly describe the abstract framework required.

Consider a real Hilbert space B and let $A_0 : D(A_0) \longrightarrow B$ be an unbounded self-adjoint operator in B with

$$\sigma_{ess}(A_0) = (-\infty, \mu^-] \cup [\mu^+, +\infty), \tag{48}$$

for some $\mu^- < \mu^+$. Let H denote the real Hilbert space obtained from the domain of A_0 equipped with the graph topology and let us consider the nonlinear problem

$$A_0 u + M(u) = \lambda u, \quad (\lambda, u) \in \mathbb{R} \times H, \tag{49}$$

where $M : H \longrightarrow B$ is a continuous and compact map such that

$$M(u) = o(\|u\|_H), \quad u \to 0.$$

Finally, let Σ be the closure of nontrivial solutions to (49). Hence, the following result holds true:

Theorem 7.1. *Let $\mu \in (\mu^-, \mu^+)$ be an eigenvalue of A_0 of odd multiplicity and let C_μ denote the component of Σ containing $(\mu, 0)$. Then, C_μ has one of the following properties:*

(1) C_μ is unbounded in $(\mu^-, \mu^+) \times H$.
(2) $\sup\{\lambda : (\lambda, u) \in C_\mu\} \geq \mu^+$ or $\inf\{\lambda : (\lambda, u) \in C_\mu\} \leq \mu^-$.
(3) C_μ contains an element $(\mu^*, 0) \in \Sigma$ with $\mu^* \neq \mu$.

The original version of Stuart deals with a linear operator A_0 satisfying the condition

$$\sigma_{ess}(A_0) = [\mu^*, +\infty),$$

for some $\mu^* \in \mathbb{R}$, instead of (48). The situation when the essential spectrum of A_0 is bounded from below arises e.g. in the study of the one-dimensional Schrödinger operators, which was the subject of the paper [49].

In our situation, B is the space $L^2(1, +\infty)$; moreover, it is possible to show that the operator ι defined by (44) has a self-adjoint extension A_0 with

$$D(A_0) = D_0$$

(see (43)) and such that (48) holds true, with $\mu^\pm$ given in (45).

The nonlinear term M is, as usual, the Nemitskii operator associated to Q, given by

$$M(u)(t) = Q(t, u(t))u(t), \quad \forall\, t \geq 1,$$

for every $u \in D_0$. The following result illustrates some assumptions on Q which ensure that M has the properties required in Theorem 7.1.

Proposition 7.2. *Assume that $Q \in \mathcal{Q}$ and that*

$$\lim_{t \to +\infty} \alpha(t) = 0, \tag{50}$$

where α is given in (46). Then $M : D_0 \longrightarrow L^2(1, +\infty)$ is a continuous compact map and satifies $M(u) = o(\|u\|_{D_0}), u \to 0$.

Two problems are now in order: the existence of (simple) eigenvalues of A_0 in $\Lambda := (\mu^-, \mu^+)$ and the definition (and continuity) of an index associated to solutions to (42).

As far as the first problem is concerned, it is possible to show that the existence of eigenvalues of A_0 (and their number) in Λ is related to the oscillatory behaviour of the linear problem

$$\begin{cases} Jz' + P(t)z = \lambda z, \ z = (u, v) \\ v(1) = 0 \end{cases} \tag{51}$$

when $\lambda = \mu^\pm$. This fact is well known in the case of the second-order differential operator (see e.g. [20, Theorems 53–55]). In our context, if we assume that the coefficients P_{11} and P_{22} of P are increasing in $[1, +\infty)$, then if (51) is oscillatory (i.e. the number of rotations of the nontrivial solutions is infinite) an infinite sequence of eigenvalues accumulating to μ^+ does exist; on the contrary, when the solutions of (51) for $\lambda = \mu^+$ have a finite number of rotations in the phase-plane, then only finitely many eigenvalues fall in (μ^-, μ^+).

In the case of a system of the form (42) where τ is the radial Dirac operator with a Coulomb-like potential, i.e. where P has the form

$$P(t) = \begin{pmatrix} -1 + V(t) & k/t \\ k/t & 1 + V(t) \end{pmatrix}, \quad k \in \mathbb{N},$$

then the following result holds true:

Proposition 7.3. *Assume that* $V \in C(1, +\infty)$ *is a strictly increasing negative potential such that*

$$V(t) \sim \frac{c}{t^\alpha}, \quad t \to +\infty,$$

with $\alpha \in (0, 1]$. *Then,* A_0 *has a sequence of simple eigenvalues in* $(-1, 1)$ *converging to* 1.

Observe that Dirac operators of the above form (which are the classical Dirac operators when $V(t) = c/t$) have been considered, among others, in [46]. In this paper, the authors study the eigenvalue problem on $(0, +\infty)$, and, as a consequence, the operator τ is singular also at $x = 0$. More precisely, in [46] it is proved a result similar to Proposition 7.3 (under more restrictive conditions on α) in the case when V is singular at zero; however, no information on the nodal properties of the eigenfunctions is provided.

Let us now focus on the definition of the index; we start considering solutions of the linear problem (51). Let us fix a nontrivial solution (λ, z) of (51) and write $z = (u, v)$ in polar coordinates as $u = \rho \cos \theta, v = \rho \sin \theta$; the angular coordinate θ is normalized in such a way that

$$\theta(1, \lambda) = 0, \quad \forall \lambda \in \Lambda,$$

(according to the boundary condition $v(1) = 0$). Let us observe that we cannot a priori ensure that $\theta(\cdot, \lambda)$ is bounded, since we are considering an unbounded interval $[1, +\infty)$; nevertheless, the fact that $\lambda \in \Lambda$ guarantees that there exist $t_\lambda \geq 1$ and $k_\lambda \in \mathbb{Z}$ such that

$$k_\lambda \pi \leq \theta(t, \lambda) < (k_\lambda + 1)\pi, \quad \forall t \geq t_\lambda.$$

As a consequence, we can give the following definition:

Definition 7.4. Assume that $P \in \mathscr{P}$, $\lambda \in \Lambda$ and let (λ, z) be a nontrivial solution of (51). We define

$$i(\lambda, z) = \left[\frac{\theta(t_\lambda, \lambda)}{\pi} \right].$$

From the properties of $\theta(t_\lambda, \lambda)$ it follows that the above defined index is independent of z. Indeed, it is related to $j(\lambda \mathrm{Id} - P)$ in the interval $[1, t_\lambda]$.

When the nonlinear system (42), together with the boundary condition $v(1) = 0$, is considered, we can define the index of a solution by means of a standard linearization procedure. Indeed, we have the following:

Definition 7.5. Assume that $P \in \mathscr{P}$ and $Q \in \mathscr{Q}$ and let (μ, w) be a solution of (42), with $w \in D_0$.

If $(\mu, w) \neq (\mu, 0)$, then the index of (μ, w) is defined as the index $i(\mu, w)$ of (μ, w) as a solution of the linear problem

$$\begin{cases} Jz' + P(t)z = \mu z + Q(t, w(t))z, \\ \\ v(1) = 0. \end{cases}$$

If $(\mu, w) = (\mu, 0)$ and the linear problem

$$\begin{cases} Jz' + P(t)z = \mu z, \\ \\ v(1) = 0 \end{cases}$$

has a nontrivial solution $z_\mu \in D_0$, then the index of (μ, w) is defined as the index $i(\mu, z_\mu)$ of (μ, z_μ) as a solution of the above problem.

According to the discussion of Sect. 1, the possibility of excluding (3) in Theorem 7.1 relies on some continuity property of the index i defined above and on the fact that indeces associated to different eigenfunctions of the linear problem are different. In our situation continuity can be proved by carefully studying the properties of θ, while in general it is not possible to guarantee that different eigenfunctions have different indeces. Indeed, we have the following:

Proposition 7.6. *Let* $\{\lambda_k\}_{k \in K}$ *be the set of simple eigenvalues of* A_0 *in* (μ^-, μ^+), *for some* $K \subset \mathbb{N}$. *Then, there exist at most two indeces* k_1 *and* $k_2 \in K$ *such that*

$$i(\lambda_{k_1}) = i(\lambda_{k_2}) \ \text{ and } \ i(\lambda_j) \neq i(\lambda_m), \quad \forall \ j \neq m, \ j, m \in K \setminus \{k_1, k_2\},$$

where $i(\lambda_k)$ *denotes the index of* $(\lambda_k, z_{\lambda_k})$, *being* z_{λ_k} *an eigenfunction associated to* λ_k.

The consequence of Proposition 7.6 is that for branches bifurcating from $(\lambda_j, 0)$, with $j \neq k_1$ and $j \neq k_2$, condition (3) in Theorem 7.1 cannot hold; however, we cannot exclude that the two branches bifurcating from $(\lambda_{k_1}, 0)$ and $(\lambda_{k_2}, 0)$ coincide.

We are now in position to state a global bifurcation result for the nonlinear problem (42).

Theorem 7.7. *Assume that* $P \in \mathscr{P}$, $Q \in \mathscr{Q}$ *and* (50) *holds true. Then for every* $k \in K \setminus \{k_1, k_2\}$ *there exists a continuum* C_k *of nontrivial solutions of* (42) *in* $(\mu^-, \mu^+) \times D_0$ *bifurcating from* $(\lambda_k, 0)$ *and such that*

(1) C_k *is unbounded in* $(\mu^-, \mu^+) \times D_0$, *or*
(2) $\sup\{\lambda : \ (\lambda, u) \in C_k\} \geq \mu^+$ *or* $\inf\{\lambda : \ (\lambda, u) \in C_k\} \leq \mu^-$.

Moreover, we have

$$i(\lambda, z) = i(\lambda_k), \quad \forall \ (\lambda, z) \in C_k.$$

Acknowledgements We wish to thank J. Pejsachowicz for introducing us to the study of the Maslov index and for many fruitful conversations. The second author wishes to thank the C.I.M.E. foundation and the course directors R. Johnson and M.P. Pera for the kind invitation to deliver these lectures.

References

1. A. Abbondandolo, *Morse Theory for Hamiltonian systems*. Research Notes in Mathematics (Chapman & Hall, CRC, Boca Raton, 2001)
2. V.I. Arnold, On a characteristic class entering in a quantum condition. Funct. Anal. Appl. **1**, 1–14 (1967)
3. F.V. Atkinson, *Discrete and Continuous Boundary Problems* (Academic Press, London, 1964)
4. M. Audin, A. Cannas da Silva, E. Lerman, in *Symplectic Geometry of Integrable Hamiltonian Systems*. Lectures delivered at the Euro Summer School held in Barcelona, July 10–15, 2001. Advanced Courses in Mathematics. CRM Barcelona (Birkhäuser, Basel, 2003)
5. M. Balabane, T. Cazenave, L. Vázquez, Existence of standing waves for Dirac fields with singular nonlinearities. Comm. Math. Phys. **133**, 53–74 (1990)
6. C. Bereanu, On a multiplicity result of J. R. Ward for superlinear planar systems. Topological Meth. Nonlinear Anal. **27**, 289–298 (2006)
7. A. Boscaggin, Global bifurcation and topological invariants for nonlinear boundary value problems, Thesis, University of Torino, 2008

8. A. Boscaggin, A. Capietto, Infinitely many solutions to superquadratic planar Dirac-type systems. Discrete Contin. Dynam. Syst. Differential Equations and Applications. 7th AIMS Conference, Supplement 72–81 (2009)
9. A. Boscaggin, M. Garrione, A note on a linear spectral theorem for a class of first order systems in $\mathbb{R}^{2N}$. Electron. J. Qual. Theor. Differ. Equat. **75**, 1–22 (2010)
10. A. Capietto, W. Dambrosio, Preservation of the Maslov index along bifurcating branches of solutions of first order systems in $\mathbb{R}^N$. J. Differ. Equat. **227**, 692–713 (2006)
11. A. Capietto, W. Dambrosio, Planar Dirac-type systems: the eigenvalue problem and a global bifurcation result. J. Lond. Math. Soc. **81**, 477–498 (2010)
12. A. Capietto, W. Dambrosio, D. Papini, Global bifurcation for singular planar Dirac-type systems. Submitted
13. S.E. Cappell, R. Lee, E.Y. Miller, On the Maslov index. Comm. Pure Appl. Math. **47**, 121–186 (1994)
14. F. Chardard, Stabilité des ondes solitaires, Ph. D. Thesis, Ecole Normale Supérieure de Cachan, 2009
15. C.N. Chen, X. Hu, Maslov index for homoclinic orbits of Hamiltonian systems. Ann. Inst. H. Poincaré Anal. Non Linéaire **24**, 589–603 (2007)
16. C. Conley, An oscillation theorem for linear systems with more than one degree of freedom, IBM Technical Report 18004, IBM Watson Research Center, New York, 1972
17. C. Conley, E. Zehnder, Morse-type index theory for flows and periodic solutions for Hamiltonian equations. Comm. Pure Appl. Math. **37**, 207–253 (1984)
18. Y. Dong, Maslov type index theory for linear Hamiltonian systems with Bolza boundary value conditions and multiple solutions for nonlinear Hamiltonian systems. Pac. J. Math. **221**, 253–280 (2005)
19. J.J. Duistermaat, On the Morse index in variational calculus. Adv. Math. **21**, 173–195 (1976)
20. N. Dunford, J. Schwartz, *Linear Operators - Part II: spectral theory* Interscience Publishers, New York, 1963
21. M.S.P. Eastham, *The Asymptotic Solution of Linear Differential Systems*. London Mathematical Society Monographs. New Series, 4. Oxford Science Publications. The Clarendon Press, Oxford University Press, New York, 1989
22. I. Ekeland, *Convexity Methods in Hamiltonian Mechanics* (Springer, Berlin, 1990)
23. M.J. Esteban, An overview on linear and nonlinear Dirac equations. Discrete Contin. Dynam. Syst. **8**, 381–397 (2002)
24. R. Fabbri, R. Johnson, C. Nunez, Rotation number for non-autonomous linear Hamiltonian systems. I: Basic properties. Z. Angew. Math. Phys. **54**, 484–501 (2003)
25. M. Fitzpatrick, J. Pejsachowicz, C. Stuart, Spectral flow for paths of unbounded operators and bifurcation of critical points (in preparation)
26. L. Greenberg, A Prüfer method for calculating eigenvalues of self-adjoint systems of ordinary differential equations. Part I. Technical Report, Department of Mathematics, University of Maryland, 1991. Available from the authors
27. R. Johnson, J. Moser, The rotation number for almost periodic potentials. Comm. Math. Phys. **84**, 403–438 (1982)
28. R. Johnson, M. Nerurkar, Exponential dichotomy and rotation number for linear hamiltonian systems. J. Differ. Equat. **108**, 201–216 (1994)
29. T. Kato, *Perturbation Theory for Linear Operators* (Springer, Berlin, 1995)
30. J. Kellendonk, S. Richard, The topological meaning of Levinson's theorem, half-bound states included. J. Phys. A **41**, 7 (2008)
31. C.G. Liu, Maslov-type index theory for symplectic paths with Lagrangian boundry conditions. Adv. Nonlinear Stud. **7**, 131–161 (2007)
32. V.B. Lidskiy, Oscillation theorems for canonical systems of differential equations. NASA Technical Translation **TT-F-14696** (1973). (Russian) Dokl. Akad. Nauk SSSR (N.S.) **102**, 877–880 (1955)
33. Y. Long, *Index Theory for Symplectic Paths with Applications* (Birkhäuser, Basel, 2002)
34. Z.-Q. Ma, The Levinson theorem. J. Phys. A **39**, 625–659 (2006)

35. A. Margheri, C. Rebelo, F. Zanolin, Maslov index, Poincaré-Birkhoff theorem and periodic solutions of asymptotically linear planar Hamiltonian systems. J. Differ. Equat. **183**, 342–367 (2002)

36. V.P. Maslov, *Théorie des perturbations et méthodes asymptotiques* (Dunod, Paris, 1972)

37. J. Mawhin, M. Willem, *Critical Point Theory and Hamiltonian Systems* (Springer, New York, 1989)

38. D. McDuff, D. Salamon, *Introduction to Symplectic Topology* (Oxford University Press, New York, 1998)

39. M. Musso, J. Pejsachowicz, A. Portaluri, Morse index and bifurcation of p-geodesics on semi Riemannian manifolds. ESAIM Contr. Optim. Calc. Var. **13**, 598–621 (2007)

40. P. Piccione, D.V. Tausk, An index theorem for non-periodic solutions of Hamiltonian systems. Proc. Lond. Math. Soc. (3) **83**, 351–389 (2001)

41. A. Portaluri, Maslov index for Hamiltonian systems. Electron. J. Differ. Equat. **237**, 85–124 (2001)

42. P.J. Rabier, C. Stuart, Global bifurcation for quasilinear elliptic equations on $\mathbb{R}^N$. Math. Z. **237**, 85–124 (2001)

43. P.H. Rabinowitz, Some aspects of nonlinear eigenvalue problems. Rocky Mt. J. Math. **3**, 161–202 (1973)

44. J. Robbin, D. Salamon, The Maslov index for paths. Topology **32**, 827–844 (1993)

45. J. Robbin, D. Salamon, The spectral flow and the Maslov index. Bull. Lond. Math. Soc. **27**, 1–33 (1995)

46. H. Schmid, C. Tretter, Singular Dirac systems and Sturm-Liouville problems nonlinear in the spectral parameter. J. Differ. Equat. **181**, 511–542 (2002)

47. S. Secchi, C. Stuart, Global bifurcation of homoclinic solutions of Hamiltonian systems. Discrete Contin. Dynam. Syst. **9**, 1493–1518 (2003)

48. G.R. Sell, *Topological Dynamics and Ordinary Differential Equations* (Van Nostrand Reinhold Co., London, 1971)

49. Stuart C., Global properties of components of solutions of non-linear second order differential equations on the half-line. Ann. Scuola Norm. Sup. Pisa Cl. Sci. (4) **2**, 265–286 (1975)

50. B. Thaller, *The Dirac Equation* (Springer, Berlin, 1992)

51. J. Ward Jr., Rotation numbers and global bifurcation in systems of ordinary differential equations. Adv. Nonlinear Stud. **5**, 375–392 (2005)

52. J. Ward Jr., Existence, multiplicity, and bifurcation in systems of ordinary differential equations. Electron. J. Differ. Equat. Conf. **15**, 399–415 (2007)

53. J. Weidmann, *Linear Operators in Hilbert Spaces* (Springer, New York, 1980)

54. J. Weidmann, *Spectral Theory of Ordinary Differential Equations*. Lectures Notes in Mathematics, 1258 Springer-Verlag, Berlin (1987)

55. V.A. Yakubovich, V.M. Starzhinskii, *Linear Differential Equations with Periodic Coefficients* (Wiley, New York, 1975)

Discrete-Time Nonautonomous Dynamical Systems

P.E. Kloeden, C. Pötzsche, and M. Rasmussen

Abstract These notes present and discuss various aspects of the recent theory for time-dependent difference equations giving rise to nonautonomous dynamical systems on general metric spaces:

First, basic concepts of autonomous difference equations and discrete-time (semi-) dynamical systems are reviewed for later contrast in the nonautonomous case. Then time-dependent difference equations or discrete-time nonautonomous dynamical systems are formulated as processes and as skew products. Their attractors including invariants sets, entire solutions, as well as the concepts of pullback attraction and pullback absorbing sets are introduced for both formulations. In particular, the limitations of pullback attractors for processes is highlighted. Beyond that Lyapunov functions for pullback attractors are discussed.

Two bifurcation concepts for nonautonomous difference equations will be introduced, namely attractor and solution bifurcations.

Finally, random difference equations and discrete-time random dynamical systems are investigated using random attractors and invariant measures.

P.E. Kloeden (✉)
Institut für Mathematik, Goethe-Universität, Postfach 11 19 32, 60054 Frankfurt a.M., Germany
e-mail: kloeden@math.uni-frankfurt.de

C. Pötzsche
Institut für Mathematik, Universität Klagenfurt, Universitätsstraße 65–67, 9020 Klagenfurt, Austria
e-mail: christian.poetzsche@aau.at

M. Rasmussen
Department of Mathematics, Imperial College, London SW7 2AZ, UK
e-mail: m.rasmussen@imperial.ac.uk

A. Capietto et al., *Stability and Bifurcation Theory for Non-Autonomous Differential Equations*, Lecture Notes in Mathematics 2065, DOI 10.1007/978-3-642-32906-7_2, © Springer-Verlag Berlin Heidelberg 2013

1 Introduction

The qualitative theory of dynamical systems has seen an enormous development since the groundbreaking contributions of Poincaré and Lyapunov over a century ago. Meanwhile it provides a successful framework to describe and understand a large variety of phenomena in areas as diverse as physics, life science, engineering or sociology.

Such a success benefits, in part, from the fact that the law of evolution in various problems from the above areas is static and does not change with time (or chance). Thus a description with autonomous evolutionary equations is appropriate. Nevertheless, many real world problems involve time-dependent parameters and, furthermore, one wants to understand control, modulation or other effects. In doing so, periodically or almost periodically driven systems are special cases, but, in principle, a theory for arbitrary time-dependence is desirable. This led to the observation that many of the meanwhile well-established concepts, methods and results for autonomous systems are not applicable and require an appropriate extension—the theory of *nonautonomous dynamical systems*.

The goal of these notes is to give a solid foundation to describe the long-term behaviour of nonautonomous evolutionary equations. Here we restrict to the discrete-time case in form of nonautonomous difference equations. This has the didactical advantage to feature many aspects of infinite-dimensional continuous-time problems (namely nonexistence and uniqueness of backward solutions) without an involved theory to guarantee the existence of a semiflow. Moreover, even in low dimensions, discrete dynamics can be quite complex.

Beyond that a time-discrete theory is strongly motivated from applications e.g., in population biology. In addition, it serves as a basic tool to understand numerical temporal discretization and is often essential for the analysis of continuous-time problems thorough concepts like time-1- or Poincaré mappings.

The focus of our presentation is on two formulations of time-discrete non-autonomous dynamical systems, namely processes (two-parameter semigroups) and skew-product systems. For both we construct, discuss and compare the so-called pullback attractor in Chaps. 4–6. A pullback attractor serves as nonautonomous counterpart to the global attractor, i.e., the object capturing the essential dynamics of a system. Furthermore, in Chap. 7 we sketch two approaches to a bifurcation theory for time-dependent problems to illuminate a current field of research. The final Chap. 8 on random dynamical systems emphasises similarities to the corresponding nonautonomous theory and provides results on random Markov chains and the approximation of invariant measures.

To conclude this introduction we point out that a significantly more comprehensive approach is given in the up-coming monograph [25] (see also the lecture notes [38, 44]). In particular, we neglect various contributions to the discrete-time nonautonomous theory: An appropriate spectral notion for linear difference equations (cf. [6, 35, 46, 47]) substitutes the dynamical role of eigenvalues from the autonomous special case. Gaps in this spectrum enable to construct nonautonomous invariant manifolds (so-called invariant fiber bundles, see [5, 42]). As special case they include centre fiber bundles and therefore allow one to deduce a time-dependent version of Pliss's reduction principle [33, 41]. The pullback attractors constructed in these notes are, generally, only upper semi-continuous in parameters. Thus, for approximation purposes it might be advantageous to embed them into a more robust dynamical object, namely a discrete counterpart to an inertial manifold [34]. Topological linearization of nonautonomous difference equations has been addressed in [7, 8], while a smooth linearization theory via normal forms was developed in [50].

2 Autonomous Difference Equations

A difference equation of the form

$$x_{n+1} = f(x_n),$$ (1)

where $f : \mathbb{R}^d \to \mathbb{R}^d$, is called a first-order *autonomous difference equation* on the state space $\mathbb{R}^d$. There is no loss of generality in the restriction to first-order difference equations (1), since higher-order difference equations can be reformulated as (1) by the use of an appropriate higher dimensional state space.

Successive iteration of an autonomous difference equation (1) generates the forwards solution mapping $\pi : \mathbb{Z}^+ \times \mathbb{R}^d \to \mathbb{R}^d$ defined by

$$x_n = \pi(n, x_0) = f^n(x_0) := \underbrace{f \circ f \circ \cdots \circ f}_{n \text{ times}}(x_0),$$

which satisfies the *initial condition* $\pi(0, x_0) = x_0$ and the *semigroup property*

$$\pi(n, \pi(m, x_0)) = f^n(\pi(m, x_0)) = f^n \circ f^m(x_0) = f^{n+m}(x_0)$$
$$= \pi(n + m, x_0) \quad \text{for all } n, m \in \mathbb{Z}^+, x_0 \in \mathbb{R}^d.$$ (2)

Here, and later,

$$\mathbb{Z}^+ := \{0, 1, 2, 3, \ldots\}, \qquad \mathbb{Z}^- := \{\ldots, -3, -2, -1, 0\}$$

denote the nonnegative and nonpositive integers, respectively, and a *discrete interval* is the intersection of a real interval with the set of integers $\mathbb{Z}$.

Property (2) says that the solution mapping π forms a semigroup under composition; it is typically only a semigroup rather than a group since the mapping f need not be invertible. It will be assumed here that the mapping f in the difference

Fig. 1 Semigroup property
(ii) of a discrete-time
semidynamical system
$\pi : \mathbb{Z}^+ \times X \to X$

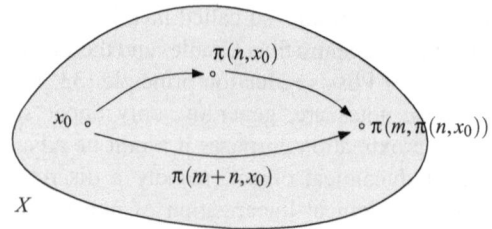

equation (1) is at least continuous, from which it follows that the mappings $\pi(n, \cdot)$ are continuous for every $n \in \mathbb{Z}^+$. The solution mapping π then generates a discrete-time *semidynamical system* on $\mathbb{R}^d$.

More generally, the state space could be a metric space (X, d).

Definition 2.1. A mapping $\pi : \mathbb{Z}^+ \times X \to X$ satisfying

(i) $\pi(0, x_0) = x_0$ for all $x_0 \in X$,
(ii) $\pi(m + n, x_0) = \pi(m, \pi(n, x_0))$ for all $m, n \in \mathbb{Z}^+$ and $x_0 \in X$,
(iii) The mapping $x_0 \mapsto \pi(n, x_0)$ is continuous for each $n \in \mathbb{Z}^+$,

is called a (discrete-time) *autonomous semidynamical system* or a *semigroup* on the *state space* X.

The semigroup property (ii) is illustrated in Fig. 1 below. Note that such an autonomous semidynamical system π on X is equivalent to a first-order autonomous difference equation on X with the right-hand side f defined by $f(x) := \pi(1, x)$ for all $x \in X$.

If $\mathbb{Z}^+$ in Definition 2.1 is replaced by $\mathbb{Z}$, then π is called a (discrete-time) autonomous *dynamical system* or *group* on the state space X. See [10, 49]

Autonomous dynamical systems need not be generated by autonomous difference equations as above.

Example 2.2. Consider the space $X = \{1, \cdots, r\}^{\mathbb{Z}}$ of bi-infinite sequences $x = \{k_n\}_{n \in \mathbb{Z}}$ with $k_n \in \{1, \cdots, r\}$ w.r.t. the group of left shift operators $\theta_n := \theta^n$ for $n \in \mathbb{Z}$, where the mapping $\theta : X \to X$ is defined by $\theta(\{k_n\}_{n \in \mathbb{Z}}) = \{k_{n+1}\}_{n \in \mathbb{Z}}$. This forms an autonomous dynamical system on X, which is a compact metric space with the metric

$$d\left(x, x'\right) = \sum_{n \in \mathbb{Z}} (r + 1)^{-|n|} \left|k_n - k'_n\right|.$$

The proximity and convergence of sets is given in terms of the *Hausdorff separation* $\mathrm{dist}_X(A, B)$ of nonempty compact subsets $A, B \subseteq X$ as

$$\mathrm{dist}_X(A, B) := \max_{a \in A} \mathrm{dist}(a, B) = \max_{a \in A} \min_{b \in B} d(a, b)$$

and the *Hausdorff metric* $H_X(A, B) = \max\{\mathrm{dist}_X(A, B), \mathrm{dist}_X(B, A)\}$ on the space $\mathscr{H}(X)$ of nonempty compact subsets of X. In absence of possible confusion we simply write dist or H for the Hausdorff separation resp. metric.

2.1 Autonomous Semidynamical Systems

The dynamical behaviour of a semidynamical system π on a state space X is characterised by its invariant sets and what happens in neighbourhoods of such sets. A nonempty subset A of X is called *invariant* under π, or π-invariant, if

$$\pi(n, A) = A \quad \text{for all } n \in \mathbb{Z}^+ \tag{3}$$

or, equivalently, if $f(A) = \pi(1, A) = A$.

Simple examples are equilibria (steady state solutions) and periodic solutions; in the first case A consists of a single point, which must thus be a fixed point of the mapping f, whereas for a solution with period r it consists of a finite set of r distinct points $\{p_1, \ldots, p_r\}$ which are fixed point of the composite mapping f^r (but not for an f^j with j smaller than r).

Invariant sets can also be much more complicated, for example fractal sets. Many are the ω-limit sets of some trajectory, i.e., defined by

$$\omega^+(x_0) = \{y \in X \ : \ \exists n_j \to \infty, \ \pi(n_j, x_0) \to y\},$$

which is nonempty, compact and π-invariant when the forwards trajectory $\{\pi(n, x_0); n \in \mathbb{Z}^+\}$ is a precompact subset of X and the metric space (X, d) is complete. However, $\omega(x_0)$ needs not to be connected.

The asymptotic behaviour of a semidynamical system is characterised by its ω-limit sets, in general, and by its attractors and their associated absorbing sets, in particular. An *attractor* is a nonempty π-invariant compact set A^* that attracts all trajectories starting in some neighbourhood $\mathcal{U}$ of A^*, that is with $\omega^+(x_0) \subset A^*$ for all $x_0 \in \mathcal{U}$ or, equivalently, with

$$\lim_{n \to \infty} \text{dist}\left(\pi(n, x_0), A^*\right) = 0 \quad \text{for all } x_0 \in \mathcal{U}.$$

A^* is called a maximal or *global attractor* when $\mathcal{U}$ is the entire state space X. Note that a global attractor, if it exists, must be unique. For later comparison the formal definition follow.

Definition 2.3. A nonempty compact subset A^* of X is a *global attractor* of the semidynamical system π on X if it is π-invariant and attracts bounded sets, i.e.,

$$\lim_{n \to \infty} \text{dist}\left(\pi(n, D), A^*\right) = 0 \quad \text{for any bounded subset } D \subset X. \tag{4}$$

As simple example consider the autonomous difference equation (1) on $X = \mathbb{R}$ with the map $f(x) := \max\{0, 4x(1 - x)\}$ for $x \in \mathbb{R}$. Then $A^* = [0, 1]$ is invariant and $f(x_0) \in A^*$ for all $x_0 \in \mathbb{R}$, so A^* is the maximal attractor. The dynamics are very simple outside of the attractor, but chaotic within it.

The existence and approximate location of a global attractor follow from that of more easily found absorbing sets, which typically have a convenient simpler shape such as a ball or ellipsoid.

Definition 2.4. A nonempty compact subset B of X is called an *absorbing set* of a semidynamical system π on X if for every bounded subset D of X there exists a $N_D \in \mathbb{Z}^+$ such that $\pi(n, D) \subset B$ for all $n \geq N_D$ in $\mathbb{Z}^+$.

Absorbing sets are often called attracting sets when they are also *positively invariant* in the sense that $\pi(n, B) \subseteq B$ holds for all $n \in \mathbb{Z}^+$, i.e., if one has the inclusion $f(B) = \pi(1, B) \subseteq B$. Attractors differ from attracting sets in that they consist entirely of limit points of the system and are thus strictly invariant in the sense of (3).

Theorem 2.5 (Existence of global attractors). *Suppose that a semidynamical system π on X has an absorbing set B. Then π has a unique global attractor $A^* \subset B$ given by*

$$A^* = \bigcap_{m \geq 0} \overline{\bigcup_{n \geq m} \pi(n, B)}, \tag{5}$$

or simply by $A^* = \bigcap_{m \geq 0} \pi(n, B)$ when B is positively invariant.

For a proof we refer to the more general situation of Theorem 4.11.

Similar results hold if the absorbing set is assumed to be only closed and bounded and the mapping π to be compact or asymptotically compact.

For later comparison note that, in view of the invariance of A^*, the attraction (4) can be written equivalently as the *forwards convergence*

$$\text{dist}\left(\pi(n, D), \pi(n, A^*)\right) \to 0 \qquad \text{as} \quad n \to \infty. \tag{6}$$

A global attractor is, in fact, uniformly Lyapunov asymptotically stable. The asymptotic stability of attractors and that of attracting sets in general can be characterised by Lyapunov functions. Such Lyapunov functions can be used to establish the existence of an absorbing set and hence that of a nearby global attractor in a perturbed system.

2.2 Lyapunov Functions for Autonomous Attractors

Consider an autonomous semidynamical system π on a compact metric space (X, d) which is generated by an autonomous difference equation

$$x_{n+1} = f(x_n), \tag{7}$$

where $f : X \to X$ is globally Lipschitz continuous with Lipschitz constant $L > 0$, i.e.,

$$d(f(x), f(y)) \leq L d(x, y), \quad \text{for all } x, y \in X.$$

Definition 2.6. A nonempty compact subset $A \subsetneq X$ is called *globally uniformly asymptotically stable* if it is both

(i) *Lyapunov stable*, i.e., for all $\epsilon > 0$, there exists a $\delta = \delta(\epsilon) > 0$ with

$$\operatorname{dist}(x, A) < \delta \implies \operatorname{dist}(f^n(x), A) < \epsilon \quad \text{for all } n \in \mathbb{Z}^+, \tag{8}$$

(ii) *Globally uniformly attracting*, i.e., for all $\epsilon > 0$, there exists an integer $N = N(\epsilon) > 1$ such that

$$\operatorname{dist}(f^n(x), A) < \epsilon \quad \text{for all } x \in X, \, n \geq N. \tag{9}$$

Note that such a set A is the global attractor for the semidynamical system generated by an autonomous difference equation (7). In particular, it is invariant, i.e., $f(A) = A$.

Global uniform asymptotical stability is characterized in terms of a Lyapunov function by the following necessary and sufficient conditions. The following theorem is taken from Diamond and Kloeden [13]. See also [53].

Theorem 2.7. *Let $f : X \to X$ be globally Lipschitz continuous, and let A be a nonempty compact subset of X. Then A is globally uniformly asymptotically stable w.r.t. the dynamical system generated by (7) if and only if there exist*

(i) *A Lyapunov function $V : X \to \mathbb{R}^+$,*
(ii) *Monotone increasing continuous functions $\alpha, \beta : \mathbb{R}^+ \to \mathbb{R}^+$ with $\alpha(0) = \beta(0) = 0$ and $0 < \alpha(r) < \beta(r)$ for all $r > 0$, and*
(iii) *Constants $K > 0, 0 \leq q < 1$ such that for all $x, y \in X$, it holds that*

1. $|V(x) - V(y)| \leq K d(x, y)$,
2. $\alpha(\operatorname{dist}(x, A)) \leq V(x) \leq \beta(\operatorname{dist}(x, A))$ *and*
3. $V(f(x)) \leq q V(x)$.

Proof. Sufficiency. Let V be a Lyapunov function as described in the theorem. Choose $\epsilon > 0$ arbitrarily and define $\delta := \beta^{-1}(\alpha(\epsilon)/q)$, which means that $\alpha(\epsilon) = q\beta(\delta)$. This implies that

$$\alpha(\operatorname{dist}(f^n(x), A)) \leq V(f^n(x)) \leq q^n V(x) \leq q V(x) \leq q\beta(\operatorname{dist}(x, A)),$$

so that

$$\operatorname{dist}(f^n(x), A) \leq \alpha^{-1}(q\beta(\operatorname{dist}(x, A))) \leq \alpha^{-1}(\alpha(\epsilon)) \leq \epsilon \quad \text{for all } n \in \mathbb{N},$$

when $\operatorname{dist}(x, A) < \delta$. Thus, A is Lyapunov stable. Now define

$$N := \max \left\{ 1, 1 + \left\lfloor \frac{\ln(\alpha(\epsilon)/V_0)}{\ln q} \right\rfloor \right\},$$

where $V_0 := \max_{x \in X} V(x)$ is finite by continuity of V and compactness of X. For $n \geq N$, one has $q^n \leq q^N$, since $0 \leq q < 1$. Since, from above,

$$\alpha(\mathrm{dist}(f^n(x), A)) \leq q^n V(x) \leq q^n V_0 \leq q^N V_0 \leq \alpha(\epsilon) \quad \text{for all } n \geq N,$$

one has $\mathrm{dist}(f^n(x), A) < \epsilon$ for $n \geq N$, $x \in X$. This means that A is globally uniformly attracting and hence globally uniformly asymptotically stable.

Necessity. This will just be sketched here; the details can be found in [13]. Let A be globally uniformly asymptotically stable, i.e., for given $\epsilon > 0$, there exists $\delta = \delta(\epsilon)$ such that (8) holds, and for given $\epsilon > 0$, there exists $N = N(\epsilon)$ such that (9) holds. Define $G_k : \mathbb{R}_0^+ \to \mathbb{R}_0^+$ for $k \in \mathbb{N}$ by

$$G_k(r) := \begin{cases} r - \frac{1}{k} : r \geq \frac{1}{k}, \\ 0 \quad : 0 \leq r < \frac{1}{k}, \end{cases} \quad \text{for all } r \geq 0.$$

Then

$$|G_k(r) - G_k(s)| \leq |r - s| \quad \text{for all } r, s \geq 0.$$

Now choose q so that $0 < q < \min\{1, L\}$, where L is the Lipschitz constant of the mapping f, and define

$$g_k := \left(\frac{q}{L}\right)^{N(1/k)} \quad \text{for all } k \in \mathbb{N}$$

and

$$V_k(x) = g_k \sup_{n \in \mathbb{Z}^+} q^{-n} G_k(\mathrm{dist}(f^n(x), A)) \quad \text{for all } k \in \mathbb{N}.$$

Then

(i) $V_k(x) = 0$ if and only if $\mathrm{dist}(x, A) < \delta(1/k)$, due to Lyapunov stability.
(ii) Since $|\mathrm{dist}(x, A) - \mathrm{dist}(y, A)| \leq d(x, y)$ and

$$d(f^n(x), f^n(y)) \leq L d\left(f^{n-1}(x), f^{n-1}(y)\right) \leq \cdots \leq L^n d(x, y),$$

it follows that

$$|V_k(x) - V_k(y)|$$
$$\leq g_k \sup_{n \geq 0} q^{-n} |G_k(\mathrm{dist}(f^n(x), A)) - G_k(\mathrm{dist}(f^n(y), A))|$$
$$\leq g_k \sup_{0 \leq n \leq N(1/k)} q^{-n} |G_k(\mathrm{dist}(f^n(x), A)) - G_k(\mathrm{dist}(f^n(y), A))|$$
$$\leq g_k \sup_{0 \leq n \leq N(1/k)} q^{-n} d\left(f^n(x), f^n(y)\right)$$
$$\leq g_k \sup_{0 \leq n \leq N(1/k)} q^{-n} L^n d(x, y) = d(x, y).$$

(iii) From above, it holds that $V_k(x) \leq V_k(y) + d(x, y)$. For all $y \in A$, one obtains that $V_k(y) = 0$ and $V_k(x) \leq d(x, y)$, and since A is compact, the minimum over all $y \in A$ is attained and $V_k(x) \leq \text{dist}(x, A)$.

(iv) $V_k(f(x)) \leq q V_k(x)$, since

$$V_k(f(x)) \leq g_k \sup_{n \geq 0} q^{-n} G_k(\text{dist}(f^n(f(x)), A))$$

$$= q g_k \sup_{n \geq 0} q^{-n-1} G_k(\text{dist}(f^{n+1}(x), A))$$

$$= q g_k \sup_{n \geq 1} q^{-n} G_k(\text{dist}(f^n(x), A))$$

$$\leq q g_k \sup_{n \geq 0} q^{-n} G_k(\text{dist}(f^n(x), A)) = q V_k(x)$$

Finally, define

$$V(x) = \sum_{k=1}^{\infty} 2^{-k} V_k(x).$$

The main difficulty is to show the existence of the lower bound function α. This is systematically built up via the component functions V_k, which vanish successively on a closed $\frac{1}{k}$-neighbourhood of the set A. $\qquad\square$

Remarks. For a more comprehensive introduction to discrete dynamical systems and their attractors we refer to e.g. [32, 51]. In particular, for the case of infinite-dimensional state spaces see [14] and [48, Chap. 2], where also connectedness issues of attractors or compactness properties for the semigroup π are addressed.

3 Nonautonomous Difference Equations

Difference equations on $\mathbb{R}^d$ of the form

$$x_{n+1} = f_n(x_n), \qquad (\Delta)$$

in which continuous mappings $f_n : \mathbb{R}^d \to \mathbb{R}^d$ on the right-hand side are allowed to vary with the time n, are called *nonautonomous difference equations*.

Such nonautonomous difference equations arise quite naturally in many different ways. The mappings f_n in (Δ) may of course vary completely arbitrarily, but often there is some relationship between them or some regularity in the way in which they are given.

For example, the mappings may all be the same as in the very special autonomous subcase (1) or they may vary periodically within, or be chosen irregularly from, a finite family $\{g_1, \cdots, g_r\}$, in which case (Δ) can be rewritten as

$$x_{n+1} = g_{k_n}(x_n), \qquad (10)$$

with the $k_n \in \{1, \ldots, r\}$ and $f_n = g_{k_n}$.

As another example, the difference equation (Δ) may represent a variable time-step discretization method for a differential equation $\dot{x} = f(x)$, the simplest of which being the Euler method with a variable time-step $h_n > 0$,

$$x_{n+1} = x_n + h_n f(x_n), \tag{11}$$

in which case $f_n(x) = x + h_n f(x)$. More generally, a difference equation may involve a parameter $\lambda \in \Lambda$ which varies in time by choice or randomly, giving rise to the nonautonomous difference equation

$$x_{n+1} = g(x_n, \lambda_n), \tag{12}$$

so $f_n(x) = g(x, \lambda_n)$ here for the prescribed choice of $\lambda_n \in \Lambda$.

The nonautonomous difference equation (Δ) generates a solution mapping $\phi : \mathbb{Z}^2_\geq \times \mathbb{R}^d \to \mathbb{R}^d$, where

$$\mathbb{Z}^2_\geq := \{(n, n_0) \in \mathbb{Z}^2 \ : \ n \geq n_0\},$$

through iteration, i.e.,

$$\phi(n_0, n_0, x_0) := x_0, \qquad \phi(n, n_0, x_0) := f_{n-1} \circ \cdots \circ f_{n_0}(x_0) \quad \text{for all } n > n_0,$$

$n_0 \in \mathbb{Z}$, and each $x_0 \in \mathbb{R}^d$. This solution mapping satisfies the *two-parameter semigroup* property

$$\phi(m, n_0, x_0) = \phi(m, n, \phi(n, n_0, x_0))$$

for all $(n, n_0) \in \mathbb{Z}^2_\geq$, $(m, n) \in \mathbb{Z}^2_\geq$ and $x_0 \in \mathbb{R}^d$. In this sense, ϕ is called *general solution* of (Δ). In particular, as composition of continuous functions the mapping $x_0 \mapsto \phi(n, n_0, x_0)$ is continuous for $(n, n_0) \in \mathbb{Z}^2_\geq$.

The general nonautonomous case differs crucially from the autonomous in that the starting time n_0 is just as important as the time that has elapsed since starting, i.e., $n - n_0$, and hence many of the concepts that have been developed and extensively investigated for autonomous dynamical systems in general and autonomous difference equations in particular are either too restrictive or no longer valid or meaningful.

3.1 Processes

Solution mappings of nonautonomous difference equations (Δ) are one of the main motivations for the process formulation of an abstract nonautonomous dynamical system on a metric state space (X, d) and time set $\mathbb{Z}$.

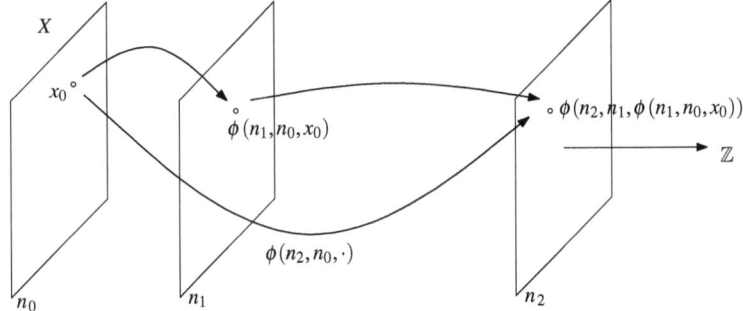

Fig. 2 Property (ii) of a discrete-time process $\phi : \mathbb{Z}_{\geq}^2 \times X \to X$

The following definition originates from Dafermos [12] and Hale [14].

Definition 3.1. A (discrete-time) *process* on a state space X is a mapping $\phi : \mathbb{Z}_{\geq}^2 \times X \to X$, which satisfies the initial value, two-parameter evolution and continuity properties:

(i) $\phi(n_0, n_0, x_0) = x_0$ for all $n_0 \in \mathbb{Z}$ and $x_0 \in X$,
(ii) $\phi(n_2, n_0, x_0) = \phi(n_2, n_1, \phi(n_1, n_0, x_0))$ for all $n_0 \leq n_1 \leq n_2$ in $\mathbb{Z}$ and $x_0 \in X$,
(iii) the mapping $x_0 \mapsto \phi(n, n_0, x_0)$ of X into itself is continuous for all $n_0 \leq n$ in $\mathbb{Z}$.

The evolution property (ii) is illustrated in Fig. 2. Given a process ϕ on X there is an associated nonautonomous difference equation like (Δ) on X with mappings defined by $f_n(x) := \phi(n + 1, n, x)$ for all $x \in X$ and $n \in \mathbb{Z}$.

A process is often called a *two-parameter semigroup* on X in contrast with the one-parameter semigroup of an autonomous semidynamical system since it depends on both the initial time n_0 and the actual time n rather than just the elapsed time $n - n_0$. This abstract formalism of a nonautonomous dynamical system is a natural and intuitive generalization of autonomous systems to nonautonomous systems.

3.2 Skew-Product Systems

The skew-product formalism of a nonautonomous dynamical system is somewhat less intuitive than the process formalism. It represents the nonautonomous system as an autonomous system on the cartesian product of the original state space and some other space such as a function or sequence space on which an autonomous dynamical system called the driving system acts. This driving system is the source of nonautonomy in the dynamics on the original state space.

Let (P, d_P) be a metric space with metric d_P and let $\theta = \{\theta_n : n \in \mathbb{Z}\}$ be a group of continuous mappings from P onto itself. Essentially, θ is an autonomous dynamical system on P that models the driving mechanism for the change in the mappings f_n on the right-hand side of a nonautonomous difference equation like (Δ), that will now be written as

$$x_{n+1} = f\left(\theta_n(p), x_n\right) \tag{13}$$

for $n \in \mathbb{Z}^+$, where $f : P \times \mathbb{R}^d \to \mathbb{R}^d$ is continuous. The corresponding solution mapping $\varphi : \mathbb{Z}^+ \times P \times \mathbb{R}^d \to \mathbb{R}^d$ is now defined by

$$\varphi(0, p, x) := x, \qquad \varphi(n, p, x) := f\left(\theta_{n-1}(p), \cdot\right) \circ \cdots \circ f\left(p, x\right) \quad \text{for all } n \in \mathbb{N}$$

and $p \in P$, $x \in \mathbb{R}^d$. The mapping φ satisfies the *cocycle property* w.r.t. the driving system θ on P, i.e.,

$$\varphi(0, p, x) := x, \qquad \varphi(m + n, p, x) := \varphi\left(m, \theta_n(p), \varphi\left(n, p, x\right)\right) \tag{14}$$

for all $m, n \in \mathbb{Z}^+$, $p \in P$ and $x \in \mathbb{R}^d$.

3.2.1 Definition

Consider now a state space X instead of $\mathbb{R}^d$, where (X, d) is a metric space with metric d. The above considerations lead to the following definition of a skew-product system, which is an alternative abstract formulation of a discrete nonautonomous dynamical system on the state space X.

Definition 3.2. A (discrete-time) *skew-product system* (θ, ϕ) is defined in terms of a *cocycle mapping* φ on a state space X, driven by an autonomous dynamical system θ acting on a base space P.

Specifically, the *driving system* θ on P is a group of homeomorphisms $\{\theta_n : n \in \mathbb{Z}\}$ under composition on P with the properties

(i) $\theta_0(p) = p$ for all $p \in P$,
(ii) $\theta_{m+n}(p) = \theta_m(\theta_n(p))$ for all $m, n \in \mathbb{Z}$ and $p \in P$,
(iii) The mapping $p \mapsto \theta_n(p)$ is continuous for each $n \in \mathbb{Z}$,

and the *cocycle mapping* $\phi : \mathbb{Z}^+ \times P \times X \to X$ satisfies

(I) $\varphi(0, p, x) = x$ for all $p \in P$ and $x \in X$,
(II) $\varphi(m + n, p, x) = \varphi(m, \theta_n(p), \varphi(n, p, x))$ for all $m, n \in \mathbb{Z}^+$, $p \in P$, $x \in X$,
(III) The mapping $(p, x) \mapsto \phi(n, p, x)$ is continuous for each $n \in \mathbb{Z}$.

For an illustration we refer to the subsequent Fig. 3. A difference equation of the form (13) can be obtained from a skew-product system by defining $f(p, x) := \varphi(1, p, x)$ for all $p \in P$ and $x \in X$.

A process ϕ admits a formulation as a skew-product system with $P = \mathbb{Z}$, the time shift $\theta_n(n_0) := n + n_0$ and the cocycle mapping

$$\varphi(n, n_0, x) := \phi(n + n_0, n_0, x) \quad \text{for all } n \in \mathbb{Z}^+, x \in X.$$

Fig. 3 A discrete-time
skew-product system (θ, φ)
over the base space P

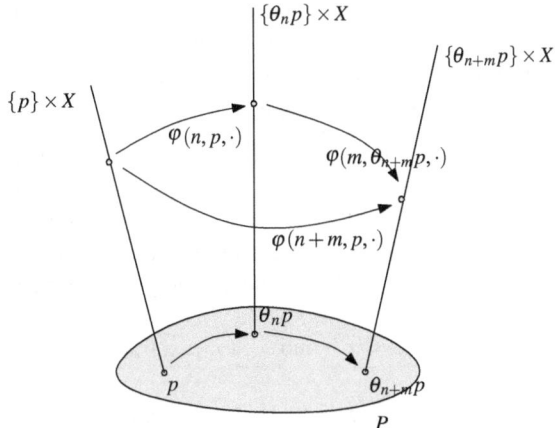

The real advantage of the somewhat more complicated skew-product system formulation of nonautonomous dynamical systems occurs when P is compact. This never happens for a process reformulated as a skew-product system as above since the parameter space P then is $\mathbb{Z}$, which is only locally compact and not compact.

3.2.2 Examples

The examples above can be reformulated as skew-product systems with appropriate choices of parameter space P and the driving system θ.

Example 3.3. A nonautonomous difference equation (Δ) with continuous right-hand sides $f_n : \mathbb{R}^d \to \mathbb{R}^d$ generates a cocycle mapping φ over the parameter set $P = \mathbb{Z}$ w.r.t. the group of left shift mappings $\theta_j := \theta^j$ for $j \in \mathbb{Z}$, where $\theta(n) := n + 1$ for $n \in \mathbb{Z}$. Here φ is defined by

$$\varphi(0, n, x) := x \quad \text{and} \quad \varphi(j, n, x) := f_{n+j-1} \circ \cdots \circ f_n(x) \quad \text{for all } j \in \mathbb{N}$$

and $n \in \mathbb{Z}$, $x \in \mathbb{R}^d$. The mappings $\varphi(j, n, \cdot) : \mathbb{R}^d \to \mathbb{R}^d$ are all continuous.

Example 3.4. Let $f : \mathbb{R}^d \to \mathbb{R}^d$ be a continuous mapping used in an autonomous difference equation (1). The solution mapping φ defined by

$$\varphi(0, x) := x \quad \text{and} \quad \varphi(j, x) = f^j(x) := \underbrace{f \circ \cdots \circ f}_{j \text{ times}}(x) \quad \text{for all } j \in \mathbb{N}$$

and $x \in \mathbb{R}^d$ generates a semigroup on $\mathbb{R}^d$. It can be considered as a cocycle mapping w.r.t. a singleton parameter set $P = \{p_0\}$ and the singleton group consisting only of identity mapping $\theta := \text{id}_P$ on P. Since the driving system just sits at p_0, the dependence on the parameter in φ can be suppressed.

While the integers $\mathbb{Z}$ appears to be the natural choice for the parameter set in Example 3.3 and the choice is trivial in the autonomous case of Example 3.4, in the remaining examples the use of sequence spaces is more advantageous because such spaces are often compact.

Example 3.5. The nonautonomous difference equation (10) with continuous mappings $g_k : \mathbb{R}^d \rightarrow \mathbb{R}^d$ for $k \in \{1, \cdots, r\}$ generates a cocycle mapping over the parameter set $P = \{1, \cdots, r\}^{\mathbb{Z}}$ of bi-infinite sequences $p = \{k_n\}_{n \in \mathbb{Z}}$ with $k_n \in \{1, \cdots, r\}$ w.r.t. the group of left shift operators $\theta_n := \theta^n$ for $n \in \mathbb{Z}$, where $\theta(\{k_n\}_{n \in \mathbb{Z}}) = \{k_{n+1}\}_{n \in \mathbb{Z}}$. The mapping φ is defined by

$$\varphi(0, p, x) := x \quad \text{and} \quad \varphi(j, p, x) := g_{k_{j-1}} \circ \cdots \circ g_{k_0}(x) \quad \text{for all } j \in \mathbb{N}$$

and $x \in \mathbb{R}^d$, where $p = \{k_n\}_{n \in \mathbb{Z}}$, is a cocycle mapping. Note that the parameter space $\{1, \cdots, r\}^{\mathbb{Z}}$ here is a compact metric space with the metric

$$d(p, p') = \sum_{n \in \mathbb{Z}} (r + 1)^{-|n|} |k_n - k'_n|.$$

In addition, $\theta_n : P \rightarrow P$ and $\varphi(j, \cdot, \cdot) : P \times \mathbb{R}^d \rightarrow \mathbb{R}^d$ are all continuous.

We omit a reformulation of the numerical scheme (11) as it is similar to the next example, but with a bi-infinite sequence $p = \{h_n\}_{n \in \mathbb{Z}}$ of stepsizes satisfying a constraint such as $\frac{1}{2}\delta \leq h_n \leq \delta$ for $n \in \mathbb{Z}$ with appropriate $\delta > 0$.

Example 3.6. As an example of a parametrically perturbed difference equation (12), consider the mapping $g : \mathbb{R}^1 \times \left[\frac{1}{2}, 1\right] \mapsto \mathbb{R}^1$ defined by

$$g(x, \lambda) = \frac{|x| + \lambda^2}{1 + \lambda},$$

which is continuous in $x \in \mathbb{R}^1$ and $\lambda \in \left[\frac{1}{2}, 1\right]$. Let $P = \left[\frac{1}{2}, 1\right]^{\mathbb{Z}}$ be the space of bi-infinite sequences $p = \{\lambda_n\}_{n \in \mathbb{Z}}$ taking values in $\left[\frac{1}{2}, 1\right]$, which is a compact metric space with the metric

$$d(p, p') = \sum_{n \in \mathbb{Z}} 2^{-|n|} |\lambda_n - \lambda'_n|,$$

and let $\{\theta_n, n \in \mathbb{Z}\}$ be the group generated by the left shift operator θ on this sequence space (analogously to Example 3.5). The mapping φ is defined by

$$\varphi(0, p, x) := x \quad \text{and} \quad \varphi(j, p, x) := g(q_{j-1}, \cdot) \circ \cdots \circ g(q_0, x) \quad \text{for all } j \in \mathbb{N}$$

and $x \in \mathbb{R}^1$, where $p = \{\lambda_n\}_{n \in \mathbb{Z}}$, is a cocycle mapping on $\mathbb{R}^1$ with parameter space $\left[\frac{1}{2}, 1\right]^{\mathbb{Z}}$ and the above shift operators θ_n. The mappings $\theta_n : P \rightarrow P$ and $\varphi(j, \cdot, \cdot) : P \times \mathbb{R}^d \rightarrow \mathbb{R}^d$ are all continuous here.

3.2.3 Skew-Product Systems as Autonomous Semidynamical Systems

A skew-product system (θ, φ) can be reformulated as autonomous semidynamical system on the *extended state space* $\mathbb{X} := P \times X$. Define a mapping $\pi : \mathbb{Z}^+ \times \mathbb{X} \to \mathbb{X}$ by

$$\pi (n, (p, x_0)) := \big(\theta_n(p), \phi(n, p, x_0)\big) \quad \text{for all } n \in \mathbb{Z}^+, (p, x_0) \in \mathbb{X}.$$

Note that the variable n in $\pi (n, (p, x_0))$ is the time that has elapsed since starting at state (p, x_0).

Theorem 3.7. π *is an autonomous semidynamical system on* $\mathbb{X}$.

Proof. It is obvious that $\pi(n, \cdot)$ is continuous in its variables (p, x_0) for every $n \in \mathbb{Z}^+$ and satisfies the initial condition

$$\pi(0, (p, x_0)) = (p, \varphi(0, p, x_0)) = (p, x_0) \quad \text{for all } p \in P, x_0 \in X.$$

It also satisfies the one-parameter semigroup property

$$\pi(m + n, (p, x_0)) = \pi (m, \pi(n, (p, x_0))) \quad \text{for all } m, n \in \mathbb{Z}^+, \ p \in P, \ x_0 \in X$$

since, by the group property of the driving system and the cocycle property of the skew-product,

$$\pi(m + n, (p, x_0)) = (\theta_{m+n}(p), \varphi(m + n, p, x_0))$$

$$= \big(\theta_m (\theta_n(p)), \varphi(m, \theta_n(p), \varphi(n, p, x_0))\big)$$

$$= \pi\big(m, (\theta_n(p), \varphi(n, p, x_0))\big) = \pi\big(m, \pi(n, (p, x_0))\big).$$

$\square$

As seen in Example 3.3, a process ϕ on the state space X is also a skew-product on X with the shift operator θ on $P := \mathbb{Z}$ and thus generates an autonomous semidynamical system π on the extended state space $\mathbb{X} := \mathbb{Z} \times X$. This semidynamical system has some unusual properties. In particular, π has no nonempty ω-limit sets and, indeed, no compact subset of $\mathbb{X}$ can be π-invariant. This is a direct consequence of the fact that the initial time is a component of the extended state space.

Remarks. An early reference to the description of nonautonomous discrete dynamics via processes or skew-product flows, is given in [32, pp. 45–56, Chap. 4].

4   Nonautonomous Invariant Sets and Attractors of Processes

Invariant sets and attractors are important regions of state space that characterize the long-term behaviour of a dynamical system.

Let $\phi : \mathbb{Z}_{\geq}^2 \times X \to X$ be a process on a metric state space (X, d). This generates a solution $x_n = \phi(n, n_0, x_0)$ to (Δ) that depends on the starting time n_0 as well as the current time n and not just on the time $n - n_0$ that has elapsed since starting as in an autonomous system. This has some profound consequences in terms of definitions and the interpretation of dynamical behaviour. As pointed out above, many concepts and results from the autonomous case are no longer valid or are too restrictive and exclude many interesting types of possible behaviour.

For example, it is too great a restriction of generality to consider a single subset A of X to be invariant under ϕ in the sense that

$$\phi(n, n_0, A) = A \quad \text{for all } n \geq n_0, \ n_0 \in \mathbb{Z},$$

which is equivalent to $f_n(A) = A$ for every $n \in \mathbb{Z}$, where the f_n are mappings in the corresponding nonautonomous difference equation (Δ). Then, in general, neither the trajectory $\{\chi_n^* : n \in \mathbb{Z}\}$ of a solution χ^* that exists on all of $\mathbb{Z}$ nor a nonautonomous ω-limit set defined by

$$\omega^+(n_0, x_0) = \left\{ y \in X \ : \ \exists n_j \to \infty, \ \phi\left(n_j, n_0, x_0\right) \to y \right\},$$

will be invariant in such a sense.

Moreover, such nonautonomous ω-limit sets exist in the infinite future in absolute time rather than in current time like autonomous ω-limit sets, so it is not so clear how useful or even meaningful dynamically they are. Hence, the appropriate formulation of asymptotic behaviour of a nonautonomous dynamical system needs some careful consideration. Lyapunov asymptotical stability of a solution of a nonautonomous system provides a clue. This requires the definition of an entire solution.

Definition 4.1. An *entire solution* of a process ϕ on X is a sequence $\{\chi_k : k \in \mathbb{Z}\}$ in X such that

$$\phi(n, n_0, \chi_{n_0}) = \chi_n \quad \text{for all } n \geq n_0 \text{ and all } n_0 \in \mathbb{Z},$$

or equivalently, $\chi_{n+1} = f_n(\chi_n)$ for all $n \in \mathbb{Z}$ in terms of the nonautonomous difference equation (Δ) corresponding to the process ϕ.

Definition 4.2. An entire solution χ^* of a process ϕ on X is said to be (globally) *Lyapunov asymptotically stable* if it is *Lyapunov stable*, i.e., for every $\epsilon > 0$ and $n_0 \in \mathbb{Z}$ there exists a $\delta = \delta(\epsilon, n_0) > 0$ such that

$$d\left(\phi(n, n_0, x_0), \chi_n^*\right) < \epsilon \quad \text{for all } n \geq n_0 \text{ whenever } d\left(x_0, \chi_{n_0}^*\right) < \delta,$$

and *attracting* in the sense that

$$d\left(\phi\left(n, n_0, x_0\right), \chi_n^*\right) \to 0 \quad \text{as } n \to \infty \tag{15}$$

for all $x_0 \in X$ and $n_0 \in \mathbb{Z}$.

Note, in particular, that the limiting "target" χ_n^* exists for all time and is, in general, also changing in time as the limit is taken.

4.1 Nonautonomous Invariant Sets

Let χ^* be an entire solution of a process ϕ on a metric space (X, d) and consider the family $\mathscr{A} = \{A_n : n \in \mathbb{Z}\}$ of singleton subsets $A_n := \{\chi_n^*\}$ of X. Then by the definition of an entire solution it follows that

$$\phi\left(n, n_0, A_{n_0}\right) = A_n \quad \text{for all } n \geq n_0, \, n_0 \in \mathbb{Z}.$$

This suggests the following generalization of invariance for nonautonomous dynamical systems.

Definition 4.3. A family $\mathscr{A} = \{A_n : n \in \mathbb{Z}\}$ of nonempty subsets of X is *invariant* under a process ϕ on X, or ϕ-*invariant*, if

$$\phi\left(n, n_0, A_{n_0}\right) = A_n \quad \text{for all } n \geq n_0 \text{ and all } n_0 \in \mathbb{Z},$$

or, equivalently, if $f_n(A_n) = A_{n+1}$ for all $n \in \mathbb{Z}$ in terms of the corresponding nonautonomous difference equation (Δ).

A ϕ-invariant family consists of entire solutions. This is essentially due to have in fact a process is onto between the component subsets. The backward solutions, however, need not to be uniquely determined, since the mappings f_n are usually not assumed to be one-to-one.

Proposition 4.4 (Characterization of invariant sets). *A family $\mathscr{A} = \{A_n : n \in \mathbb{Z}\}$ is ϕ-invariant if and only if for every pair $n_0 \in \mathbb{Z}$ and $x_0 \in A_{n_0}$ there exists an entire solution χ such that $\chi_{n_0} = x_0$ and $\chi_n \in A_n$ for all $n \in \mathbb{Z}$.*

Moreover, the entire solution χ is uniquely determined provided the mapping $f_n(\cdot) := \phi(n + 1, n, \cdot) : X \to X$ is one-to-one for every $n \in \mathbb{Z}$.

Proof. Sufficiency. Let $\mathscr{A}$ be ϕ-invariant and pick an arbitrary $x_0 \in A_{n_0}$. For $n \geq n_0$ define the sequence $\chi_n := \phi(n, n_0, x_0)$. Then the ϕ-invariance of $\mathscr{A}$ yields $\chi_n \in A_n$. On the other hand, $A_{n_0} = \phi(n_0, n, A_n)$ for $n \leq n_0$, so there exists a sequence $x_n \in A_n$ with $x_0 = \phi(n_0, n, x_n)$ and $x_n = \phi(n, n - 1, x_{n-1})$ for all $n < n_0$. Hence define $\chi_n := x_n$ for $n < n_0$ and χ becomes an entire solution with the desired properties. If the mappings f_n are all one-to-one, then the sequence $\{x_n\}$ is uniquely determined.

Fig. 4 Forward convergence
$n \to \infty$

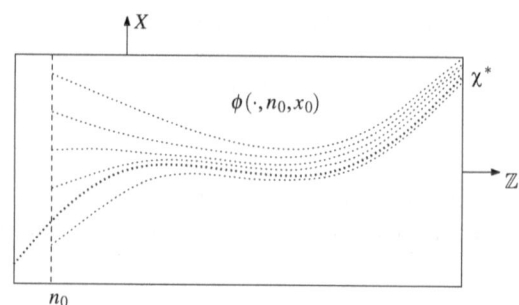

Necessity. Suppose for an arbitrary $n_0 \in \mathbb{Z}$ and $x_0 \in A_{n_0}$ that there is an entire solution χ with $\chi_{n_0} = x_0$ and $\chi_n \in A_n$ for all $n \in \mathbb{Z}$. Hence $\phi(n, n_0, x_0) = \phi(n, n_0, \chi_{n_0}) = \chi_n \in A_n$ for $n \geq n_0$. From this it follows that $f_n(A_n) \subseteq A_{n+1}$. The remaining inclusion $f_n(A_n) \supseteq A_{n+1}$ follows from the fact that $x_0 = \phi(n_0, n, \chi_n) \in \phi(n_0, n, A_n)$ for $n \leq n_0$. $\qquad\square$

4.2 Forwards and Pullback Convergence

The convergence

$$d\left(\phi(n, n_0, x_0), \chi_n^*\right) \to 0 \quad \text{as } n \to \infty \quad (n_0 \text{ fixed})$$

in the attraction property (15) in the definition of a Lyapunov asymptotically stable entire solution χ^* of a process ϕ will be called *forwards convergence* (cf. Fig. 4) to distinguish it from another kind of convergence that is useful for nonautonomous systems.

Forwards convergence does not, however, provide convergence to a particular point $\chi_{n^*}^*$ for a fixed $n^* \in \mathbb{Z}$, which is important in many practical situations because the actual solution χ^* may not be known and thus needs to be determined. To obtain such convergence one has to start progressively earlier. This leads to the concept of *pullback convergence*, defined by

$$d\left(\phi(n, n_0, x_0), \chi_n^*\right) \to 0 \quad \text{as } n_0 \to -\infty \quad (n \text{ fixed})$$

and illustrated in Fig. 5.

In terms of the elapsed time j, forwards convergence can be rewritten as

$$d\left(\phi(n_0 + j, n_0, x_0), \chi_{n_0+j}^*\right) \to 0 \quad \text{as } j \to \infty \tag{16}$$

for all $x_0 \in X$ and $n_0 \in \mathbb{Z}$, while pullback convergence becomes

$$d\left(\phi(n, n - j, x_0), \chi_n^*\right) \to 0 \quad \text{as } j \to \infty$$

for all $x_0 \in X$ and $n \in \mathbb{Z}$.

Fig. 5 Pullback convergence
$n_0 \to -\infty$

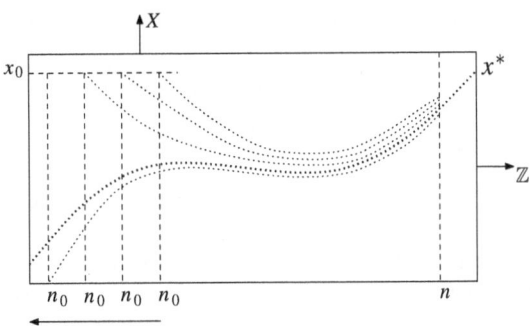

Example 4.5. The nonautonomous difference equation $x_{n+1} = \frac{1}{2}x_n + g_n$ on $\mathbb{R}$ has the solution mapping $\phi(j + n_0, n_0, x_0) = 2^{-j}x_0 + \sum_{k=0}^{j} 2^{-j+k} g_{n_0+n}$, for which pullback convergence gives

$$\phi(n_0, n_0 - j, x_0) = 2^{-j}x_0 + \sum_{k=0}^{j} 2^{-k} g_{n_0-k} \to \sum_{k=0}^{\infty} 2^{-k} g_{n_0-k} \quad \text{as } j \to \infty,$$

provided the infinite series here converges. The limiting solution χ^* is given by $\chi_{n_0}^* := \sum_{k=0}^{\infty} 2^{-k} g_{n_0-k}$ for each $n_0 \in \mathbb{Z}$. It is an entire solution of the nonautonomous difference equation.

Pullback convergence makes use of information about the nonautonomous dynamical system from the past, while forwards convergence uses information about the future.

In autonomous dynamical systems, forwards and pullback convergence are equivalent since the elapsed time $n - n_0 \to \infty$ if either $n \to \infty$ with n_0 fixed or $n_0 \to -\infty$ with n fixed. In nonautonomous dynamical systems pullback convergence and forwards convergence do not necessarily imply each other.

Example 4.6. Consider the process ϕ on $\mathbb{R}$ generated $f_n = g_1$ for $n \leq 0$ and $f_n = g_2$ for $n \geq 1$ where the mappings $g_1, g_2 : \mathbb{R} \to \mathbb{R}$ are given by $g_1(x) := \frac{1}{2}x$ and $g_2(x) := \max\{0, 4x(1 - x)\}$ for all $x \in \mathbb{R}$. Then ϕ is pullback convergent to the entire solution χ^* defined by $\chi_n^* \equiv 0$ for $n \in \mathbb{Z}$, but is not forwards convergent to χ^*. In particular, χ^* is not Lyapunov stable.

4.3 Forwards and Pullback Attractors

Forwards and pullback convergence can be used to define two distinct types of nonautonomous attractors for a process ϕ on a state space X. Instead of a family $\mathscr{A} = \{A_n : n \in \mathbb{Z}\}$ of singleton subsets $A_n := \{\chi_n^*\}$ for an entire solution χ^* of the process consider a ϕ-invariant family of $\mathscr{A} = \{A_n : n \in \mathbb{Z}\}$ of nonempty subsets A_n of X.

In this context forwards convergence generalizes to

$$\text{dist}\left(\phi(n_0 + j, n_0, x_0), A_{n_0+j}\right) \to 0 \quad \text{as } j \to \infty \quad (n_0 \text{ fixed}) \qquad (17)$$

and pullback convergence to

$$\text{dist}\left(\phi(n, n - j, x_0), A_n\right) \to 0 \quad \text{as } j \to \infty \quad (n \text{ fixed}). \qquad (18)$$

More generally, $\mathscr{A}$ is said to forwards (resp. pullback) attract bounded subsets of X if x_0 is replaced by an arbitrary bounded subset D of X in (17) (resp. (18)).

Definition 4.7. A ϕ–invariant family $\mathscr{A} = \{A_n : n \in \mathbb{Z}\}$ of nonempty compact subsets of X is called a *forward attractor* if it forward attracts bounded subsets of X and a *pullback attractor* if it pullback attracts bounded subsets of X.

As a ϕ-invariant family $\mathscr{A}$ of nonempty compact subsets of X, by Proposition 4.4, both pullback and forwards attractors consist of entire solutions.

In fact when the component subsets of a pullback attractor are uniformly bounded, i.e., if there exists a bounded subset B of X such that $A_n \subset B$ for all $n \in \mathbb{Z}$, then pullback attractors are characterized by the bounded entire solutions of the process.

Proposition 4.8 (Dynamical characterization of pullback attractors). *A uniformly bounded pullback attractor* $\mathscr{A} = \{A_n : n \in \mathbb{Z}\}$ *admits the dynamical characterization: for each* $n_0 \in \mathbb{Z}$

$$x_0 \in A_{n_0} \quad \Leftrightarrow \quad \text{there exists a bounded entire solution } \chi \text{ with } \chi_{n_0} = x_0.$$

Such a pullback attractor is therefore uniquely determined.

Proof. Sufficiency. Pick $n_0 \in \mathbb{Z}$ and $x_0 \in A_{n_0}$ arbitrarily. Then, due to the ϕ-invariance of the pullback attractor $\mathscr{A}$, by Proposition 4.4 there exists an entire solution χ with $\chi_{n_0} = x_0$ and $\chi_n \in A_n$ for each $n \in \mathbb{Z}$. Moreover, χ is bounded since the component sets of the pullback attractor are uniformly bounded.

Necessity. If there exists a bounded entire solution χ of the process ϕ, then the set of points $D_\chi := \{\chi_n : n \in \mathbb{Z}\}$ is bounded in X. Since $\mathscr{A}$ pullback attracts bounded subsets of X, for each $n \in \mathbb{Z}$,

$$0 \le \text{dist}\left(\chi_n, A_n\right) \le \lim_{j \to \infty} \text{dist}\left(\phi(n, n - j, D_\chi), A_n\right) = 0,$$

so $\chi_n \in A_n$. $\square$

4.4 Existence of Pullback Attractors

Absorbing sets can also be defined for pullback attraction. Wider applicability can be attained if they are also allowed to depend on time.

Definition 4.9. A family $\mathscr{B} = \{B_n : n \in \mathbb{Z}\}$ of nonempty compact subsets of X is called a *pullback absorbing family* for a process ϕ on X if for each $n \in \mathbb{Z}$ and every bounded subset D of X there exists an $N_{n,D} \in \mathbb{Z}^+$ such that

$$\phi(n, n - j, D) \subseteq B_n \quad \text{for all } j \geq N_{n,D}, \, n \in \mathbb{Z}.$$

The existence of a pullback attractor follows from that of a pullback absorbing family in the following generalization of Theorem 2.5 for autonomous global attractors. The proof is simpler if the pullback absorbing family is assumed to be ϕ-positive invariant.

Definition 4.10. A family $\mathscr{B} = \{B_n : n \in \mathbb{Z}\}$ of nonempty compact subsets of X is said to be ϕ-*positive invariant* if

$$\phi(n, n_0, B_{n_0}) \subseteq B_n \quad \text{for all } n \geq n_0.$$

Theorem 4.11 (Existence of pullback attractors). *Suppose that a process ϕ on a complete metric space (X, d) has a ϕ-positive invariant pullback absorbing family $\mathscr{B} = \{B_n : n \in \mathbb{Z}\}$. Then there exists a global pullback attractor $\mathscr{A} = \{A_n : n \in \mathbb{Z}\}$ with component sets determined by*

$$A_n = \bigcap_{j \geq 0} \phi(n, n - j, B_{n-j}) \quad \text{for all } n \in \mathbb{Z}. \tag{19}$$

Moreover, if $\mathscr{A}$ is uniformly bounded then it is unique.

Proof. Let $\mathscr{B}$ be a pullback absorbing family and let A_n be defined as in (19). Clearly $A_n \subset B_n$ for each $n \in \mathbb{Z}$.

(i) First, it will be shown for any $n \in \mathbb{Z}$ that

$$\lim_{j \to \infty} \text{dist}\left(\phi(n, n - j, B_{n-j}), A_n\right) = 0. \tag{20}$$

Assume to the contrary that there exist sequences $x_{j_k} \in \phi\left(n, n - j_k, B_{n-j_k}\right) \subset B_n$ and $j_k \to \infty$ such that $\text{dist}(x_{j_k}, A_n) > \epsilon$ for all $k \in \mathbb{N}$. The set $\{x_{j_k} : k \in \mathbb{N}\} \subset B_n$ is relatively compact, so there is a point $x_0 \in B_n$ and an index subsequence $k' \to \infty$ such that $x_{j_{k'}} \to x_0$. Now

$$x_{j_{k'}} \in \phi\left(n, n - j_{k'}, B_{n-j_{k'}}\right) \subset \phi(n, n - k, B_{n-k})$$

for all $k_{j'} \geq k$ and each $k \geq 0$. This implies that

$$x_0 \in \phi(n, n - k, B_{n-k}) \quad \text{for all } k \geq 0.$$

Hence, $x_0 \in A_n$, which is a contradiction. This proves the assertion (20).

(ii) By (20), for every $\epsilon > 0, n \in \mathbb{Z}$, there exists an $N = N_{\epsilon,n} \geq 0$ such that

$$\text{dist}\,(\phi(n, n - N, B_{n-N}), A_n) < \epsilon\,.$$

Let D be a bounded subset of X. The fact that $\mathscr{B}$ is a pullback absorbing family implies that $\phi\,(n, n - j, D) \subset B_n$ for all sufficiently large j. Hence, by the cocycle property,

$$\phi\,(n, n - N - j, D) = \phi\,(n, n - N, \phi(n - N, n - N - j, D))$$
$$\subset \phi\,(n, n - N, B_{n-N})\,.$$

(iii) The ϕ-invariance of the family $\mathscr{A}$ will now be shown. By (19), the set $F_m(n) := \phi\,(n, n - m, B_{n-m})$ is contained in B_n for every $m \geq 0$, and by definition, $A_{n-j} = \bigcap_{m\geq0} F_m\,(n - j)$. First, it will be shown that

$$\phi\left(n, n - j, \bigcap_{m\geq0} F_m(n - j)\right) = \bigcap_{m\geq0} \phi\,(n, n - j, F_m(n - j))\,. \qquad (21)$$

One sees directly that "$\subset$" holds. To prove "$\supset$", let x be contained in the set on the right side. Then for any $n \geq 0$, there exists an $x^m \in F_m(n - j) \subset B_{n-j}$ such that $x = \phi\,(n, n - j, x^m)$. Since the sets $F_m(n - j)$ are compact and monotonically decreasing with increasing m, the set $\{x^m : m \geq 0\}$ has a limit point $\hat{x} \in \bigcap_{m\geq0} F_m(n - j)$. By the continuity of $\phi\,(n, n - j, \cdot)$, it follows that $x = \phi\,(n, n - j, \hat{x})$. Thus,

$$x \in \phi\left(n, n - j, \bigcap_{m\geq0} F_m(n - j)\right) = \phi\left(n, n - j, A_{n-j}\right)\,.$$

Hence, equation (21), the compactness of $F_m(n - j)$ and the continuity of $\phi\,(n, n - j, \cdot)$ imply that

$$\phi\left(n, n - j, A_{n-j}\right) = \bigcap_{m\geq0} \phi\,(n, n - j, F_m(n - j))$$

$$= \bigcap_{m\geq0} \phi\left(n, n - j, \phi\left(n - j, n - j - m, B_{n-j-m}\right)\right)$$

$$= \bigcap_{m\geq0} \phi\left(n, n - j - m, B_{n-j-m}\right)$$

$$= \bigcap_{m\geq j} \phi\,(n, n - m, B_{n-m})\,,$$

$$\supset A_n$$

which means that

$$A_n \subset \phi\left(n, n - j, A_{n-j}\right), \quad j \in \mathbb{Z}^+ \quad \text{for all } n \in \mathbb{Z}. \tag{22}$$

Replacing n by $m - m$ in (22) and using the cocycle property gives

$$\phi\left(n, n - m, A_{n-m}\right) \subset \phi\left(n, n - m, \phi\left(m - n, n - m - j, A_{n-m-j}\right)\right)$$

$$= \phi\left(n, n - j, \phi\left(n - j, n - m - j, A_{n-m-j}\right)\right)$$

$$\subset \phi\left(n, n - j, \phi(n - j, n - m - j, B_{n-m-j})\right)$$

$$\subset \phi\left(n, n - j, B_{n-j}\right) \subset U_\epsilon(A_n)$$

for all ϵ-neighborhoods $U_\epsilon(A_n)$ of A_n, where $\epsilon > 0$, provided that $j = J(\epsilon)$ is sufficiently large. Hence, $\phi\left(n, n - m, A_{n-m}\right) \subset A_n$ For all $m \in \mathbb{Z}^+, n \in \mathbb{Z}$. With m replaced by j, this yields with (22) the ϕ-invariance of the family $\{A_n : n \in \mathbb{Z}\}$.

(iv) It remains to observe that if the sets in $\mathscr{A} = \{A_n : n \in \mathbb{Z}\}$ are uniformly bounded, then the pullback attractor $\mathscr{A}$ is unique by Proposition 4.8.   □

Remark 4.12. There is no counterpart of Theorem 4.11 for nonautonomous forwards attractors.

If the pullback absorbing family $\mathscr{B}$ is not ϕ-positive invariant, then the proof is somewhat more complicated and the component subsets of the pullback attractor of $\mathscr{A}$ are given by

$$A_n = \bigcap_{k \geq 0} \overline{\bigcup_{j \geq k} \phi\left(n, n - j, B_{n-j}\right)}.$$

However, the assumption in Theorem 4.11 that ϕ-positively invariant pullback absorbing systems is not a serious restriction.

Proposition 4.13. *If $\mathscr{B} = \{B_n : n \in \mathbb{Z}\}$ is a pullback absorbing system for a process ϕ fulfilling $B_n \subset C$ for $n \in \mathbb{Z}$, where C is a bounded subset of X, then there exists a ϕ-positively invariant pullback absorbing system $\widehat{\mathscr{B}} = \{\widehat{B}_n : n \in \mathbb{Z}\}$ containing $\mathscr{B} = \{B_n : n \in \mathbb{Z}\}$ component set-wise.*

Proof. For each $n \in \mathbb{Z}$ define

$$\widehat{B}_n := \overline{\bigcup_{j \geq 0} \phi(n, n - j, B_{n-j})}.$$

Obviously $B_n \subset \widehat{B}_n$ for every $n \in \mathbb{Z}$.

To show positive invariance, the cocycle property is used in what follows.

$$\phi(n+1, n, \widehat{B}_n) = \overline{\bigcup_{j \geq 0} \phi(n+1, n, \phi(n, n-j, B_{n-j}))}$$

$$= \overline{\bigcup_{j \geq 0} \phi(n+1, n-j, B_{n-j})}$$

$$= \overline{\bigcup_{i \geq 1} \phi(n+1, n+1-i, B_{n+1-i})}$$

$$\subseteq \overline{\bigcup_{i \geq 0} \phi(n+1, n+1-i, B_{n+1-i})} = \widehat{B}_{n+1},$$

so $\phi(n+1, n, \widehat{B}_n) \subseteq \widehat{B}_{n+1}$. By this and the cocycle property again

$$\phi(n+2, n, \widehat{B}_n) = \phi\left(n+2, n+1, \phi(n+1, n, \widehat{B}_n)\right)$$

$$\subseteq \phi(n+2, n+1, \widehat{B}_{n+1}) \subseteq \widehat{B}_{n+2}.$$

The general positive invariance assertion then follows by induction.

Now by the continuity of $\phi(n, n-j, \cdot)$ and the compactness of B_{n-j}, the set $\phi(n, n-j, B_{n-j})$ is compact for each $j \geq 0$ and $n \in \mathbb{Z}$. Moreover, $B_{n-j} \subset C$ for each $j \geq 0$ and $n \in \mathbb{Z}$, so by the pullback absorbing property of $\mathscr{B}$ there exists an $N = N_{n,C} \in \mathbb{N}$ such that

$$\phi(n, n-j, B_{n-j}) \subset \phi(n, n-j, C) \subset B_n$$

for all $j \geq N$. Hence

$$\widehat{B}_n = \overline{\bigcup_{j \geq 0} \phi(n, n-j, B_{n-j})}$$

$$\subseteq \overline{B_n \cup \bigcup_{0 \leq j < N} \phi(n, n-j, B_{n-j})}$$

$$= \overline{\bigcup_{0 \leq j < N} \phi(n, n-j, B_{n-j})},$$

which is compact as a finite union of compact sets, so $\widehat{B}_n$ is compact.

To see that $\widehat{\mathscr{B}}$ so constructed is pullback absorbing, let D be a bounded subset of X and fix $n \in \mathbb{Z}$. Since $\mathscr{B}$ is pullback absorbing, there exists an $N_{n,D} \in \mathbb{N}$ such that $\phi(n, n - j, D) \subset B_n$ for all $j \geq N_{n,D}$. But $B_n \subset \widehat{B}_n$, so

$$\phi(n, n - j, D) \subset \widehat{B}_n \quad \text{for all } j \geq N_{n,D}.$$

Hence $\widehat{\mathscr{B}}$ is pullback absorbing as required.   □

4.5 Limitations of Pullback Attractors

Pullback attractors are based on the behaviour of a nonautonomous system in the past and may not capture the complete dynamics of a system when it is formulated in terms of a process. This was already indicated by Example 4.6 and will be illustrated here through some simpler examples. See [29].

First consider the autonomous scalar difference equation

$$x_{n+1} = \frac{\lambda x_n}{1 + |x_n|} \tag{23}$$

depending on a real parameter $\lambda > 0$. Its zero solution $x^* = 0$ exhibits a pitchfork bifurcation at $\lambda = 1$. Its global dynamical behavior can be summarized as follows (see Fig. 6):

- If $\lambda \leq 1$, then $x^* = 0$ is the only constant solution and is globally asymptotically stable. Thus $\{0\}$ is the global attractor of the autonomous dynamical system generated by the difference equation (23).
- If $\lambda > 1$, then there exist two additional nontrivial constant solutions given by $x_\pm := \pm(\lambda - 1)$. The zero solution $x^* = 0$ is an unstable steady state solution and the symmetric interval $A = [x_-, x_+]$ is the global attractor.

These constant solutions are the fixed points of the mapping $f(x) = \frac{\lambda x}{1 + |x|}$.

Piecewise autonomous difference equation: Consider now the piecewise autonomous equation

$$x_{n+1} = \frac{\lambda_n x_n}{1 + |x_n|}, \qquad \lambda_n := \begin{cases} \lambda, & n \geq 0, \\ \lambda^{-1}, & n < 0 \end{cases} \tag{24}$$

for some $\lambda > 1$, which corresponds to a switch between the two autonomous problems (23) at $n = 0$.

The zero solution of the resulting nonautonomous system is the only bounded entire solution, so by Proposition 4.8 the pullback attractor $\mathscr{A}$ has component sets $A_n \equiv \{0\}$ for all $n \in \mathbb{Z}$. Note that the zero solution seems to be "asymptotically stable" for $n < 0$ and then "unstable" for $n \geq 0$. Moreover the interval $[x_-, x_+]$ is like a global attractor for the whole equation on $\mathbb{Z}$, but it is not really one since it is not invariant or minimal for $n < 0$.

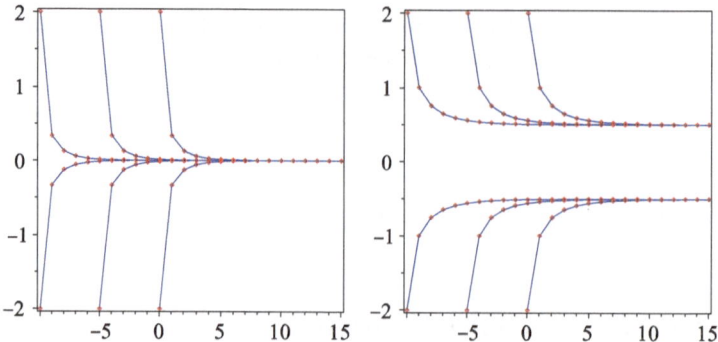

Fig. 6 Trajectories of the autonomous difference equation (23) with $\lambda = 0.5$ (*left*) and $\lambda = 1.5$ (*right*)

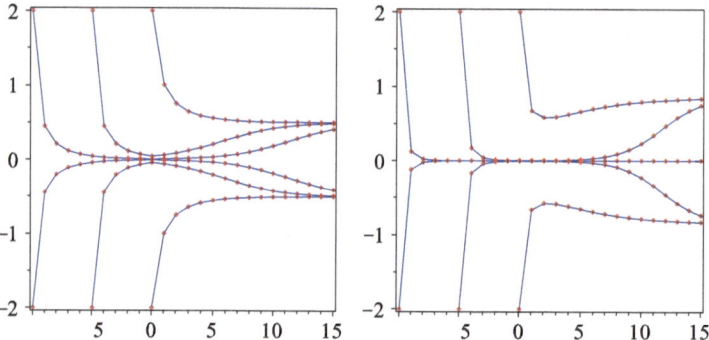

Fig. 7 Trajectories of the piecewise autonomous equation (24) with $\lambda = 1.5$ (*left*) and the asymptotically autonomous equation (25) with $\lambda_k = 1 + \frac{0.9k}{1+|k|}$ (*right*)

The nonautonomous difference equation (24) is asymptotically autonomous in both directions, but the pullback attractor does not reflect the full limiting dynamics (see Fig. 7 (left)), in particular in the forwards time direction.

Fully nonautonomous equation: If the parameters λ_n do not switch from one constant to another as above, but increase monotonically, e.g., such as $\lambda_n = 1 + \frac{0.9n}{1+|n|}$, then the dynamics is similar, although the limiting dynamics is not so obvious from the equation. See Fig. 7 (left).

Let $\{\lambda_n\}_{n \in \mathbb{Z}}$ be a monotonically increasing sequence with $\lim_{k \to \pm\infty} \lambda_n = \bar{\lambda}^{\pm 1}$ for $\bar{\lambda} > 1$. The nonautonomous problem

$$x_{n+1} = f_n(x_n) := \frac{\lambda_n x_n}{1 + |x_n|}. \tag{25}$$

is asymptotically autonomous in both directions with the limiting autonomous systems given above.

Its pullback attractor $\mathscr{A}$ has component sets $A_n \equiv \{0\}$ for all $n \in \mathbb{Z}$ corresponding to the zero entire solution, which is the only bounded entire solution. As above, the zero solution $x^* = 0$ seems to be "asymptotically stable" for $n < 0$ and then "unstable" for $n \geq 0$. However, the forward limit points for nonzero solutions are $\pm(\bar{\lambda} - 1)$, neither of which is a solution at all. In particular, they are not entire solutions, so cannot belong to an attractor, forward or pullback, since these consist of entire solutions. See Fig. 7 (right).

Remark 4.14. Pullback attraction alone does not characterize fully the bounded limiting behaviour of a nonautonomous system formulated as a process. Something in addition like nonautonomous limit sets [25, 38], limiting equations [18] or asymptotically invariant sets [19] and eventual asymptotic stability [20] or a mixture of these ideas is needed to complete the picture. However, this varies from example to example and is somewhat ad hoc. In contrast, this information is built into the skew-product system formulation of a nonautonomous dynamical system, especially when the state space P of the driving system is compact. Essentially, the skew-product system already includes the limiting dynamics and no further ad hoc methods are needed to determine it.

Remarks. Pullback attractors for nonautonomous difference equations were introduced in [23, 24] and a comparison between different attractor types is given in [11] (see also Sect. 5.2).

Without the assumption of being uniformly bounded, pullback attractors of processes need not be unique (see [38, p. 18, Example 1.3.5]). In applications, absorbing sets are frequently not compact and one has to assume ambient compactness properties of a process in order to establish the existence of a pullback attractor (see [38, pp. 12ff]).

5 Nonautonomous Invariant Sets and Attractors: Skew-Product Systems

5.1 Existence of Pullback Attractors

Consider a discrete-time skew-product system (θ, φ) on $P \times X$, where (P, d_P) and (X, d) are metric spaces. There are counterparts for skew-product systems of the concepts of invariance, forwards and pullback convergence and forwards and pullback attractors considered in the previous section for discrete-time processes.

Definition 5.1. A family $\mathscr{A} = \{A_p : p \in P\}$ of nonempty subsets of X is called φ-*invariant* for a skew-product system (θ, φ) on $P \times X$ if

$$\varphi(n, p, A_p) = A_{\theta_n(p)} \quad \text{for all } n \in \mathbb{Z}^+, \, p \in P.$$

It is called φ-*positively invariant* if

$$\varphi(n, p, A_p) \subseteq A_{\theta_n(p)} \quad \text{for all } n \in \mathbb{Z}^+, \ p \in P.$$

Definition 5.2. A family $\mathscr{A} = \{A_p : p \in P\}$ of nonempty compact subsets of X is called *pullback attractor* of a skew-product system (θ, φ) on $P \times X$ if it is φ-invariant and pullback attracts bounded sets, i.e.,

$$\text{dist}\left(\varphi(j, \theta_{-j}(p), D), A_p\right) \to 0 \quad \text{for } j \to \infty \tag{26}$$

for all $p \in P$ and all bounded subsets D of X. It is called a *forwards attractor* if it is φ-invariant and forward attracts bounded sets, i.e.,

$$\text{dist}\left(\varphi(j, p, D), A_{\theta_j(p)}\right) \to 0 \quad \text{for } j \to \infty. \tag{27}$$

As with processes, the existence of a pullback attractor for skew-product systems is ensured by that of a pullback absorbing system.

Definition 5.3. A family $\mathscr{B} = \{B_p : p \in P\}$ of nonempty compact subsets of X is called a *pullback absorbing family* for a skew-product system (θ, φ) on $P \times X$ if for each $p \in P$ and every bounded subset D of X there exists an $N_{p,D} \in \mathbb{Z}^+$ such that

$$\varphi\left(j, \theta_{-j}(p), D\right) \subseteq B_p \quad \text{for all } j \geq N_{p,D}, \ p \in P.$$

The following result generalizes Theorem 2.5 for autonomous semidynamical systems and the first half is the counterpart of Theorem 4.11 for processes. The proof is similar in the latter case, essentially with j and $\theta_{-j}(p)$ changed to n_0 and $n_0 - j$, respectively, but additional complications due to the fact that the pullback absorbing family is no longer assumed to be φ-positively invariant. See [26] for details.

Theorem 5.4 (Existence of pullback attractors). *Let (X, d) and (P, d_P) be complete metric spaces and suppose that a skew-product system (θ, φ) has a pullback absorbing set family $\mathscr{B} = \{B_p : p \in P\}$. Then there exists a pullback attractor $\mathscr{A} = \{A_p : p \in P\}$ with component sets determined by*

$$A_p = \bigcap_{n \geq 0} \overline{\bigcup_{j \geq n} \varphi\left(j, \theta_{-j}(p), B_{\theta_{-j}(p)}\right)}; \tag{28}$$

it is unique if its component sets are uniformly bounded.

The pullback attractor of a skew-product system (θ, φ) has some nice properties when its component subsets are contained in a common compact subset or if the state space P of the driving system is compact.

Proposition 5.5 (Upper semi-continuity of pullback attractors). *Suppose that $A(P) := \bigcup_{p \in P} A_p$ is compact for a pullback attractor $\mathscr{A} = \{A_p : p \in P\}$.*

Then the set-valued mapping $p \mapsto A_p$ is upper semi-continuous in the sense that

$$\operatorname{dist}\left(A_q, A_p\right) \to 0 \quad \text{as } q \to p.$$

On the other hand, if P is compact and the set-valued mapping $p \mapsto A_p$ is upper semi-continuous, then $A(P)$ is compact.

Proof. First note that, since $A(P)$ is compact, the pullback attractor is uniformly bounded by a compact set and hence is uniquely determined.

Assume that the set-valued mapping $p \mapsto A_p$ is not upper semi-continuous. Then there exist an $\epsilon_0 > 0$ and a sequence $p_n \to p_0$ in P such that $\operatorname{dist}\left(A_{p_n}, A_{p_0}\right) \geq 3\epsilon_0$ for all $n \in \mathbb{N}$. Since the sets A_{p_n} are compact, there exists an $a_n \in A_{p_n}$ such that

$$\operatorname{dist}\left(a_n, A_{p_0}\right) = \operatorname{dist}\left(A_{p_n}, A_{p_0}\right) \geq 3\epsilon_0 \quad \text{for each } n \in \mathbb{N}. \tag{29}$$

By pullback attraction, $\operatorname{dist}\left(\varphi\left(m, \theta_{-m}(p_0), B\right), A_{p_0}\right) \leq \epsilon_0$ for $m \geq M_{B,\epsilon_0}$ for any bounded subset B of X; in particular, below $A(P)$ will be used for the set B. By the φ-invariance of the pullback attractor, there exist $b_n \in A_{\theta_{-m}(p_n)} \subset A(P)$ for $n \in \mathbb{N}$ such that $\varphi\left(m, \theta_{-m}(p_n), b_n\right) = a_n$. Since $A(P)$ is compact, there is a convergent subsequence $b_{n'} \to \bar{b} \in A(P)$. Finally, by the continuity of $\theta_{-m}(\cdot)$ and of the cocycle mapping $\varphi(n, \cdot, \cdot)$,

$$d\left(\varphi(m, \theta_{-m}(p_{n'}), b_{n'}), \varphi(m, \theta_{-m}(p_0), \bar{b})\right) \leq \epsilon_0 \quad \text{for } n' \text{ large enough.}$$

Thus,

$$\operatorname{dist}\left(a_{n'}, A_{p_0}\right) = \operatorname{dist}\left(\varphi(m, \theta_{-m}(p_{n'}), b_{n'}), A_{p_0}\right)$$

$$\leq d\left(\varphi(m, \theta_{-m}(p_{n'}), b_{n'}), \varphi(m, \theta_{-m}(p_0), \bar{b})\right)$$

$$+ \operatorname{dist}\left(\varphi(m, \theta_{-m}(p_0), \bar{b}), A_{p_0}\right) \leq 2\epsilon_0,$$

which contradicts (29). Hence, $p \mapsto A_p$ must be upper semi-continuous.

The remaining assertion follows since the image of a compact subset under an upper semi-continuous compact set-valued mapping is compact (cf. [4]).   □

Pullback attractors are in general not forwards attractors. When, however, the state space P of the driving system is compact, then one has the following partial forwards convergence result for the pullback attractor.

Theorem 5.6. *In addition to the assumptions of Theorem 5.4, suppose that P is compact and suppose that the pullback absorbing family $\mathscr{B}$ is uniformly bounded by a compact subset C of X. Then*

$$\lim_{n \to \infty} \sup_{p \in P} \operatorname{dist}\left(\varphi(n, p, D), A(P)\right) = 0 \tag{30}$$

for every bounded subset D of X, where $A(P) := \overline{\bigcup_{p \in P} A_p}$.

Proof. First note that $A(P)$ is compact since the component subsets A_p are all contained in the common compact set C. This means also that the pullback attractor is unique.

Suppose to the contrary that the convergence (30) does not hold. Then there exist an $\epsilon_0 > 0$ and sequences $n_j \to \infty$, $\hat{p}_j \in P$ and $x_j \in C$ such that

$$\text{dist}\left(\varphi(n_j, \hat{p}_j, x_j), A(P)\right) > \epsilon_0. \tag{31}$$

Set $p_j = \theta_{n_j}(\hat{p}_j)$. By the compactness of P, there exists a convergent subsequence $p_{j'} \to p_0 \in P$. From the pullback attraction, there exists an $n > 0$ such that

$$\text{dist}\left(\varphi(n, \theta_{-n}(p_0), C), A_{p_0}\right) < \frac{\epsilon_0}{2}.$$

The cocycle property then gives

$$\varphi\left(n_j, \theta_{-n_j}(p_j), x_j\right) = \varphi\left(n, \theta_{-n}(p_j), \varphi\left(n_j - n, \theta_{-n_j}(p_j), x_j\right)\right)$$

for any $n_j > n$. By the pullback absorption of $\mathcal{B}$, it follows that

$$\varphi\left(n_j - n, \theta_{-n_j}(p_j), x_j\right) \subset B_{\theta_{-n}(p_j)} \subset C,$$

and since C is compact, there is a further index subsequence j'' of j' (depending on n) such that

$$z_{n_{j''}} := \varphi\left(n_{j''} - n, \theta_{-n_{j''}}(p_{j''}), x_{j''}\right) \to z_0 \in C.$$

The continuity of the skew-product mappings in the p and x variables implies

$$\text{dist}\left(\varphi(n, \theta_{-n}(p_{j''}), z_{n_{j''}}), \varphi(n, \theta_{-n}(p_0), z_0)\right) < \frac{\epsilon_0}{2}, \quad \text{when } n_{j''} > n(\epsilon_0).$$

Therefore,

$$\epsilon_0 > \text{dist}\left(\varphi(n_{j''}, \theta_{-n_{j''}}(p_0), x_{j''}), A_{p_0}\right)$$

$$= \text{dist}\left(\varphi\left(n_{j''}, \hat{p}_{j''}, x_{j''}\right), A_{p_0}\right) \geq \text{dist}\left(\varphi\left(n_{j''}, \hat{p}_{j''}, x_{j''}\right), A(P)\right),$$

which contradicts (31). Thus, the asserted convergence (30) must hold.   □

5.2   Comparison of Nonautonomous Attractors

Recall from Theorem 3.7 that the mapping $\pi : \mathbb{Z}^+ \times \mathbb{X} \to \mathbb{X}$ defined by

$$\pi(n, (p, x)) := (\theta_n(p), \varphi(n, p, x))$$

for all $j \in \mathbb{Z}^+$ and $(p, x) \in \mathbb{X} := P \times X$ forms an autonomous semidynamical system on the extended state space $\mathbb{X}$ with the metric

$$\text{dist}_{\mathbb{X}} ((p_1, x_1), (p_2, x_2)) = d_P(p_1, p_2) + d(x_1, x_2).$$

Proposition 5.7 (Uniform and global attractors). *Suppose that $\mathscr{A}$ is a uniform attractor (i.e., uniformly attracting in both the forward and pullback senses) of a skew-product system (θ, φ) and that $\bigcup_{p \in P} A_p$ is precompact in X. Then the union $\mathbb{A} := \bigcup_{p \in P} \{p\} \times A_p$ is the global attractor of the autonomous semidynamical system π.*

Proof. The π-invariance of $\mathbb{A}$ follows from the φ-invariance of $\mathscr{A}$, and the θ-invariance of P via

$$\pi(n, \mathbb{A}) = \bigcup_{p \in P} \{\theta_n(p)\} \times \varphi(n, p, A_p) = \bigcup_{p \in P} \{\theta_n(p)\} \times A_{\theta_n(p)} = \bigcup_{q \in P} \{q\} \times A_q = \mathbb{A}.$$

Since $\mathscr{A}$ is also a pullback attractor and $\bigcup_{p \in P} A_p$ is precompact in X (and P is compact too), the set-valued mapping $p \mapsto A_p$ is upper semi-continuous, which means that $p \mapsto F(p) := \{p\} \times A_p$ is also upper semi-continuous. Hence, $F(P) = \mathbb{A}$ is a compact subset of $\mathbb{X}$. Moreover, the definition of the metric $\text{dist}_{\mathbb{X}}$ on $\mathbb{X}$ implies that

$$\text{dist}_{\mathbb{X}} (\pi(n, (p, x)), \mathbb{A}) = \text{dist}_{\mathbb{X}} ((\theta_n(p), \varphi(n, p, x)), \mathbb{A})$$

$$\leq \text{dist}_{\mathbb{X}} ((\theta_n(p), \varphi(n, p, x)), \{\theta_n(p)\} \times A_{\theta_n(p)})$$

$$= \text{dist}_P (\theta_n(p), \theta_n(p)) + \text{dist} (\varphi(n, p, x), A_{\theta_n(p)})$$

$$= \text{dist} (\varphi(n, p, x), A_{\theta_n(p)}),$$

where $\pi(n, (p, x)) = (\theta_n(p), \varphi(n, p, x))$. The desired attraction to $\mathbb{A}$ w.r.t. π then follows from the forward attraction of $\mathscr{A}$ w.r.t. φ. $\quad\square$

Without uniform attraction as in Proposition 5.7 a pullback attractor need not give a global attractor, but the following result does hold.

Proposition 5.8. *If $\mathscr{A}$ is a pullback attractor for a skew-product system (θ, φ) and $\bigcup_{p \in P} A_p$ is precompact in X, then $\mathbb{A} := \bigcup_{p \in P} \{p\} \times A_p$ is the maximal invariant compact set of the autonomous semidynamical system π.*

Proof. The compactness and π-invariance of $\mathbb{A}$ are proved in the same way as in first part of the proof of Proposition 5.7. To prove that the compact invariant set $\mathbb{A}$ is maximal, let $\mathbb{C}$ be any other compact invariant set of the autonomous semidynamical system π. Then $\mathbb{A}$ is a compact and φ-invariant family of compact sets, and by pullback attraction,

$$\text{dist} (\varphi (n, \theta_{-n}(p), C_{\theta_{-n}(p)}), A_p) \leq \text{dist} (\varphi (n, \theta_{-n}(p), K), A_p) \to 0$$

as $n \to \infty$, where $K := \overline{\bigcup_{p \in P} C_p}$ is compact. Hence, $C_p \subseteq A_p$ for all $p \in P$, i.e., $\mathbb{C} := \bigcup_{p \in P}\{p\} \times C_p \subseteq \mathbb{A}$, which finally means that $\mathbb{A}$ is a maximal π-invariant set.   □

The set $\mathbb{A}$ here need not be the global attractor of π. In the opposite direction, the global attractor of the associated autonomous semidynamical system always forms a pullback attractor of the skew-product system.

Proposition 5.9 (Global and pullback attractors). *If an autonomous semidynamical system π has a global attractor*

$$\mathbb{A} = \bigcup_{p \in P}\{p\} \times A_p,$$

then $\mathscr{A} = \{A_p : p \in P\}$ is a pullback attractor for the skew-product system (θ, φ).

Proof. The sets P and $K := \bigcup_{p \in P} A_p$ are compact by the compactness of $\mathbb{A}$. Moreover, $\mathbb{A} \subset P \times K$, which is a compact set. Now

$$\text{dist}\,(\varphi(n, p, x), K) = \text{dist}_P\,(\theta_n(p), P) + \text{dist}\,(\varphi(n, p, x), K)$$

$$= \text{dist}_{\mathbb{X}}\,((\theta_n(p), \varphi(n, p, x)), P \times K)$$

$$\le \text{dist}_{\mathbb{X}}\,(\pi(n, (p, x)), P \times K)$$

$$\le \text{dist}_{\mathbb{X}}\,(\pi(n, P \times D), \mathbb{A}) \to 0 \text{ as } n \dot\to \infty$$

for all $(p, x) \in P \times D$ and every arbitrary bounded subset D of X, since $\mathbb{A}$ is the global attractor of π.

Hence, replacing p by $\theta_{-n}(p)$ implies

$$\lim_{n \to \infty} \text{dist}\,(\varphi(n, \theta_{-n}(p), D), K) = 0.$$

Then the system is pullback asymptotic compact (see the definition in Chapter 12 of [25]) and by Theorem 12.12 in [25] this is a sufficient condition for the existence of a pullback attractor $\mathscr{A}' = \{A'_p : p \in P\}$ with $\bigcup_{p \in P} A'_p \subset K$. From Proposition 5.8, $\mathbb{A}' := \bigcup_{p \in P}\{p\} \times A'_p$ is the maximal π-invariant subset of $\mathbb{X}$, but so is the global attractor $\mathbb{A}$. This means that $\mathbb{A}' = \mathbb{A}$. Thus, $\mathscr{A}$ is a pullback attractor of the skew-product system (θ, φ).   □

5.3 Limitations of Pullback Attractors Revisited

The limitations of pullback attraction for processes were illustrated in Sect. 4.5 through the scalar nonautonomous difference equation

$$x_{n+1} = f_n(x_n) := \frac{\lambda_n x_n}{1 + |x_n|}, \tag{32}$$

where $\{\lambda_n\}_{n \in \mathbb{Z}}$ is an increasing sequence with $\lim_{n \to \pm\infty} \lambda_n = \bar{\lambda}^{\pm 1}$ for $\bar{\lambda} > 1$.

The pullback attractor $\mathscr{A}$ of the corresponding process has component sets $A_n \equiv \{0\}$ for all $n \in \mathbb{Z}$ corresponding to the zero entire solution, which is the only bounded entire solution. The zero solution $x^* = 0$ seems to be "asymptotically stable" for $n < 0$ and then "unstable" for $n \geq 0$. However the forward limit points for nonzero solutions are $\pm(\bar{\lambda} - 1)$, which both are not solutions at all. In particular, they are not entire solutions.

An elegant way to resolve the problem is to consider the skew-product system formulation of a nonautonomous dynamical system. This includes an autonomous dynamical system as a driving mechanism, which is responsible for the temporal change in the dynamics of the nonautonomous difference equation. It also includes the dynamics of the asymptotically autonomous difference equations above and their limiting autonomous systems.

The nonautonomous difference equation (32) can be formulated as a skew-product system with the diving system defined in terms of the shift operator θ on the space of bi-infinite sequences

$$\Lambda_L = \{\lambda = \{\lambda_n\}_{n \in \mathbb{Z}} : \lambda_n \in [0, L], \quad n \in \mathbb{Z}\}$$

for some $L > \bar{\lambda} > 1$. It yields a compact metric space with the metric

$$d_{\Lambda_L}(\lambda, \lambda') := \sum_{n \in \mathbb{Z}} (L + 1)^{-|n|} |\lambda_n - \lambda'_n|.$$

This is coupled with a cocycle mapping with values $x_n = \varphi(n, \lambda, x_0)$ on $\mathbb{R}$ generated by the difference equation (32) with a given coefficient sequence λ.

For the sequence λ from (32), the limit of the shifted sequences $\theta_n(\lambda)$ in the above metric as $n \to \infty$ is the constant sequence λ_+^* equal to $\bar{\lambda}$, while the limit as $n \to -\infty$ is the sequence λ_-^* with all components equal to $\bar{\lambda}^{-1}$.

The pullback attractor of the corresponding skew-product system (θ, φ) on $\Lambda \times \mathbb{R}$ consists of compact subsets A_λ of $\mathbb{R}$ for each $\lambda \in \Lambda_L$. It is easy to see that $A_\lambda = \{0\}$ for any λ with components $\lambda_n < 1$ for $n \leq 0$, which includes the constant sequence λ_-^* as well as the switched sequence in (32). On the other hand, $A_{\lambda_+^*} = [-\bar{\lambda}, \bar{\lambda}]$. Here $\cup_{\lambda \in \Lambda_L} A_\lambda$ is precompact, so contains all future limiting dynamics.

The pullback attractor of the skew-product system includes that of the process for a given bi-infinite coefficient sequence, but also includes its forward asymptotic limits and much more. The coefficient sequence set Λ_L includes all possibilities, in fact, far more than may be of interest in particular situation.

If one is interested in the dynamics of a process corresponding to a specific $\hat{\lambda} \in \Lambda_L$, then it would suffice to consider the skew-product system w.r.t. the driving system on the smaller space $\Lambda_{\hat{\lambda}}$ defined as the hull of this sequence, i.e., the set of accumulation points of the set $\{\theta_n(\hat{\lambda}) : n \in \mathbb{Z}\}$ in the metric space $(\Lambda_L, d_{\Lambda_L})$. In particular, if $\hat{\lambda}$ is the specific sequence in (32), then the union $\cup_{\lambda \in \Lambda_{\hat{\lambda}}} A_\lambda = A_{\lambda_+^*} = [-\bar{\lambda}, \bar{\lambda}]$ contains all future limiting dynamics, i.e.,

$$\lim_{n \to \infty} \text{dist}\left(\varphi(n, \lambda, x), [-\bar{\lambda}, \bar{\lambda}]\right) = 0 \quad \text{for all } x \in \mathbb{R}.$$

The example described by nonautonomous difference equation (32) is asymptotically autonomous with $\Lambda_\lambda = \{\lambda_\pm^*\} \cup \{\theta_n(\lambda) : n \in \mathbb{Z}\}$. The forward limit points $\pm(\bar{\lambda} - 1)$ of the process generated by (25), which were not steady states of the process, are now locally asymptotic steady states of the skew product flow with base space $P = \bar{\Lambda}$ consisting of the single constant sequence $\lambda_k \equiv \bar{\lambda}$, when the skew product system is interpreted as an autonomous semidynamical system on the product space $P \times X$. More generally, unlike the process formulation, the skew-product system formulation and its pullback attractor include the forwards limiting dynamics.

5.4 Local Pullback Attractors

Less uniform behaviour such as parameter dependent domains of definition and local pullback attractors can be handled by introducing the concept of a basin of attraction system.

Let $\text{Dom}_p \subset X$ be the domain of definition of $f(p, \cdot)$ in the nonautonomous equation (13), which requires $f(p, \text{Dom}_p) \subset \text{Dom}_{\theta(p)}$. Then the corresponding cocycle mapping φ has the domain of definition $\mathbb{Z}^+ \times \cup_{p \in P} (\{p\} \times \text{Dom}_p)$. Consequently one needs to restrict the admissible families of bounded sets in the pullback convergence to subsets of Dom_p for each $p \in P$.

Definition 5.10. An ensemble $\mathfrak{D}_{ad}$ of families $\mathscr{D} = \{D_p : p \in P\}$ of nonempty subsets X is called *admissible* if

(i) D_p is bounded and $D_p \subset \text{Dom}_p$ for each $p \in P$ and every $\mathscr{D} = \{D_p : p \in P\}$ $\in \mathfrak{D}_{ad}$; and
(ii) $\widehat{D}^{(1)} = \{D_p^{(1)} : p \in P\} \in \mathfrak{D}_{ad}$ whenever $\widehat{D}^{(2)} = \{D_p^{(2)} : p \in P\} \in \mathfrak{D}_{ad}$ and $D_p^{(1)} \subseteq D_p^{(2)}$ for all $p \in P$.

Further restrictions will allow one to consider local or otherwise restricted form of pullback attraction.

Definition 5.11. A φ-invariant family $\mathscr{A} = \{A_p : p \in P\}$ of nonempty compact subsets of X with $A_p \subset \text{Dom}_p$ for each $p \in P$ is called a *pullback attractor* w.r.t. the basin of attraction system $\mathfrak{D}_{att}$ if $\mathfrak{D}_{att}$ is an admissible ensemble of families of subsets such that

$$\lim_{j \to \infty} \text{dist}\left(\varphi(j, \theta_{-j}(p), D_{\theta_{-j}(p)}), A_p\right) = 0 \tag{33}$$

for every $\mathscr{D} = \{D_p : p \in P\} \in \mathfrak{D}_{att}$.

In this case a pullback absorbing set system $\mathscr{B} = \{B_p : p \in P\}$ should also satisfy $\mathscr{B} \in \mathfrak{D}_{att}$ and the pullback absorbing property should be modified to

$$\varphi\left(j, \theta_{-j}(p), D_{\theta_{-j}(p)}\right) \subseteq B_p$$

for all $j \geq N_{p,\mathscr{D}}$, $p \in P$ and $\mathscr{D} = \{D_p; \ p \in P\} \in \mathfrak{D}_{att}$.

A counterpart of Theorem 5.4 then holds here. In this case the pullback attractor is unique within the basin of attraction system, but the skew-product system may have other pullback attractors within other basin of attraction systems, which may be either disjoint from or a proper sub-ensemble of the original basin of attraction system.

Example 5.12. Consider the scalar nonautonomous difference equation

$$x_{n+1} = f_n(x_n) := x_n + \gamma_n x_n \left(1 - x_n^2\right) \tag{34}$$

for given parameters $\gamma_n > 0$, $n \in \mathbb{Z}$.

First let $\gamma_n \equiv \bar{\gamma}$ for all $n \in \mathbb{Z}$, so the system is autonomous. It has the attractor $A^* = [-1, 1]$ for the maximal basin of attraction $\left(-1 - \bar{\gamma}^{-1}, 1 + \bar{\gamma}^{-1}\right)$, but if one restricts attention further to the basin of attraction $\left(0, 1 + \bar{\gamma}^{-1}\right)$ then the attractor is only $A^{**} = \{1\}$.

Now let γ_n be variable with $\gamma_n \in \left[\frac{1}{2}\bar{\gamma}, \bar{\gamma}\right]$ for each $n \in \mathbb{Z}$, so the system is now nonautonomous and representable as a skew-product on the state space $\mathbb{X} = \mathbb{Z} \times \mathbb{R}$ with the parameter set $P = \mathbb{Z}$. Then $\mathscr{A}^* = \{A_n^* : n \in \mathbb{Z}\}$ with $A_n^* = [-1, 1]$ for all $n \in \mathbb{Z}$ is the pullback attractor for the basin of attraction system $\mathfrak{D}_{att}$ consisting of all families $\mathscr{D} = \{D_n : n \in \mathbb{Z}\}$ satisfying $D_n \subset \left(-1 - \bar{\gamma}^{-1}, 1 + \bar{\gamma}^{-1}\right)$, whereas $\mathscr{A}^{**} = \{A_n^{**} : n \in \mathbb{Z}\}$ with $A_n^{**} = \{1\}$ for all $n \in \mathbb{Z}$ is the pullback attractor for the basin of attraction system $\mathfrak{D}_{att}$ consisting of all families $\mathscr{D} = \{D_n : n \subset \mathbb{Z}\}$ with $D_n \subset \left(0, 1 + \bar{\gamma}^{-1}\right)$.

6   Lyapunov Functions for Pullback Attractors

A Lyapunov function characterizing pullback attraction and pullback attractors for a discrete-time process in $\mathbb{R}^d$ will be constructed here. Consider a nonautonomous difference equation

$$x_{n+1} = f_n(x_n) \tag{Δ}$$

on $\mathbb{R}^d$, where the $f_n : \mathbb{R}^d \to \mathbb{R}^d$ are Lipschitz continuous mappings. This generates a process $\phi : \mathbb{Z}_{\geq}^2 \times \mathbb{R}^d \to \mathbb{R}^d$ through iteration by

$$\phi(n, n_0, x_0) = f_{n-1} \circ \cdots \circ f_{n_0}(x_0) \quad \text{for all } n \geq n_0$$

and each $x_0 \in \mathbb{R}^d$, which in particular satisfies the *continuity property*

$$x_0 \mapsto \phi(n, n_0, x_0) \quad \text{is Lipschitz continuous for all } n \geq n_0.$$

The pullback attraction is taken w.r.t. a basin of attraction system, which is defined as follows for a process.

Definition 6.1. A *basin of attraction system* $\mathfrak{D}_{att}$ consists of families $\mathscr{D} = \{D_n : n \in \mathbb{Z}\}$ of nonempty bounded subsets of $\mathbb{R}^d$ with the property that $\mathscr{D}^{(1)} = \{D_n^{(1)} : n \in \mathbb{Z}\} \in \mathfrak{D}_{att}$ if $\mathscr{D}^{(2)} = \{D_n^{(2)} : n \in \mathbb{Z}\} \in \mathfrak{D}_{att}$ and $D_n^{(1)} \subseteq D_n^{(2)}$ for all $n \in \mathbb{Z}$.

Although somewhat complicated, the use of such a basin of attraction system allows both nonuniform and local attraction regions, which are typical in nonautonomous systems, to be handled.

Definition 6.2. A ϕ-invariant family of nonempty compact subsets $\mathscr{A} = \{A_n : n \in \mathbb{Z}\}$ is called a *pullback attractor* w.r.t. a basin of attraction system $\mathfrak{D}_{att}$ if it is pullback attracting

$$\lim_{j \to \infty} \operatorname{dist} \big(\phi(n, n - j, D_{n-j}), A_n\big) = 0 \tag{35}$$

for all $n \in \mathbb{Z}$ and all $\mathscr{D} = \{D_n : n \in \mathbb{Z}\} \in \mathfrak{D}_{att}$.

Obviously $\mathscr{A} \in \mathfrak{D}_{att}$.

The construction of the Lyapunov function requires the existence of a pullback absorbing neighbourhood family.

6.1   Existence of a Pullback Absorbing Neighbourhood System

The following lemma shows that there always exists such a pullback absorbing neighbourhood system for any given pullback attractor. This will be required for the construction of the Lyapunov function for the proof of Theorem 6.4. The proof is very similar to that of Proposition 4.13.

Lemma 6.3. *If $\mathscr{A}$ is a pullback attractor with a basin of attraction system $\mathfrak{D}_{att}$ for a process ϕ, then there exists a pullback absorbing neighbourhood system $\mathscr{B} \subset \mathfrak{D}_{att}$ of $\mathscr{A}$ w.r.t. ϕ. Moreover, $\mathscr{B}$ is ϕ-positive invariant.*

Proof. For each $n_0 \in \mathbb{Z}$ pick $\delta_{n_0} > 0$ such that

$$B[A_{n_0}; \delta_{n_0}] := \{x \in \mathbb{R}^d : \operatorname{dist}(x, A_{n_0}) \leq \delta_{n_0}\}$$

satisfies $\{B[A_{n_0}; \delta_{n_0}] : n_0 \in \mathbb{Z}\} \in \mathfrak{D}_{att}$ and define

$$B_{n_0} := \overline{\bigcup_{j \geq 0} \phi(n_0, n_0 - j, B[A_{n_0-j}; \delta_{n_0-j}])}.$$

Obviously $A_{n_0} \subset \operatorname{int} B[A_{n_0}; \delta_{n_0}] \subset B_{n_0}$. To show positive invariance the two-parameter semigroup property will be used in what follows.

$$\phi(n_0 + 1, n_0, B_{n_0}) = \overline{\bigcup_{j \geq 0} \phi(n_0 + 1, n_0, \phi(n_0, n_0 - j, B[A_{n_0-j}; \delta_{n_0-j}]))}$$

$$= \overline{\bigcup_{j \geq 0} \phi(n_0 + 1, n_0 - j, B[A_{n_0-j}; \delta_{n_0-j}])}$$

$$= \overline{\bigcup_{i \geq 1} \phi(n_0 + 1, n_0 + 1 - i, B[A_{n_0+1-i}; \delta_{n_0+1-i}])}$$

$$\subseteq \overline{\bigcup_{i \geq 0} \phi(n_0 + 1, n_0 + 1 - i, B[A_{n_0+1-i}; \delta_{n_0+1-i}])} = B_{n_0+1},$$

so $\phi(n_0 + 1, n_0, B_{n_0}) \subseteq B_{n_0+1}$. This and the two-parameter semigroup property again gives

$$\phi(n_0 + 2, n_0, B_{n_0}) = \phi(n_0 + 2, n_0 + 1, \phi(n_0 + 1, n_0, B_{n_0}))$$

$$\subseteq \phi(n_0 + 2, n_0 + 1, B_{n_0+1}) \subseteq B_{n_0+2}.$$

The general positive invariance assertion then follows by induction.

Now referring to the continuity of $\phi(n_0, n_0 - j, \cdot)$ and the compactness of $B[A_{n_0-j}; \delta_{n_0-j}]$, the set $\phi(n_0, n_0 - j, B[A_{n_0-j}; \delta_{n_0-j}])$ is compact for each $j \geq 0$ and $n_0 \in \mathbb{Z}$. Moreover, by pullback convergence, there exists an $N = N(n_0, \delta_{n_0}) \in \mathbb{N}$ such that

$$\psi(n_0, n_0 - j, B[A_{n_0-j}; \delta_{n_0-j}]) \subseteq B[A_{n_0}; \delta_{n_0}] \subset B_{n_0}$$

for all $j \geq N$. Hence

$$B_{n_0} = \overline{\bigcup_{j \geq 0} \phi(n_0, n_0 - j, B[A_{n_0-j}; \delta_{n_0-j}])}$$

$$\subseteq B[A_{n_0}; \delta_{n_0}] \bigcup \overline{\bigcup_{0 \leq j < N} \phi(n_0, n_0 - j, B[A_{n_0-j}; \delta_{n_0-j}])}$$

$$= \overline{\bigcup_{0 \leq j < N} \phi(n_0, n_0 - j, B[A_{n_0-j}; \delta_{n_0-j}])},$$

which is compact, so B_{n_0} is compact.

To see that $\mathscr{B}$ so constructed is pullback absorbing w.r.t. $\mathfrak{D}_{att}$, let $\mathscr{D} \in \mathfrak{D}_{att}$. Fix $n_0 \in \mathbb{Z}$. Since $\mathscr{A}$ is pullback attracting, there exists an $N(\mathscr{D}, \delta_{n_0}, n_0) \in \mathbb{N}$ such that

$$\text{dist}\left(\phi(n_0, n_0 - j, D_{n_0-j}), A_{n_0}\right) < \delta_{n_0}$$

for all $j \geq N(\mathcal{D}, \delta_{n_0}, n_0)$. But $(\phi(n_0, n_0 - j, D_{n_0-j}) \subset \text{int} B[A_{n_0}; \delta_{n_0}]$ and $B[A_{n_0}; \delta_{n_0}] \subset B_{n_0}$, so

$$\phi(n_0, n_0 - j, D_{n_0-j}) \subset \text{int} B_{n_0}$$

for all $j \geq N(\mathcal{D}, \delta_{n_0}, n_0)$. Hence $\mathcal{B}$ is pullback absorbing as required. $\square$

6.2   Necessary and Sufficient Conditions

The main result is the construction of a Lyapunov function that characterizes this pullback attraction. See [21, 22].

Theorem 6.4. *Let the f_n be uniformly Lipschitz continuous on $\mathbb{R}^d$ for each $n \in \mathbb{Z}$ and let ϕ be the process that they generate. In addition, let $\mathscr{A}$ be a ϕ-invariant family of nonempty compact sets that is pullback attracting with respect to ϕ with a basin of attraction system $\mathcal{D}_{att}$. Then there exists a Lipschitz continuous function $V : \mathbb{Z} \times \mathbb{R}^d \to \mathbb{R}$ such that*

Property 1 (upper bound).   For all $n_0 \in \mathbb{Z}$ and $x_0 \in \mathbb{R}^d$

$$V(n_0, x_0) \leq \text{dist}(x_0, A_{n_0}); \tag{36}$$

Property 2 (lower bound).   For each $n_0 \in \mathbb{Z}$ there exists a function $a(n_0, \cdot) : \mathbb{R}^+ \to \mathbb{R}^+$ with $a(n_0, 0) = 0$ and $a(n_0, r) > 0$ for all $r > 0$ which is monotonically increasing in r such that

$$a(n_0, \text{dist}(x_0, A_{n_0})) \leq V(n_0, x_0) \quad \text{for all } x_0 \in \mathbb{R}^d; \tag{37}$$

Property 3 (Lipschitz condition).   For all $n_0 \in \mathbb{Z}$ and $x_0, y_0 \in \mathbb{R}^d$

$$|V(n_0, x_0) - V(n_0, y_0)| \leq \|x_0 - y_0\|; \tag{38}$$

Property 4 (pullback convergence).   For all $n_0 \in \mathbb{Z}$ and any $\mathcal{D} \in \mathcal{D}_{att}$

$$\limsup_{n \to \infty} \sup_{z_{n_0-n} \in D_{n_0-n}} V(n_0, \phi(n_0, n_0 - n, z_{n_0-n})) = 0. \tag{39}$$

In addition,

Property 5 (forwards convergence).   There exists $\mathscr{N} \in \mathcal{D}_{att}$. which is positively invariant under ϕ and consists of nonempty compact sets N_{n_0} with $A_{n_0} \subset \text{int} N_{n_0}$ for each $n_0 \in \mathbb{Z}$ such that

$$V(n_0 + 1, \phi(n_0 + 1, n_0, x_0)) \le e^{-1} V(n_0, x_0) \tag{40}$$

for all $x_0 \in N_{n_0}$ and hence

$$V(n_0 + j, \phi(j, n_0, x_0)) \le e^{-j} V(n_0, x_0) \quad \text{for all } x_0 \in N_{n_0}, \ j \in \mathbb{N}. \tag{41}$$

Proof. The aim is to construct a Lyapunov function $V(n_0, x_0)$ that characterizes a pullback attractor $\mathscr{A}$ and satisfies properties 1–5 of Theorem 6.4. For this define

$$V(n_0, x_0) := \sup_{n \in \mathbb{N}} e^{-T_{n_0,n}} \operatorname{dist}(x_0, \phi(n_0, n_0 - n, B_{n_0-n}))$$

for all $n_0 \in \mathbb{Z}$ and $x_0 \in \mathbb{R}^d$, where

$$T_{n_0,n} = n + \sum_{j=1}^{n} \alpha_{n_0-j}^{+}$$

with $T_{n_0,0} = 0$. Here $\alpha_n = \log L_n$, where L_n is the uniform Lipschitz constant of f_n on $\mathbb{R}^d$, and $a^+ = (a + |a|)/2$, i.e., the positive part of a real number a.
Note 4: $T_{n_0,n} \ge n$ and $T_{n_0,n+m} = T_{n_0,n} + T_{n_0-n,m}$ for $n, m \in \mathbb{N}, n_0 \in \mathbb{Z}$.

Proof of property 1

Since $e^{-T_{n_0,n}} \le 1$ for all $n \in \mathbb{N}$ and $\operatorname{dist}(x_0, \phi(n_0, n_0 - n, B_{n_0-n}))$ is monotonically increasing from $0 \le \operatorname{dist}(x_0, \phi(n_0, n_0, B_{n_0}))$ at $n = 0$ to $\operatorname{dist}(x_0, A_{n_0})$ as $n \to \infty$,

$$V(n_0, x_0) = \sup_{n \in \mathbb{N}} e^{-T_{n_0,n}} \operatorname{dist}(x_0, \phi(n_0, n_0 - n, B_{n_0-n})) \le 1 \cdot \operatorname{dist}(x_0, A_{n_0}).$$

Proof of property 2

If $x_0 \in A_{n_0}$, then $V(n_0, x_0) = 0$ by Property 1, so assume that $x_0 \in \mathbb{R}^d \setminus A_{n_0}$. Now in

$$V(n_0, x_0) = \sup_{n \ge 0} e^{-T_{n_0,n}} \operatorname{dist}(x_0, \phi(n_0, n_0 - n, B_{n_0-n}))$$

the supremum involves the product of an exponentially decreasing quantity bounded below by zero and a bounded increasing function, since the sets $\phi(n_0, n_0 - n, B_{n_0-n})$ are a nested family of compact sets decreasing to A_{n_0} with increasing n. In particular,

$$\operatorname{dist}(x_0, A_{n_0}) \ge \operatorname{dist}(x_0, \phi(n_0, n_0 - n, B_{n_0-n})) \quad \text{for all } n \in \mathbb{N}.$$

Hence there exists an $N^* = N^*(n_0, x_0) \in \mathbb{N}$ such that

$$\frac{1}{2}\text{dist}(x_0, A_{n_0}) \leq \text{dist}(x_0, \phi(n_o, n_0 - n, B_{n_0-n})) \leq \text{dist}(x_0, A_{n_0})$$

for all $n \geq N^*$, but not for $n = N^* - 1$. Then, from above,

$$V(n_0, x_0) \geq e^{-T_{n_0, N^*}} \text{dist}\left(x_0, \phi(n_0, n_0 - N^*, B_{n_0-N^*})\right)$$

$$\geq \frac{1}{2} e^{-T_{n_0, N^*}} \text{dist}\left(x_0, A_{n_0}\right).$$

Define
$$N^*(n_0, r) := \sup\{N^*(n_0, x_0) : \text{dist}\,(x_0, A_{n_0}) = r\}$$

Now $N^*(n_0, r) < \infty$ for $x_0 \notin A_{n_0}$ with $\text{dist}\,(x_0, A_{n_0}) = r$ and $N^*(n_0, r)$ is nondecreasing with $r \to 0$. To see this note that by the triangle rule

$$\text{dist}(x_0, A_{n_0}) \leq \text{dist}(x_0, \phi(n_0, n_0 - n, B_{n_0-n})) + \text{dist}(\phi(n_0, n_0 - n, B_{n_0-n}), A_{n_0}).$$

Also by pullback convergence there exists an $N(n_0, r/2)$ such that

$$\text{dist}(\phi(n_0, n_0 - n, B_{n_0-n}), A_{n_0}) < \frac{1}{2}r$$

for all $n \geq N(n_0, r/2)$. Hence for $\text{dist}(x_0, A_{n_0}) = r$ and $n \geq N(n_0, r/2)$,

$$r \leq \text{dist}(x_0, \phi(n_0, n_0 - n, B_{n_0-n})) + \frac{1}{2}r,$$

that is
$$\frac{1}{2}r \leq \text{dist}(x_0, \phi(n_0, n_0 - n, B_{n_0-n})).$$

Obviously $N^*(n_0, r) \leq N^*(n_0, r/2)$.

Finally, define

$$a(n_0, r) := \frac{1}{2}r\, e^{-T_{n_0, N^*(n_0, r)}}. \tag{42}$$

Note that there is no guarantee here (without further assumptions) that $a(n_0, r)$ does not converge to 0 for fixed $r \neq 0$ as $n_0 \to \infty$.

Proof of property 3

$$|V(n_0, x_0) - V(n_0, y_0)|$$

$$= \left| \sup_{n \in \mathbb{N}} e^{-T_{n_0, n}} \text{dist}\,(x_0, \phi(n_0, n_0 - n, B_{n_0-n})) \right.$$

$$\left. - \sup_{n \in \mathbb{N}} e^{-T_{n_0, n}} \text{dist}\,(y_0, \phi(n_0, n_0 - n, B_{n_0-n})) \right|$$

$$\leq \sup_{n \in \mathbb{N}} e^{-T_{n_0,n}} |\text{dist}(x_0, \phi(n_0, n_0 - n, B_{n_0-n})) - \text{dist}(y_0, \phi(n_0, n_0 - n, B_{n_0-n}))|$$

$$\leq \sup_{n \in \mathbb{N}} e^{-T_{n_0,n}} \|x_0 - y_0\| \leq \|x_0 - y_0\|$$

since

$$|\text{dist}(x_0, C) - \text{dist}(y_0, C)| \leq \|x_0 - y_0\|$$

for any $x_0, y_0 \in \mathbb{R}^d$ and nonempty compact subset C of $\mathbb{R}^d$.

Proof of property 4

Assume the opposite. Then there exists an $\varepsilon_0 > 0$, a sequence $n_j \to \infty$ in $\mathbb{N}$ and points $x_j \in \phi(n_0, n_0 - n_j, D_{n_0-n_j})$ such that $V(n_0, x_j) \geq \varepsilon_0$ for all $j \in \mathbb{N}$. Since $\mathscr{D} \in \mathfrak{D}_{att}$ and $\mathscr{B}$ is pullback absorbing, there exists an $N = N(\mathscr{D}, n_0) \in \mathbb{N}$ such that

$$\phi(n_0, n_0 - n_j, D_{n_0-n_j}) \subset B_{n_0} \quad \text{for all } n_j \geq N.$$

Hence, for all j such that $n_j \geq N$, it holds $x_j \in B_{n_0}$, which is a compact set, so there exists a convergent subsequence $x_{j'} \to x^* \in B_{n_0}$. But also

$$x_{j'} \in \overline{\bigcup_{n \geq n_{j'}} \phi(n_0, n_0 - n, D_{n_0-n})}$$

and

$$\bigcap_{n_{j'}} \overline{\bigcup_{n \geq n_{j'}} \phi(n_0, n_0 - n, D_{n_0-n})} \subseteq A_{n_0}$$

by the definition of a pullback attractor. Hence $x^* \in A_{n_0}$ and $V(n_0, x^*) = 0$. But V is Lipschitz continuous in its second variable by property 3, so

$$\varepsilon_0 \leq V(n_0, x_{j'}) = \|V(n_0, x_{j'}) - V(n_0, x^*)\| \leq \|x_{j'} - x^*\|,$$

which contradicts the convergence $x_{j'} \to x^*$. Hence property 4 must hold.

Proof of property 5

Define

$$N_{n_0} := \{x_0 \in B[B_{n_0}; 1] : \phi(n_0 + 1, n_0, x_0) \in B_{n_0+1}\},$$

where $B[B_{n_0}; 1] = \{x_0 : \text{dist}(x_0, B_{n_0}) \leq 1\}$ is bounded because B_{n_0} is compact and $\mathbb{R}^d$ is locally compact, so N_{n_0} is bounded. It is also closed, hence compact, since $\phi(n_0 + 1, n_0, \cdot)$ is continuous and B_{n_0+1} is compact. Now $A_{n_0} \subset \text{int} B_{n_0}$ and $B_{n_0} \subset N_{n_0}$, so $A_{n_0} \subset \text{int} N_{n_0}$. In addition,

$$\phi(n_0 + 1, n_0, N_{n_0}) \subset B_{n_0+1} \subset N_{n_0+1},$$

so $\mathcal{N}$ is positive invariant.

It remains to establish the exponential decay inequality (40). This needs the following Lipschitz condition on $\phi(n_0 + 1, n_0, \cdot) \equiv f_{n_0}(\cdot)$:

$$\|\phi(n_0 + 1, n_0, x_0) - \phi(n_0 + 1, n_0, y_0)\| \le e^{\alpha_{n_0}} \|x_0 - y_0\|$$

for all $x_0, y_0 \in D_{n_0}$. It follows from this that

$$\mathrm{dist}(\phi(n_0 + 1, n_0, x_0), \phi(n_0 + 1, n_0, C_{n_0})) \le e^{\alpha_{n_0}} \mathrm{dist}(x_0, C_{n_0})$$

for any compact subset $C_{n_0} \subset \mathbb{R}^d$. From the definition of V,

$V(n_0 + 1, \phi(n_0 + 1, n_0, x_0))$

$$= \sup_{n \ge 0} e^{-T_{n_0+1,n}} \mathrm{dist}(\phi(n_0 + 1, n_0, x_0), \phi(n_0, n_0 - n, B_{n_0-n}))$$

$$= \sup_{n \ge 1} e^{-T_{n_0+1,n}} \mathrm{dist}(\phi(n_0 + 1, n_0, x_0), \phi(n_0, n_0 - n, B_{n_0-n}))$$

since $\phi(n_0 + 1, n_0, x_0) \in B_{n_0+1}$ when $x_0 \in N_{n_0}$. Hence re-indexing and then using the two-parameter semigroup property and the Lipschitz condition on $\phi(1, n_0, \cdot)$

$V(n_0 + 1, \phi(n_0 + 1, n_0, x_0))$

$$= \sup_{j \ge 0} e^{-T_{n_0+1,j+1}} \mathrm{dist}(\phi(n_0 + 1, n_0, x_0), \phi(n_0, n_0 - j - 1, B_{n_0-j-1}))$$

$$= \sup_{j \ge 0} e^{-T_{n_0+1,j+1}} \mathrm{dist}(\phi(n_0 + 1, n_0, x_0), \phi(n_0 + 1, n_0, \phi(n_0, n_0 - j, B_{n_0-j})))$$

$$\le \sup_{j \ge 0} e^{-T_{n_0+1,j+1}} e^{\alpha_{n_0}} \mathrm{dist}(x_0, \phi(n_0, n_0 - j, B_{n_0-j}))$$

Now $T_{n_0+1,j+1} = T_{n_0,j} + 1 - \alpha_{n_0}^+$, so

$$V(n_0 + 1, \phi(n_0 + 1, n_0, x_0))$$

$$\le \sup_{j \ge 0} e^{-T_{n_0+1,j+1} + \alpha_{n_0}} \mathrm{dist}(x_0, \phi(n_0 + j, n_0 - j, B_{n_0-j}))$$

$$= \sup_{j \ge 0} e^{-T_{n_0,j} - 1 - \alpha_{n_0}^+ + \alpha_{n_0}} \mathrm{dist}(x_0, \phi(n_0, n_0 - j, B_{n_0-j}))$$

$$\le e^{-1} \sup_{j \ge 0} e^{-T_{n_0,j}} \mathrm{dist}(x_0, \phi(n_0, n_0 - j, B_{n_0-j})) \le e^{-1} V(n_0, x_0),$$

which is the desired inequality. Moreover, since $\phi(1, n_0, x_0) \in B_{n_0+1} \subset N_{n_0+1}$, the proof continues inductively to give

$$V(n_0 + j, \phi(n_0 + j, n_0, x_0)) \leq e^{-j} V(n_0, x_0) \quad \text{for all } j \in \mathbb{N}.$$

This completes the proof of Theorem 6.4.   □

6.2.1   Comments on Theorem 6.4

Note 1:   It would be nice to use $\phi(n_0, n_0 - n, x_0)$ for a fixed x_0 in the pullback convergence property (39), but this may not always be possible due to nonuniformity of the attraction region, i.e., there may not be a $\mathcal{D} \in \mathfrak{D}_{att}$ with $x_0 \in D_{n_0-n}$ for all $n \in \mathbb{N}$.

Note 2:   The forwards convergence inequality (41) does not imply forwards Lyapunov stability or Lyapunov asymptotical stability. Although

$$a(n_0 + j, \text{dist}(\phi(n_0 + j, n_0, x_0), A_{n_0+j})) \leq e^{-j} V(n_0, x_0)$$

there is no guarantee (without additional assumptions) that

$$\inf_{j \geq 0} a(n_0 + j, r) > 0$$

for $r > 0$, so $\text{dist}(\phi(n_0 + j, n_0, x_0), A_{n_0+j})$ need not become small as $j \to \infty$.

As a counterexample consider Example 4.6 of the process ϕ on $\mathbb{R}$ generated by (Δ) with $f_n = g_1$ for $n \leq 0$ and $f_n = g_2$ for $n \geq 1$ where the mappings $g_1, g_2 : \mathbb{R} \to \mathbb{R}$ are given by $g_1(x) := \frac{1}{2}x$ and $g_2(x) := \max\{0, 4x(1-x)\}$ for all $x \in \mathbb{R}$. Then $\mathscr{A}$ with $A_{n_0} = \{0\}$ for all $n_0 \subset \mathbb{Z}$ is pullback attracting for ϕ but is not forwards Lyapunov asymptotically stable. (Note one can restrict g_1, g_2 to $[-R, R] \to [-R, R]$ for any fixed $R > 1$ to ensure the required uniform Lipschitz continuity of the f_n).

Note 3:   The forwards convergence inequality (41) can be rewritten as

$$V(n_0, \phi(n_0, n_0 - j, x_{n_0-j})) \leq e^{-j} V(n_0 - j, x_{n_0-j}) \leq e^{-j} \text{dist}(x_{n_0-j}, A_{n_0-j})$$

for all $x_{n_0-j} \in N_{n_0-j}$ and $j \in \mathbb{N}$.

Definition 6.6.   A family $\mathcal{D} \in \mathfrak{D}_{att}$ is called *past–tempered* w.r.t. $\mathscr{A}$ if

$$\lim_{j \to \infty} \frac{1}{j} \log^+ \text{dist}(D_{n_0-j}, A_{n_0-j}) = 0 \quad \text{for all } n_0 \in \mathbb{Z},$$

or equivalently if

$$\lim_{j \to \infty} e^{-\gamma j} \text{dist}(D_{n_0-j}, A_{n_0-j}) = 0 \quad \text{for all } n_0 \in \mathbb{Z}, \gamma > 0.$$

This says that there is at most subexponential growth backwards in time of the starting sets. It is reasonable to restrict attention to such sets.

For a past-tempered family $\mathscr{D} \subset \mathscr{N}$ it follows that

$$V(n_0, \phi(n_0, n_0 - j, x_{n_0-j})) \leq e^{-j} \operatorname{dist}(D_{n_0-j}, A_{n_0-j}) \longrightarrow 0$$

as $j \to \infty$. Hence

$$a\left(n_0, \operatorname{dist}(\phi(n_0, n_0 - j, x_{n_0-j}), A_{n_0})\right) \leq e^{-j} \operatorname{dist}(D_{n_0-j}, A_{n_0-j}) \longrightarrow 0$$

as $j \to \infty$. Since n_0 is fixed in the lower expression, this implies the pullback convergence

$$\lim_{j \to \infty} \operatorname{dist}(\phi(n_0, n_0 - j, D_{n_0-j}), A_{n_0}) = 0.$$

A rate of pullback convergence for more general sets $\mathscr{D} \in \mathfrak{D}_{att}$ will be considered in the next subsection.

6.2.2 Rate of Pullback Convergence

Since $\mathscr{B}$ is a pullback absorbing neighbourhood system, then for every $n_0 \in \mathbb{Z}$, $n \in \mathbb{N}$ and $\mathscr{D} \in \mathfrak{D}_{att}$ there exists an $N(\mathscr{D}, n_0, n) \in \mathbb{N}$ such that

$$\phi(n_0 - n, n_0 - n - m, D_{n_0-n-m}) \subseteq B_{n_0-n} \quad \text{for all } m \geq N.$$

Hence, by the two-parameter semigroup property,

$$\phi(n_0, n_0 - n - m, D_{n_0-n-m})$$
$$= \phi(n_0, n_0 - n, \phi(n_0 - n, n_0 - n - m, D_{n_0-n-m}))$$
$$\subseteq \phi(n_0, n_0 - n, B_{n_0-n})$$
$$= \phi(n_0, n_0 - i, \phi(n_0 - i, n_0 - n, B_{n_0-n}))$$
$$\subseteq \phi(n_0, n_0 - i, B_{n_0-i}) \quad \text{for all } m \geq N, \, 0 \leq i \leq n,$$

where the positive invariance of $\mathscr{B}$ was used in the last line. Hence

$$\phi(n_0, n_0 - n - m, D_{n_0-n-m}) \subseteq \phi(n_0, n_0 - i, B_{n_0-i})$$

for all $m \geq N(\mathscr{D}, n_0, n)$ and $0 \leq i \leq n$, or equivalently

$$\phi(n_0, n_0 - m, D_{n_0-m}) \subseteq \phi(n_0, n_0 - i, B_{n_0-i}) \quad \text{for all } m \geq n + N(\mathscr{D}, n_0, n)$$

and $0 \leq i \leq n$. This means that for any $z_{n_0-m} \in D_{n_0-m}$ the supremum in

$$V(n_0, \phi(n_0, n_0 - m, z_{n_0-m}))$$
$$= \sup_{i \geq 0} e^{-T_{n_0,i}} \operatorname{dist}(\phi(n_0, n_0 - m, z_{n_0-m}), \phi(n_0, n_0 - i, B_{n_0-i}))$$

need only be considered over $i \geq n$. Hence

$$V(n_0, \phi(n_0, n_0 - m, z_{n_0-m}))$$
$$= \sup_{i \geq n} e^{-T_{n_0,i}} \operatorname{dist}(\phi(n_0, n_0 - m, z_{n_0-m}), \phi(n_0, n_0 - i, B_{n_0-i}))$$

$$\leq e^{-T_{n_0,n}} \sup_{j \geq 0} e^{-T_{n_0-n,j}} \operatorname{dist}(\phi(n_0, n_0 - m, z_{n_0-m}), \phi(n_0, n_0 - n - j, B_{n_0-n-j}))$$

$$\leq e^{-T_{n_0,n}} \operatorname{dist}(\phi(n_0, n_0 - m, z_{n_0-m}), A_{n_0})$$

$$\leq e^{-T_{n_0,n}} \operatorname{dist}(B_{n_0}, A_{n_0})$$

since $A_{n_0} \subseteq \phi(n_0, n_0 - n - j, B_{n_0-n-j})$ and $\phi(n_0, n_0 - m, z_{n_0-m}) \in B_{n_0}$. Thus

$$V(n_0, \phi(n_0, n_0 - m, z_{n_0-m})) \leq e^{-T_{n_0,n}} \operatorname{dist}(B_{n_0}, A_{n_0})$$

for all $z_{n_0-m} \in D_{n_0-m}$, $m \geq n + N(\mathscr{D}, n_0, n)$ and $n \geq 0$.

It can be assumed that the mapping $n \mapsto n + N(\mathscr{D}, n_0, n)$ is monotonic increasing in n (by taking a larger $N(\mathscr{D}, n_0, n)$ if necessary), and is hence invertible. Let the inverse of $m = n + N(\mathscr{D}, n_0, n)$ be $n = M(m) = M(\mathscr{D}, n_0, m)$. Then

$$V(n_0, \phi(n_0, n_0 - m, z_{n_0-m})) \leq e^{-T_{n_0,M(m)}} \operatorname{dist}(B_{n_0}, A_{n_0})$$

for all $m \geq N(\mathscr{D}, n_0, 0) \geq 0$. Usually $N(\mathscr{D}, n_0, 0) > 0$. This expression can be modified to hold for all $m \geq 0$ by replacing $M(m)$ by $M^*(m)$ defined for all $m \geq 0$ and introducing a constant $K_{\mathscr{D}, n_0} \geq 1$ to account for the behaviour over the finite time set $0 \leq m < N(\mathscr{D}, n_0, 0)$. For all $m \geq 0$ this gives

$$V(n_0, \phi(n_0, n_0 - m, z_{n_0-m})) \leq K_{\mathscr{D}, n_0} e^{-T_{n_0,M^*(m)}} \operatorname{dist}(B_{n_0}, A_{n_0}) .$$

7 Bifurcations

The classical theory of dynamical bifurcation focusses on autonomous difference equations

$$x_{n+1} = g(x_n, \lambda) \tag{43}$$

with a right-hand side $g : \mathbb{R}^d \times \Lambda \to \mathbb{R}^d$ depending on a parameter λ from some parameter space Λ, which is typically a subset of $\mathbb{R}^n$ (cf., e.g., [30] or [17]). A central question is how stability and multiplicity of invariant sets for (43) changes when the parameter λ is varied. In the simplest, and most often considered situation, these invariant sets are fixed points or periodic solutions to (43).

Given some parameter value λ^*, a fixed point $x^* = g(x^*, \lambda^*)$ of (43) is called *hyperbolic*, if the derivative $D_1 g(x^*, \lambda^*)$ has no eigenvalue on the complex unit circle $\mathbb{S}^1$. Then it is an easy consequence of the implicit function theorem (cf. [31, p. 365, Theorem 2.1]) that x^* allows a unique continuation $x(\lambda) \equiv g(x(\lambda), \lambda)$ in a neighborhood of λ^*. In particular, hyperbolicity rules out bifurcations understood as topological changes in the set $\{x \in \mathbb{R}^d : g(x, \lambda) = x\}$ near (x^*, λ^*) or a stability change of x^*.

On the other hand, eigenvalues on the complex unit circle give rise to various well-understood autonomous bifurcation scenarios. Examples include fold, transcritical or pitchfork bifurcations (eigenvalue 1), flip bifurcations (eigenvalue -1) or the Sacker–Neimark bifurcation (a pair of complex conjugate eigenvalues for $d \geq 2$).

7.1 Hyperbolicity and Simple Examples

Even in the autonomous set-up of (43) one easily encounters intrinsically nonautonomous problems, where the classical methods of, for instance, [17, 30] do not apply:

1. Investigate the behaviour of (43) along an entire reference solution $(\chi_n)_{n \in \mathbb{Z}}$, which is not constant or periodic. This is typically done using the (obviously nonautonomous) *equation of perturbed motion*

$$x_{n+1} = g(x_n + \chi_n, \lambda) - g(\chi_n, \lambda).$$

2. Replace the constant parameter λ in (43) by a sequence $(\lambda_n)_{n \in \mathbb{Z}}$ in Λ, which varies in time. Also the resulting *parametrically perturbed* equation

$$x_{n+1} = g(x_n, \lambda_n)$$

becomes nonautonomous. This situation is highly relevant from an applied point of view, since parameters in real world problems are typically subject to random perturbations or an intrinsic background noise.

Both the above problems fit into the framework of general nonautonomous difference equations

$$x_{n+1} = f_n(x_n, \lambda) \tag{Δ_λ}$$

with a sufficiently smooth right-hand side $f_n : \mathbb{R}^d \times \Lambda \to \mathbb{R}^d$, $n \in \mathbb{Z}$. In addition, suppose that f_n and its derivatives map bounded subsets of $\mathbb{R}^d \times \Lambda$ into bounded sets uniformly in $n \in \mathbb{Z}$.

Generically, nonautonomous equations (Δ_λ) do not have constant solutions, and the fixed point sequences $x_n^* = f_n(x_n^*, \lambda^*)$ are usually not solutions to (Δ_λ). This gives rise to the following question:

If there are no equilibria, what should bifurcate in a nonautonomous set-up?

Before suggesting an answer, a criterion to exclude bifurcations is proposed. For motivational purposes consider again the autonomous case (43) and the problem of parametric perturbations.

Example 7.1. The autonomous difference equation $x_{n+1} = \frac{1}{2}x_n + \lambda$ has the unique fixed point $x^*(\lambda) = 2\lambda$ for all $\lambda \in \mathbb{R}$. Replace λ by a bounded sequence $(\lambda_n)_{n\in\mathbb{Z}}$ and observe as in Example 4.5 that the nonautonomous counterpart

$$x_{n+1} = \frac{1}{2}x_n + \lambda_n$$

has a unique bounded entire solution $\chi_n^* := \sum_{k=-\infty}^{n-1} \left(\frac{1}{2}\right)^{n-k-1} \lambda_k$. For the special case $\lambda_n \equiv \lambda$, this solution reduces to the known fixed point $\chi_n^* \equiv 2\lambda$.

This simple example yields the conjecture that equilibria of autonomous equations persist as bounded entire solutions under parametric perturbations. It will be shown below in Theorem 7.5 (or in [37, Theorem 3.4]) that this conjecture is generically true in the sense that the fixed point of (43) has to be hyperbolic in order to persist under parametric perturbations.

Example 7.2. The linear difference equation $x_{n+1} = x_n + \lambda_n$ has the forward solution $x_n = x_0 + \sum_{k=0}^{n-1} \lambda_n$, whose boundedness requires the assumption that the real sequence $(\lambda_n)_{n\geq 0}$ is summable. Thus, the nonhyperbolic equilibria x^* of $x_{n+1} = x_n$ do not necessarily persist as bounded entire solutions under arbitrary bounded parametric perturbations.

Typical examples of nonautonomous equations having an equilibrium, given by the trivial solution are equations of perturbed motion. Their variational equation along $(\chi_n)_{n\in\mathbb{Z}}$ is given by $x_{n+1} = D_1 g(\chi_n, \lambda)x_n$ and investigating the behaviour of its trivial solution under variation of λ requires an appropriate nonautonomous notion of hyperbolicity.

Suppose that $A_n \in \mathbb{R}^{d\times d}$, $n \in \mathbb{Z}$, is a sequence of invertible matrices, and consider a linear difference equation

$$x_{n+1} = A_n x_n \tag{44}$$

with the *transition matrix*

$$\Phi(n,l) := \begin{cases} A_{n-1}\cdots A_l, & l < n, \\ I, & n = l, \\ A_n^{-1}\cdots A_{l-1}^{-1}, & n < l. \end{cases}$$

Let $\mathbb{I}$ be a discrete interval and define $\mathbb{I}' := \{k \in \mathbb{I} : k + 1 \in \mathbb{I}\}$. An *invariant projector* for (44) is a sequence $P_n \in \mathbb{R}^{d \times d}$, $n \in \mathbb{I}$, of projections $P_n = P_n^2$ such that

$$A_{n+1} P_n = P_n A_n \quad \text{for all } n \in \mathbb{I}'.$$

Definition 7.3. A linear difference equation (44) is said to admit an *exponential dichotomy* on $\mathbb{I}$, if there exist an invariant projector P_n and real numbers $K \geq 0$, $\alpha \in (0, 1)$ such that for all $n, l \in \mathbb{I}$ one has

$$\|\Phi(n, l) P_l\| \leq K\alpha^{n-l} \quad \text{if } l \leq n,$$

$$\|\Phi(n, l)[\text{id} - P_l]\| \leq K\alpha^{l-n} \quad \text{if } n \leq l.$$

Remark 7.4. An autonomous difference equation $x_{n+1} = Ax_n$ has an exponential dichotomy, if and only if the coefficient matrix $A \in \mathbb{R}^{d \times d}$ has no eigenvalues on the complex unit circle.

In terms of this terminology an entire solution $(\chi_n)_{n \in \mathbb{Z}}$ of (Δ_λ) is called *hyperbolic*, if the variational equation

$$x_{n+1} = D_1 f_n(\chi_n, \lambda) x_n \tag{V_λ}$$

has an exponential dichotomy on $\mathbb{Z}$.

Let ℓ^∞ denote the space of bounded sequences in $\mathbb{R}^d$.

Theorem 7.5 (Continuation of bounded entire solutions). *If $\chi^* = (\chi_n^*)_{n \in \mathbb{Z}}$ is an entire bounded and hyperbolic solution of (Δ_{λ^*}), then there exists an open neighborhood $\Lambda_0 \subseteq \Lambda$ of λ^* and a unique function $\chi : \Lambda_0 \to \ell^\infty$ such that*

(i) $\chi(\lambda^*) = \chi^*$,
(ii) *Each $\chi(\lambda)$ is a bounded entire and hyperbolic solution of (Δ_λ),*
(iii) $\chi : \Lambda_0 \to \ell^\infty$ *is as smooth as the functions f_n.*

Proof. The proof is based on the idea to formulate a nonautonomous difference equation (Δ_λ) as an abstract equation $F(\chi, \lambda) = 0$ in the space ℓ^∞. This is solved using the implicit mapping theorem, where the invertibility of the Fréchet derivative $D_1 F(\chi^*, \lambda^*)$ is characterised by the hyperbolicity assumption on χ^*. For details, see [40, Theorem 2.11].   □

Consequently, in order to deduce sufficient conditions for bifurcations, one must violate the hyperbolicity of χ^*. For this purpose, the following characterisation of an exponential dichotomy is useful.

Theorem 7.6 (Characterization of exponential dichotomies). *A variational equation (V_λ) has an exponential dichotomy on $\mathbb{Z}$, if and only if the following conditions are fulfilled:*

(i) *(V_λ) has an exponential dichotomy on $\mathbb{Z}^+$ with projector P_n^+, as well as an exponential dichotomy on $\mathbb{Z}^-$ with projector P_n^-,*

(ii) $R(P_0^+) \oplus N(P_0^-) = \mathbb{R}^d$.

Proof. See [9, Lemma 2.4].   □

The subsequent examples illustrate various scenarios that can arise, if a condition stated in Theorem 7.6 is violated.

Example 7.7 (Pitchfork bifurcation). Consider the difference equation

$$x_{n+1} = f_n(x_n, \lambda), \qquad\qquad f_n(x, \lambda) := \frac{\lambda x}{1 + |x|}$$

from Sect. 4.5. It is a prototypical example of a supercritical autonomous pitchfork bifurcation (cf., e.g. [30, pp. 119ff, Sect. 4.4]), where the unique asymptotically stable equilibrium $x^* = 0$ for $\lambda \in (0, 1)$ bifurcates into two asymptotically stable equilibria $x_\pm := \pm(\lambda - 1)$ for $\lambda > 1$.

Along the trivial solution the variational equation $x_{n+1} = \lambda x_n$ becomes nonhyperbolic for $\lambda = 1$. Indeed, criterion (i) of Theorem 7.6 is violated, since the variational equation does not admit a dichotomy on $\mathbb{Z}^+$ or on $\mathbb{Z}^-$. This loss of hyperbolicity causes an attractor bifurcation, since for

- $\lambda \in (0, 1)$, the set $x^* = 0$ is the global attractor
- $\lambda > 1$, the trivial equilibrium $x^* = 0$ becomes unstable and the symmetric interval $A = [x_-, x_+]$ is the global attractor.

Bifurcations of pullback attractors can be observed as nonautonomous versions of pitchfork bifurcations.

Example 7.8 (Pullback attractor bifurcation). Consider for parameter values $\lambda > 0$ the difference equation

$$x_{n+1} = \lambda x_n - \begin{cases} \min\left\{a_n x_n^3, \frac{\lambda}{2} x_n\right\}, & x_n \geq 0, \\ \max\left\{a_n x_n^3, \frac{\lambda}{2} x_n\right\}, & x_n < 0, \end{cases}$$

where $(a_n)_{n \in \mathbb{Z}}$ is a sequence which is both bounded and bounded away from zero. Note that in a neighborhood U of 0, the difference equation is given by $x_{n+1} = \lambda x_n - a_n x_n^3$, and outside of a set $V \supset U$, the difference equation is given by $x_{n+1} = \frac{\lambda}{2} x_n$. Both U and V here can be chosen independently of λ near $\lambda = 1$. Moreover, for fixed $n \in \mathbb{Z}$, the right-hand side of this equation lies between the functions $x \mapsto \frac{\lambda}{2} x$ and $x \mapsto \lambda x$.

It is clear that for $\lambda \in (0, 1)$, the global pullback attractor is given by the trivial solution, which follows from the fact that points are contracted at each time step by the factor λ. For $\lambda > 1$, the trivial solution is no longer attractive, but there exists a (nontrivial) pullback attractor for $\lambda \in (1, 2)$. This follows from Theorem 5.4, because the family $\mathcal{B} = \{V : n \in \mathbb{Z}\}$ is pullback absorbing (the right-hand is given by $x \mapsto \frac{\lambda}{2} x$ outside of V).

At the parameter value $\lambda = 1$, the global pullback attractor changes its dimension. Thus, this difference equation provides an example of a nonautonomous pitchfork bifurcation, which will be treated below in Sect. 7.2.

While these two examples show how (autonomous) bifurcations can be understood as attractor bifurcations, the following scenario is intrinsically nonautonomous (see [36] for a deeper analysis).

Example 7.9 (Shovel bifurcation). Consider a scalar difference equation

$$x_{n+1} = a_n(\lambda)x_n, \qquad a_n(\lambda) := \begin{cases} \frac{1}{2} + \lambda, & n < 0, \\ \lambda, & n \geq 0, \end{cases} \qquad (45)$$

with parameters $\lambda > 0$. In order to understand the dynamics of (45), distinguish three cases:

(i) $\lambda \in (0, \frac{1}{2})$: The equation (45) has an exponential dichotomy on $\mathbb{Z}$ with projector $P_n \equiv 1$. The uniquely determined bounded entire solution is the trivial one, which is uniformly asymptotically stable.

(ii) $\lambda > 1$: The equation (45) has an exponential dichotomy on $\mathbb{Z}$ with projector $P_k \equiv 0$. Again, 0 is the unique bounded entire solution, but is now unstable.

(iii) $\lambda \in (\frac{1}{2}, 1)$: In this situation, (45) has an exponential dichotomy on $\mathbb{Z}^+$ with projector $P_n^+ \equiv 1$, as well as an exponential dichotomy on $\mathbb{Z}^-$ with projector $P_n^- \equiv 0$. Thus condition (ii) in Theorem 7.6 is violated and 0 is a nonhyperbolic solution. For this parameter regime, every solution of (45) is bounded. Moreover, (45) is asymptotically stable, but not uniformly asymptotically stable on the whole time axis $\mathbb{Z}$.

The parameter values $\lambda \in \{\frac{1}{2}, 1\}$ are critical. In both situations, the number of bounded entire solutions to the linear difference equation (45) changes drastically. Furthermore, there is a loss of stability in two steps: From uniformly asymptotically stable to asymptotically stable, and finally to unstable, as λ increases through the values $\frac{1}{2}$ and 1. Hence, both values can be considered as bifurcation values, since the number of bounded entire solutions changes as well as their stability properties.

The next example requires the state space to be at least two-dimensional.

Example 7.10 (Fold solution bifurcation). Consider the planar equation

$$x_{n+1} = f_n(x_n, \lambda) := \begin{pmatrix} b_n & 0 \\ 0 & c_n \end{pmatrix} x_n + \begin{pmatrix} 0 \\ (x_n^1)^2 \end{pmatrix} - \lambda \begin{pmatrix} 0 \\ 1 \end{pmatrix} \qquad (46)$$

with components $x_n = (x_n^1, x_n^2)$, depending on a parameter $\lambda \in \mathbb{R}$ and asymptotically constant sequences

$$b_n := \begin{cases} 2, & n < 0, \\ \frac{1}{2}, & n \geq 0, \end{cases} \qquad c_n := \begin{cases} \frac{1}{2}, & n < 0, \\ 2, & n \geq 0. \end{cases} \qquad (47)$$

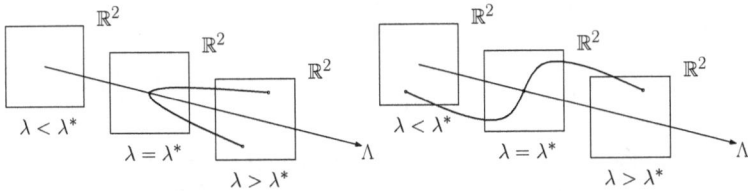

Fig. 8 *Left* (supercritical fold): Initial values $\eta \in \mathbb{R}^2$ yielding a bounded solution $\phi_\lambda(\cdot, 0, \eta)$ of (46) for different parameter values λ.
Right (cusp): Initial values $\eta \in \mathbb{R}^2$ yielding a bounded solution $\phi_\lambda(\cdot, 0, \eta)$ of (49) for different parameter values λ

The variational equation for (46) corresponding to the trivial solution and the parameter $\lambda^* = 0$ reads as

$$x_{n+1} = D_1 f_n(0,0)x_n := \begin{pmatrix} b_n & 0 \\ 0 & c_n \end{pmatrix} x_n.$$

It admits an exponential dichotomy on $\mathbb{Z}^+$, as well as on $\mathbb{Z}^-$ with corresponding invariant projectors $P_n^+ \equiv \begin{pmatrix} 1 & 0 \\ 0 & 0 \end{pmatrix}$ and $P_n^- \equiv \begin{pmatrix} 0 & 0 \\ 0 & 1 \end{pmatrix}$. This yields

$$R(P_0^+) \cap N(P_0^-) = \mathbb{R} \begin{pmatrix} 1 \\ 0 \end{pmatrix}, \qquad R(P_0^+) + N(P_0^-) = \mathbb{R} \begin{pmatrix} 1 \\ 0 \end{pmatrix}$$

and therefore condition (ii) of Theorem 7.6 is violated. Hence, the trivial solution to (46) for $\lambda = 0$ is not hyperbolic.

Let $\phi_\lambda(\cdot, 0, \eta)$ be the general solution to (46). Its first component ϕ_λ^1 is

$$\phi_\lambda^1(n, 0, \eta) = 2^{-|n|}\eta_1 \quad \text{for all } n \in \mathbb{Z}, \tag{48}$$

while the variation of constants formula (cf. [1, p. 59]) can be used to deduce the asymptotic representation

$$\phi_\lambda^2(n, 0, \eta) = \begin{cases} 2^n \left(\eta_2 + \frac{4}{7}\eta_1^2 - \lambda \right) + O(1), & n \to \infty, \\ \frac{1}{2^n} \left(\eta_2 - \frac{1}{2}\eta_1^2 + 2\lambda \right) + O(1), & n \to -\infty. \end{cases}$$

Therefore, the sequence $\phi_\lambda(\cdot, 0, \eta)$ is bounded if and only if $\eta_2 = -\frac{4}{7}\eta_1^2 + \lambda$ and $\eta_2 = \frac{1}{2}\eta_1^2 - 2\lambda$ holds, i.e., $\eta_1^2 = \frac{7}{2}\lambda$, $\eta_2 = -\lambda$. From the first relation, one sees that there exist two bounded solutions if $\lambda > 0$, the trivial solution is the unique bounded solution for $\lambda = 0$ and there are no bounded solutions for $\lambda < 0$; see Fig. 8 (left) for an illustration. For this reason, $\lambda = 0$ can be interpreted as bifurcation value, since the number of bounded entire solutions increases from 0 to 2 as λ increases through 0.

The method of explicit solutions can also be applied to the nonlinear equation

$$x_{n+1} = f_n(x_n, \lambda) := \begin{pmatrix} b_n & 0 \\ 0 & c_n \end{pmatrix} x_n + \begin{pmatrix} 0 \\ (x_n^1)^3 \end{pmatrix} - \lambda \begin{pmatrix} 0 \\ 1 \end{pmatrix}. \tag{49}$$

However, using the variation of constants formula (cf. [1, p. 59]), it is possible to show that the crucial second component of the general solution $\phi_\lambda(\cdot, 0, \eta)$ for (49) fulfills

$$\phi_\lambda^2(n, 0, \eta) = \begin{cases} 2^n\left(\eta_2 + \frac{8}{15}\eta_1^3 - \lambda\right) + O(1), & n \to \infty, \\ \frac{1}{2^n}\left(\eta_2 - \frac{2}{15}\eta_1^3 + 2\lambda\right) + O(1), & n \to -\infty. \end{cases}$$

Since the first component is given in (48), $\phi_\lambda(\cdot, 0, \eta)$ is bounded if and only if $\eta_2 = -\frac{8}{15}\eta_1^3 + \lambda$ and $\eta_2 = \frac{2}{15}\eta_1^3 - 2\lambda$, which in turn is equivalent to

$$\eta_1 = \sqrt[3]{\frac{9}{2}\lambda}, \qquad\qquad \eta_2 = -\frac{7}{5}\lambda.$$

Hence, these particular initial values $\eta \in \mathbb{R}^2$ given by the cusp shaped curve depicted in Fig. 8 (right) lead to bounded entire solutions of (49).

7.2   Attractor Bifurcation

An easy example for a bifurcation of a pullback attractor was discussed already in Example 7.8. Now a general bifurcation pattern will be derived, which ensures, under certain conditions on Taylor coefficients, that a pullback attractor changes qualitatively under variation of the parameter. This generalizes the autonomous pitchfork bifurcation pattern. Although the pullback attractor discussed in Example 7.8 is a *global* attractor, the pitchfork bifurcation only yields results for a *local* pullback attractor.

Definition 7.11. Consider a process ϕ on a metric state space X. A ϕ-invariant family $\mathscr{A} = \{A_n : n \in \mathbb{Z}\}$ of nonempty compact subsets of X is called a *local pullback attractor* if there exists an $\eta > 0$ such that

$$\lim_{k \to \infty} \text{dist}\left(\phi(n, n-k, B_\eta(A_{n-k})), A_n\right) = 0 \quad \text{for all } n \in \mathbb{Z}.$$

A local pullback attractor is a special case of a pullback attractor w.r.t. a certain basin of attraction, which was introduced in Definition 5.11. Here, the basin of attraction has to be chosen as a neighborhood of the local pullback attractor.

Suppose now that (Δ_λ) is a scalar equation $(d = 1)$ with the trivial solution for all parameters λ from an interval $\Lambda \subseteq \mathbb{R}$. The transition matrix of the corresponding variational equation

$$x_{n+1} = D_1 f_n(0, \lambda) x_n$$

is denoted by $\Phi_\lambda(n, l) \in \mathbb{R}$.

The hyperbolicity condition (i) in Theorem 7.6 will be violated when dealing with attractor bifurcations. This yields a nonautonomous counterpart to the classical pitchfork bifurcation pattern.

Theorem 7.12 (Nonautonomous pitchfork bifurcation). *Suppose that $f_n(\cdot, \lambda)$: $\mathbb{R} \to \mathbb{R}$ is invertible and of class C^4 with*

$$D_1^2 f_n(0, \lambda) = 0 \quad \text{for all } n \in \mathbb{Z} \text{ and } \lambda \in \Lambda.$$

Suppose there exists a $\lambda^* \in \mathbb{R}$ such that the following hypotheses hold.

- *Hypothesis on linear part: There exists a $K \geq 1$ and functions $\beta_1, \beta_2 : \Lambda \to (0, \infty)$ which are either both increasing or decreasing with $\lim_{\lambda \to \lambda*} b_1(\lambda) = \lim_{\lambda \to \lambda*} b_2(\lambda) = 1$ and*

$$\Phi_\lambda(n, l) \leq K\beta_1(\lambda)^{n-l} \quad \text{for all } l \leq n,$$

$$\Phi_\lambda(n, l) \leq K\beta_2(\lambda)^{n-l} \quad \text{for all } n \leq l$$

and all $\lambda \in \Lambda$.
- *Hypothesis on nonlinearity: Assume that if the functions β_1 and β_2 are increasing, then*

$$-\infty < \liminf_{\lambda \to \lambda^*} \inf_{n \in \mathbb{Z}} D_1^3 f_n(0, \lambda) \leq \limsup_{\lambda \to \lambda^*} \sup_{n \in \mathbb{Z}} D_1^3 f_n(0, \lambda) < 0,$$

and otherwise (i.e., if the functions β_1 and β_2 are decreasing), then

$$0 < \liminf_{\lambda \to \lambda^*} \inf_{n \in \mathbb{Z}} D_1^3 f_n(0, \lambda) \leq \limsup_{\lambda \to \lambda^*} \sup_{n \in \mathbb{Z}} D_1^3 f_n(0, \lambda) < \infty.$$

In addition, suppose that the remainder satisfies

$$\lim_{x \to 0} \sup_{\lambda \in (\lambda^* - x^2, \lambda^* + x^2)} \sup_{n \leq 0} x \int_0^1 (1 - t)^3 D^4 f_n(tx, \lambda)\, dt = 0,$$

$$\limsup_{\lambda \to \lambda^*} \limsup_{x \to 0} \sup_{n \leq 0} \frac{Kx^3}{1 - \min\{\beta_1(\lambda), \beta_2(\lambda)^{-1}\}} \int_0^1 (1 - t)^3 D^4 f_n(tx, \lambda)\, dt < 3.$$

Then there exist $\lambda_- < \lambda^* < \lambda_+$ so that the following statements hold:

1. *If the functions β_1, β_2 are increasing, the trivial solution is a local pullback attractor for $\lambda \in (\lambda_-, \lambda^*)$, which bifurcates to a nontrivial local pullback attractor $\{A_n^\lambda : n \in \mathbb{Z}\}$, $\lambda \in (\lambda^*, \lambda^+)$, satisfying the limit*

$$\lim_{\lambda \to \lambda^*} \sup_{n \leq 0} \text{dist}(A_n^\lambda, \{0\}) = 0.$$

2. *If the functions β_1, β_2 are decreasing, the trivial solution is a local pullback attractor for $\lambda \in (\lambda^*, \lambda^+)$, which bifurcates to a nontrivial local pullback attractor $\{A_n^\lambda : n \in \mathbb{Z}\}$, $\lambda \in (\lambda_-, \lambda^*)$, satisfying the limit*

$$\lim_{\lambda \to \lambda^*} \sup_{n \le 0} \text{dist}(A_n^\lambda, \{0\}) = 0.$$

For a proof of this theorem and extensions (to both different time domains and repellers), see [43, 45].

The next example, taken from [15], illustrates the above theorem.

Example 7.13. Consider the nonautonomous difference equation

$$x_{n+1} = \frac{\lambda x_n}{1 + \frac{b_n q}{\lambda} x_n^q}, \tag{50}$$

where $q \in \mathbb{N}$ and the sequence $(b_n)_{n \in \mathbb{N}}$ is positive and both bounded and bounded away from zero. For $q = 1$, this difference equation can be transformed into the well-known Beverton–Holt equation, which describes the density of a population in a fluctuating environment. It was shown in [15] that in this case, the system admits a nonautonomous transcritical bifurcation (the bifurcation pattern of which was derived in [43]).

For $q = 2$, a nonautonomous pitchfork bifurcation occurs. The above theorem can be applied, because the Taylor expansion of the right-hand side of (50) reads as $\lambda x_n + b_n x_n^{q+1} + O(x^{2q+1})$, and the remainder fulfills the conditions of the theorem (see [15] for details). This means that for $\lambda \in (0, 1)$, the trivial solution is a local pullback attractor, which undergoes a transition to a nontrivial local pullback attractor when $\lambda > 1$. Note that the extreme solutions of the nontrivial local pullback attractor for $\lambda > 1$ are also local pullback attractors, which gives the interpretation of this bifurcation as a bifurcation of locally pullback attractive solutions.

7.3 Solution Bifurcation

In the previous section on attractor bifurcations, the first hyperbolicity condition (i) in Theorem 7.6, given by exponential dichotomies on both semiaxes, was violated.

The present concept of solution bifurcation is based on the assumption that merely condition (ii) of Theorem 7.6 does not hold. This requires the variational difference equation (V_λ) to be intrinsically nonautonomous. Indeed, if (V_λ) is almost periodic, then an exponential dichotomy on a semiaxis extends to the whole integer axis (cf. [52, Theorem 2]) and the reference solution $\chi = (\chi_n)_{n \in \mathbb{Z}}$ becomes hyperbolic. For this reason the following bifurcation scenarios cannot occur for periodic or autonomous difference equations.

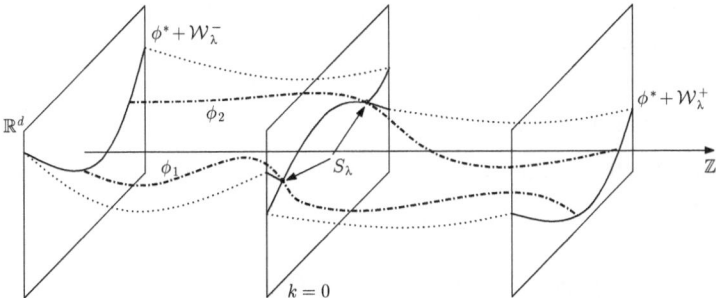

Fig. 9 Intersection $S_\lambda \subseteq \mathbb{R}^d$ of the stable fiber bundle $\phi^* + \mathcal{W}_\lambda^+ \subseteq \mathbb{Z}^+ \times \mathbb{R}^d$ with the unstable fiber bundle $\phi^* + \mathcal{W}_\lambda^- \subseteq \mathbb{Z}^- \times \mathbb{R}^d$ at time $k = 0$ yields two bounded entire solutions ϕ_1, ϕ_2 to (Δ_λ) indicated as *dotted dashed lines*

The crucial and standing assumption is the following:

Hypothesis: *The variational equation (V_λ) admits an ED both on $\mathbb{Z}^+$ (with projector P_n^+) and on $\mathbb{Z}^-$ (with projector P_n^-) such that there exists nonzero vectors $\xi_1 \in \mathbb{R}^d$, $\xi_1' \in \mathbb{R}^d$ satisfying*

$$R(P_0^+) \cap N(P_0^-) = \mathbb{R}\xi_1, \qquad (R(P_0^+) + N(P_0^-))^\perp = \mathbb{R}\xi_1'. \qquad (51)$$

Then a *solution bifurcation* is understood as follows: Suppose that for a fixed parameter $\lambda^* \in \Lambda$, the difference equation (Δ_{λ^*}) has an entire bounded reference solution $\chi^* = \chi(\lambda^*)$. One says that (Δ_λ) undergoes a *bifurcation* at $\lambda = \lambda^*$ along χ^*, or χ^* *bifurcates* at λ^*, if there exists a convergent parameter sequence $(\lambda_n)_{n\in\mathbb{N}}$ in Λ with limit λ^* so that (Δ_{λ_n}) has two distinct entire solutions $\chi_{\lambda_n}^1, \chi_{\lambda_n}^2 \in \ell^\infty$ both satisfying

$$\lim_{n\to\infty} \chi_{\lambda_n}^1 = \lim_{n\to\infty} \chi_{\lambda_n}^2 = \chi^*.$$

The above Hypothesis allows a geometrical insight into the following abstract bifurcation results using *invariant fiber bundles*, i.e., nonautonomous counterparts to invariant manifolds: Because (V_λ) has an exponential dichotomy on $\mathbb{Z}^+$, there exists a stable fiber bundle $\chi^* + \mathcal{W}_\lambda^+$ consisting of all solutions to (Δ_λ) approaching χ^* in forward time. Here, $\mathcal{W}_\lambda^+$ is locally a graph over the stable vector bundle $\{R(P_n^+) : n \in \mathbb{Z}^+\}$. Analogously, the dichotomy on $\mathbb{Z}^-$ guarantees an unstable fiber bundle $\chi^* + \mathcal{W}_\lambda^-$ consisting of solutions decaying to χ^* in backward time (cf. [40, Corollary 2.23]). Then the bounded entire solutions to (Δ_λ) are contained in the intersection $(\chi^* + \mathcal{W}_\lambda^+) \cap (\chi^* + \mathcal{W}_\lambda^-)$. In particular, the intersection of the fibers

$$S_\lambda := \left(\chi_0^* + \mathcal{W}_{\lambda,0}^+\right) \cap \left(\chi_0^* + \mathcal{W}_{\lambda,0}^-\right) \subseteq \mathbb{R}^d$$

yields initial values for bounded entire solutions (see Fig. 9).

It can be assumed without loss of generality, using the equation of perturbed motion, that $\chi^* = 0$. In addition suppose that

$$f_n(0, \lambda) \equiv 0 \quad \text{on } \mathbb{Z},$$

which means that (Δ_λ) has the trivial solution for all $\lambda \in \Lambda$. The corresponding variational equation is

$$x_{n+1} = D_1 f_n(0, \lambda) x_n$$

with transition matrix $\Phi_\lambda(n, l) \in \mathbb{R}^{d \times d}$.

Theorem 7.14 (Bifurcation from known solutions). *Let $\Lambda \subseteq \mathbb{R}$ and suppose f_n is of class C^m, $m \geq 2$. If the <u>transversality condition</u>*

$$g_{11} := \sum_{n \in \mathbb{Z}} \langle \Phi_{\lambda^*}(0, n+1)' \xi_1', D_1 D_2 f_n(0, \lambda^*) \Phi_{\lambda^*}(n, 0) \xi_1 \rangle \neq 0 \qquad (52)$$

is satisfied, then the trivial solution of a difference equation (Δ_λ) bifurcates at λ^*. In particular, there exists a $\rho > 0$, open convex neighborhoods $U \subseteq \ell^\infty(\Omega)$ of 0, $\Lambda_0 \subseteq \Lambda$ of λ^* and C^{m-1}-functions $\psi : (-\rho, \rho) \to U$, $\lambda : (-\rho, \rho) \to \Lambda_0$ with

1. *$\psi(0) = 0$, $\lambda(0) = \lambda^*$ and $\dot\psi(0) = \Phi_{\lambda^*}(\cdot, 0) \xi_1$,*
2. *Each $\psi(s)$ is a nontrivial solution of $(\Delta)_{\lambda(s)}$ homoclinic to 0, i.e.,*

$$\lim_{n \to \pm\infty} \psi(s)_n = 0.$$

Proof. See [39, Theorem 2.14].   □

Corollary 7.15 (Transcritical bifurcation). *Under the additional assumption*

$$g_{20} := \sum_{n \in \mathbb{Z}} \langle \Phi_{\lambda^*}(0, n+1)' \xi_1', D_1^2 f_n(0, \lambda^*) [\Phi_{\lambda^*}(n, 0) \xi_1]^2 \rangle \neq 0$$

one has $\dot\lambda(0) = -\frac{g_{20}}{2g_{11}}$ and the following holds locally in $U \times \Lambda_0$: The difference equation (Δ_λ) has a unique nontrivial entire bounded solution $\psi(\lambda)$ for $\lambda \neq \lambda^*$ and 0 is the unique entire bounded solution of $(\Delta)_{\lambda^*}$; moreover, $\psi(\lambda)$ is homoclinic to 0.

Proof. See [39, Corollary 2.16].   □

Example 7.16. Consider the nonlinear difference equation

$$x_{n+1} = f_n(x_n, \lambda) := \begin{pmatrix} b_n & 0 \\ \lambda & c_n \end{pmatrix} x_n + \begin{pmatrix} 0 \\ (x_n^1)^2 \end{pmatrix} \qquad (53)$$

depending on a bifurcation parameter $\lambda \in \mathbb{R}$ and sequences b_n, c_n defined in (47). As in Example 7.10, the assumptions hold with $\lambda^* = 0$ and

$$g_{11} = \frac{4}{3} \neq 0, \qquad\qquad g_{20} = \frac{12}{7} \neq 0.$$

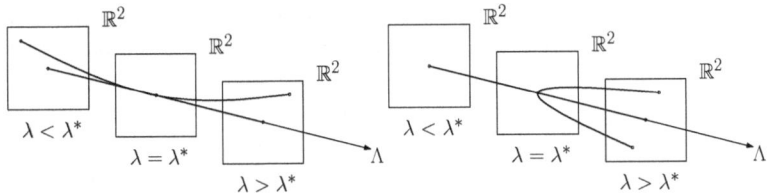

Fig. 10 *Left* (transcritical): Initial values $\eta \in \mathbb{R}^2$ yielding a homoclinic solution $\phi_\lambda(\cdot, 0, \eta)$ of (53) for different parameter values λ.
Right (supercritical pitchfork): Initial values $\eta \in \mathbb{R}^2$ yielding a homoclinic solution $\phi_\lambda(\cdot, 0, \eta)$ of (54) for different parameter values λ

Hence, Corollary 7.15 can be applied in order to see that the trivial solution of (53) has a transcritical bifurcation at $\lambda = 0$. Again, this bifurcation will be described quantitatively. While the first component of the general solution $\phi_\lambda(\cdot, 0, \eta)$ given by (48) is homoclinic, the second component satisfies

$$
\phi_\lambda^2(n, 0, \eta) = \begin{cases} 2^n \left(\eta_2 + \frac{4}{7}\eta_1^2 + \frac{2\lambda}{3}\eta_1 \right) + o(1), & n \to \infty, \\ 2^{-n} \left(\eta_2 - \frac{2}{7}\eta_1^2 - \frac{2\lambda}{3}\eta_1 \right) + o(1), & n \to -\infty. \end{cases}
$$

In conclusion, one sees that $\phi_\lambda(\cdot, 0, \eta)$ is bounded if and only if $\eta - (0, 0)$ or

$$
\eta_1 = -\frac{14}{9}\lambda, \qquad\qquad \eta_2 = \frac{28}{81}\lambda^2.
$$

Hence, besides the zero solution, there is a unique nontrivial entire solution passing through the initial point $\eta = (\eta_1, \eta_2)$ at time $n = 0$ for $\lambda \neq 0$. This means the solution bifurcation pattern sketched in Fig. 10 (left) holds.

Corollary 7.17 (Pitchfork bifurcation). *For $m \geq 3$ and under the additional assumptions*

$$
\sum_{n \in \mathbb{Z}} \langle \Phi_{\lambda^*}(0, n+1)'\xi_1', D_1^2 f_n(0, \lambda^*)[\Phi_{\lambda^*}(n, 0)\xi_1]^2 \rangle = 0,
$$

$$
g_{30} := \sum_{n \in \mathbb{Z}} \langle \Phi_{\lambda^*}(0, n+1)'\xi_1', D_1^3 f_n(0, \lambda^*)[\Phi_{\lambda^*}(n, 0)\xi_1]^3 \rangle \neq 0
$$

one has $\dot\lambda(0) = 0$, $\ddot\lambda(0) = -\frac{g_{30}}{3g_{11}}$ and the following holds locally in $U \times \Lambda_0$:

3. *Subcritical case: If $g_{30}/g_{11} > 0$, then the unique entire bounded solution of (Δ_λ) is the trivial one for $\lambda \geq \lambda^*$ and (Δ_λ) has exactly two nontrivial entire solutions for $\lambda < \lambda^*$; both are homoclinic to 0.*

4. _Supercritical case:_ If $g_{30}/g_{11} < 0$, then the unique entire bounded solution of (Δ_λ) is the trivial one for $\lambda \leq \lambda^*$ and (Δ_λ) has exactly two nontrivial entire solutions for $\lambda > \lambda^*$; both are homoclinic to 0.

Proof. See [39, Corollary 2.16]. □

Example 7.18. Let δ be a fixed nonzero real number and consider the nonlinear difference equation

$$x_{n+1} = f_n(x_n, \lambda) := \begin{pmatrix} b_n & 0 \\ \lambda & c_n \end{pmatrix} x_n + \delta \begin{pmatrix} 0 \\ (x_n^1)^3 \end{pmatrix} \tag{54}$$

depending on a bifurcation parameter $\lambda \in \mathbb{R}$ and the b_n, c_n defined in (47). As in our above Example 7.16, the assumptions of Corollary 7.17 are fulfilled with $\lambda^* = 0$. The transversality condition here reads $g_{11} = \frac{4}{3} \neq 0$. Moreover, $D_1^2 f_n(0,0) \equiv 0$ on $\mathbb{Z}$ implies $g_{20} = 0$, whereas the relation $D_1^3 f_n(0,0)\zeta^3 = \begin{pmatrix} 0 \\ 6\delta\zeta_1^3 \end{pmatrix}$ for all $n \in \mathbb{Z}$, $\zeta \in \mathbb{R}^2$ leads to $g_{30} = 4\delta \neq 0$. This gives the crucial quotient $\frac{g_{30}}{g_{11}} = 3\delta$. By Corollary 7.17, the trivial solution to (54) undergoes a subcritical (supercritical) pitchfork bifurcation at $\lambda = 0$ provided $\delta > 0$ (resp. $\delta < 0$). As before one can illustrate this result using the general solution $\phi_\lambda(\cdot, 0, \eta)$ to (54). The first component is given by (48) and helps to show for the second component that

$$\phi_\lambda^2(n, 0, \eta) = \begin{cases} 2^n \left(\eta_2 + \frac{8\delta}{15}\eta_1^3 + \frac{2\lambda}{3}\eta_1 \right) + o(1), & n \to \infty, \\ 2^{-n} \left(\eta_2 - \frac{2\delta}{15}\eta_1^3 - \frac{4\lambda}{3}\eta_1 \right) + o(1), & n \to -\infty. \end{cases}$$

This asymptotic representation shows that $\phi_\lambda(\cdot, 0, \eta)$ is homoclinic to 0 if and only if $\eta = 0$ or $\eta_1^2 = -\frac{2}{\delta}\lambda$ and $\eta_2 = \frac{4}{15}\frac{(5\delta+16\lambda)}{\delta^2}\lambda^2$. Hence, there is a correspondence to the pitchfork solution bifurcation from in Corollary 7.17. See Fig. 10 (right) for an illustration.

Remarks. In [43, Theorem 5.1] one finds a nonautonomous generalization for transcritical bifurcations.

8 Random Dynamical Systems

Random dynamical systems on a state space X are nonautonomous by the very nature of the driving noise. They can be formulated as skew-product systems with the driving system acting a probability sample space Ω rather than on a topological or metric parameter space P. A major difference is that only measurability and not continuity w.r.t. the parameter can be assumed, which changes the types of results that can be proved. In particular, the skew-product system does not form an autonomous semidynamical system on the product space $\Omega \times X$. Nevertheless, there are many interesting parallels with the theory of deterministic nonautonomous dynamical systems.

For further details see Arnold [2] and, for example, also [27], where the temporal discretization of random differential equations is also considered.

8.1 Random Difference Equations

Let $(\Omega, \mathscr{F}, \mathbb{P})$ be a probability space and let $\{\xi_n, n \in \mathbb{Z}\}$ be a discrete-time stochastic process taking values in some space Ξ, i.e., a sequence of random variables or, equivalently, $\mathscr{F}$-measurable mappings $\xi_n : \Omega \to \Xi$ for $n \in \mathbb{Z}$. Let (X, d) be a complete metric space and consider a mapping $g : \Xi \times X \to X$.

Then

$$x_{n+1}(\omega) = g\left(\xi_n(\omega), x_n(\omega)\right) \quad \text{for all } n \in \mathbb{Z}, \ \omega \in \Omega, \tag{55}$$

is a *random difference equation* on X driven by the stochastic process ξ_n.

Greater generality can be achieved by representing the driving noise process by a metrical (i.e., measure theoretic) dynamical system θ on some canonical sample space Ω, i.e., the group of $\mathscr{F}$-measurable mappings $\{\theta_n, n \in \mathbb{Z}\}$ under composition formed by iterating a measurable mapping $\theta : \Omega \to \Omega$ and its measurable inverse mapping $\theta^{-1} : \Omega \to \Omega$, i.e., with $\theta_0 = \mathrm{id}_\Omega$ and

$$\theta_{n+1} := \theta \circ \theta_n, \qquad \theta_{-n-1} := \theta^{-1} \circ \theta_{-n} \quad \text{for all } n \in \mathbb{N},$$

where $\theta_{-1} := \theta^{-1}$. It is usually assumed that θ generates an ergodic process on Ω.

Let $f : \Omega \times X \to X$ be an $\mathscr{F} \times \mathscr{B}(X)$-measurable mapping, where $\mathscr{B}(X)$ is the Borel σ-algebra on X. Then, in this context, a *random difference equation* has the form

$$x_{n+1}(\omega) = f\left(\theta^n(\omega), x_n(\omega)\right) \quad \text{for all } n \in \mathbb{Z}, \ \omega \in \Omega. \tag{56}$$

Define recursively a solution mapping $\varphi : \mathbb{Z}^+ \times \Omega \times X \to X$ for the random difference equation (56) by $\varphi(\omega, 0, x) := x$ and

$$\varphi(\omega, n+1, x) = f(\theta^n(\omega), \phi(\theta^n(\omega), n, x)) \quad \text{for all } n \in \mathbb{N}, \ x \in X$$

and $\omega \in \Omega$. Then, φ satisfies the discrete-time *cocycle property* w.r.t. θ, i.e.,

$$\varphi(n+m, \omega, x) = \varphi\left(n, \theta_m(\omega), \varphi(m, \omega, x_0)\right) \quad \text{for all } m, n \in \mathbb{Z}^+,$$

$x \in X$ and $\omega \in \Omega$. The mapping φ is called a cocycle mapping.

In terms of Arnold [2], the random difference equation (56) generates a discrete-time *random dynamical system* (θ, ϕ) on $\Omega \times X$ with the *metric dynamical system* θ on the probability space $(\Omega, \mathscr{F}, \mathbb{P})$ and the cocycle mapping φ on the state space X.

Definition 8.1. A (discrete-time) *random dynamical system* (θ, φ) on $\Omega \times X$ consists of a metrical dynamical system θ on Ω, i.e., a group of measure preserving mappings $\theta_n : \Omega \to \Omega, n \in \mathbb{Z}$, such that

(i) $\theta_0 = \mathrm{id}_\Omega$ and $\theta_n \circ \theta_m = \theta_{n+m}$ for all $n, m \in \mathbb{Z}$,
(ii) The map $\omega \mapsto \theta_n(\omega)$ is measurable and invariant w.r.t. $\mathbb{P}$ in the sense that $\theta_n(\mathbb{P}) = \mathbb{P}$ for each $n \in \mathbb{Z}$,

and a cocycle mapping $\varphi : \mathbb{Z}^+ \times \Omega \times X \to X$ such that

(a) $\varphi(0, \omega, x_0) = x_0$ for all $x_0 \in X$ and $\omega \in \Omega$,
(b) $\varphi(n + m, \omega, x_0) = \varphi(n, \theta_m(\omega), \varphi(m, \omega, x_0))$ for all $n, m \in \mathbb{Z}^+$, $x_0 \in X$ and $\omega \in \Omega$,
(c) $x_0 \mapsto \varphi(n, \omega, x_0)$ is continuous for each $(n, \omega) \in \mathbb{Z}^+ \times \Omega$,
(d) $\omega \mapsto \varphi(n, \omega, x_0)$ is $\mathscr{F}$-measurable for all $(n, x_0) \in \mathbb{Z}^+ \times X$.

The notation $\theta_n(\mathbb{P}) = \mathbb{P}$ for the measure preserving property of θ_n w.r.t. $\mathbb{P}$ is just a compact way of writing

$$\mathbb{P}(\theta_n(A)) = \mathbb{P}(A) \quad \text{for all } n \in \mathbb{Z}, A \in \mathscr{F}.$$

A systematic treatment of the random dynamical system theory, both continuous and discrete time, is propounded in Arnold [2]. Note that $\pi = (\theta, \phi)$ has a skew-product structure on $\Omega \times X$, but it is not an autonomous semidynamical system on $\Omega \times X$ since no topological structure is assumed on Ω.

8.2 Random Attractors

Unlike a deterministic skew-product system, a random dynamical system (θ, φ) on $\Omega \times X$ is not an autonomous semidynamical system on $\Omega \times X$. Nevertheless, skew-product deterministic systems and random dynamical systems have many analogous properties, and concepts and results for one can often be used with appropriate modifications for the other. The most significant modification concerns measurability and the nonautonomous sets under consideration are random sets.

Let (X, d) be a complete and separable metric space (i.e., a Polish space)

Definition 8.2. A family $\mathscr{D} = \{D_\omega, \omega \in \Omega\}$ of nonempty subsets of X is called a *random set* if the mapping $\omega \mapsto \mathrm{dist}(x, D_\omega)$ is $\mathscr{F}$-measurable for all $x \in X$. A random set $\mathscr{D}$ is called a *random closed set* if D_ω is closed for each $\omega \in \Omega$ and is called a *random compact set* if D_ω is compact for each $\omega \in \Omega$.

Random sets are called *tempered* if their growth w.r.t. the driving system θ is sub-exponential (cf. Definition 6.6).

Definition 8.3. A random set $\mathscr{D} = \{D_\omega, \omega \in \Omega\}$ in X is said to be *tempered* if there exists a $x_0 \in X$ such that

$$D_\omega \subset \{x \in X \; : \; d(x, x_0) \le r(\omega)\} \quad \text{for all } \omega \in \Omega,$$

where the random variable $r(\omega) > 0$ is tempered, i.e.,

$$\sup_{n \in \mathbb{Z}} \{ r(\theta_n(\omega)) e^{-\gamma |n|} \} < \infty \quad \text{for all } \omega \in \Omega, \gamma > 0.$$

The collection of all tempered random sets in X will be denoted by $\mathfrak{D}$.

A random attractor of a random dynamical system is a random set which is a pullback attractor in the pathwise sense w.r.t. the attracting basin of tempered random sets.

Definition 8.4. A random compact set $\mathscr{A} = (A_\omega)_{\omega \in \Omega}$ from $\mathfrak{D}$ is called a *random attractor* of a random dynamical system (θ, φ) on $\Omega \times X$ in $\mathfrak{D}$ if $\mathscr{A}$ is a φ-invariant set, i.e.,

$$\varphi(n, \omega, A_\omega) = A_{\theta_n(\omega)} \quad \text{for all } n \in \mathbb{Z}^+, \omega \in \Omega,$$

and pathwise pullback attracting in $\mathfrak{D}$, i.e.,

$$\lim_{n \to \infty} \text{dist} \left(\varphi\big(n, \theta_{-n}(\omega), D(\theta_{-n}(\omega))\big), A_\omega \right) = 0 \quad \text{for all } \omega \in \Omega, \mathscr{D} \in \mathfrak{D}.$$

If the random attractor consists of singleton sets, i.e., $A_\omega = \{Z^*(\omega)\}$ for some random variable Z^* with $Z^*(\omega) \in X$, then $\bar{Z}_n(\omega) := Z^*(\theta_n(\omega))$ is a stationary stochastic process on X.

The existence of a random attractor is ensured by that of a pullback absorbing set. The tempered random set $\mathscr{B} = \{B_\omega, \omega \in \Omega\}$ in the following theorem is called a pullback absorbing random set.

Theorem 8.5 (Existence of random attractors). *Let (θ, φ) be a random dynamical system on $\Omega \times X$ such that $\varphi(n, \omega, \cdot) : X \to X$ is a compact operator for each fixed $n > 0$ and $\omega \in \Omega$. If there exist a tempered random set $\mathscr{B} = \{B_\omega, \omega \in \Omega\}$ with closed and bounded component sets and an $N_{\mathscr{D}, \omega} \geq 0$ such that*

$$\varphi\big(n, \theta_{-n}(\omega), D(\theta_{-n}(\omega))\big) \subset B_\omega \quad \text{for all } n \geq N_{\mathscr{D}, \omega}, \tag{57}$$

and every tempered random set $\mathscr{D} = \{D_\omega, \omega \in \Omega\}$, then the random dynamical system (θ, φ) possesses a random pullback attractor $\mathscr{A} = \{A_\omega : \omega \in \Omega\}$ with component sets defined by

$$A_\omega = \bigcap_{m > 0} \overline{\bigcup_{n \geq m} \varphi(n, \theta_{-n}(\omega), B(\theta_{-n}(\omega)))} \quad \text{for all } \omega \in \Omega. \tag{58}$$

The proof of Theorem 8.5 is essentially the same as its counterparts for deterministic skew-product systems. The only new feature is that of measurability, i.e., to show that $\mathscr{A} = \{A_\omega\}, \omega \in \Omega\}$ is a random set. This follows from the fact that the set-valued mappings $\omega \mapsto \varphi\big(n, \theta_{-n}(\omega), B(\theta_{-n}(\omega))\big)$ are measurable for each $n \in \mathbb{Z}^+$.

Arnold and Schmalfuß [3] showed that a random attractor is also a *forward* attractor in the weaker sense of convergence in probability, i.e.,

$$\lim_{n\to\infty} \int_{\Omega} \text{dist}\left(\varphi(n,\omega,D_\omega), A_{\theta_n(\omega)}\right) \mathbb{P}(d\omega) = 0$$

for all $\mathscr{D} \in \mathfrak{D}$. This allows individual sample paths to have large deviations from the attractor, but for all to converge in this probabilistic sense.

8.3   Random Markov Chains

Discrete-time finite state Markov chains with a tridiagonal structure are common in biological applications. They have a transition matrix $[I_N + \Delta Q]$, where I_N is the $N \times N$ identity matrix and Q is the tridiagonal $N \times N$-matrix

$$Q = \begin{bmatrix} -q_1 & q_2 & & & \bigcirc \\ q_1 & -(q_2+q_3) & q_4 & & \\ & \ddots & \ddots & \ddots & & \ddots \\ & & & q_{2N-5} & -(q_{2N-4}+q_{2N-3}) & q_{2N-2} \\ \bigcirc & & & & q_{2N-3} & -q_{2N-2} \end{bmatrix} \tag{59}$$

where the q_j are positive constants.

Such a Markov chain is a first order linear difference equation

$$\mathbf{p}^{(n+1)} = [I_N + \Delta Q] \mathbf{p}^{(n)} \tag{60}$$

on the probability simplex Σ_N in $\mathbb{R}^N$ defined by

$$\Sigma_N = \left\{ \mathbf{p} = (p_1, \cdots, p_N)^T : \sum_{j=1}^{N} p_j = 1, \ p_1, \ldots, p_N \in [0,1] \right\}.$$

The Perron-Frobenius theorem applies to the matrix $L_\Delta := I_N + \Delta Q$ when $\Delta > 0$ is chosen sufficiently small. In particular, it has eigenvalue $\lambda = 1$ and there is a positive eigenvector $\bar{\mathbf{x}}$, which can be normalized (in the $\| \cdot \|_1$ norm) to give a probability vector $\bar{\mathbf{p}}$, i.e., $[I_N + \Delta Q]\bar{\mathbf{p}} = \bar{\mathbf{p}}$, so $Q\bar{\mathbf{p}} = 0$. Specifically, the probability vector

$$\bar{p}_1 = \frac{1}{\|\bar{\mathbf{x}}\|_1}, \quad \bar{p}_{j+1} = \frac{1}{\|\bar{\mathbf{x}}\|_1} \prod_{i=1}^{j} \frac{q_{2i-1}}{q_{2i}} \quad \text{for all } j = 1, \ldots, N-1,$$

where

$$\|\bar{\mathbf{x}}\|_1 = \sum_{j=1}^{N} \bar{x}_j = 1 + \sum_{j=1}^{N-1} \prod_{i=1}^{j} \frac{q_{2i-1}}{q_{2i}}.$$

The following result is well known.

Theorem 8.6. *The probability eigenvector $\bar{\mathbf{p}}$ is an asymptotically stable steady state of the difference equation* (60) *on the simplex Σ_N.*

In a random environment, e.g., with randomly varying food supply, the transition probabilities may be random, i.e., the band entries q_i of the matric Q may depend on the sample space parameter $\omega \in \Omega$. Thus, $q_i = q_i(\omega)$ for $i = 1, 2, \ldots, 2N - 2$, and these may vary in turn according to some metric dynamical system θ on the probability space $(\Omega, \mathcal{F}, \mathbb{P})$. The following basic assumption will be used.

Assumption 1. *There exist numbers $0 < \alpha \leq \beta < \infty$ such that the uniform estimates hold*

$$\alpha \leq q_i(\omega) \leq \beta \quad \text{for all } \omega \in \Omega, \, i = 1, 2, \ldots, 2N - 2. \tag{61}$$

Let $\mathcal{L}$ be a set of linear operators $L_\omega : \mathbb{R}^N \to \mathbb{R}^N$ parametrized by the parameter ω taking values in some set Ω and let $\{\theta_n, n \in \mathbb{Z}\}$ be a group of maps of Ω onto itself. The maps $L_\omega x$ serve as the generator of a linear cocycle $F_{\mathcal{L}}(n, \omega)$. Then $(\theta, F_{\mathcal{L}})$ is a random dynamical system on $\Omega \times \Sigma_N$.

Theorem 8.7. *Let $F_{\mathcal{L}}(n, \omega)x$ be the linear cocycle*

$$F_{\mathcal{L}}(n, \omega)x = L_{\theta_{n-1}\omega} \cdots L_{\theta_1 \omega} L_{\theta_0 \omega} x.$$

with matrices $L_\omega := I_N + \Delta Q(\omega)$, where the tridiagonal matrices $Q(\omega)$ are of the form (59) *with the entries $q_i = q_i(\omega)$ satisfying the uniform estimates* (61) *in Assumption 1. In addition, suppose that $0 < \Delta < \frac{1}{2\beta}$.*
Then, the simplex Σ_N is positively invariant under $F_{\mathcal{L}}(n, \omega)$, i.e.,

$$F_{\mathcal{L}}(n, \omega)\Sigma_N \subseteq \Sigma_N \quad \text{for all } \omega \in \Omega.$$

Moreover, for n large enough, the restriction of $F_{\mathcal{L}}(n, \omega)x$ to the set Σ_N is a uniformly dissipative and uniformly contractive cocycle (w.r.t. the Hilbert metric), which has a random attractor $\mathcal{A} = \{A_\omega, \omega \in \Omega\}$ such that each set A_ω, $\omega \in \Omega$, consists of a single point.

The proof can be found in [28]. It involves positive matrices and the Hilbert projective metric on positive cones in $\mathbb{R}^N$.

Henceforth write $A_\omega = \{a_\omega\}$ for the singleton component subsets of the random attractor $\mathcal{A}$. Then the random attractor is an entire random sequence $\{a_{\theta_n \omega}, n \in \mathbb{Z}\}$ in $\Sigma_N(\gamma) \subset \overset{\circ}{\Sigma}_N$, where

$$\Sigma_N(\gamma) = \left\{ x = (x_1, x_2, \ldots, x_N) : \sum_{i=1}^{N} x_i = 1, \, x_1, x_2, \ldots, x_N \geq \gamma^{N-1} \right\}.$$

with $\gamma := \min\{\Delta\alpha, 1 - 2\Delta\beta\} > 0$. It attracts other iterates of the random Markov chain in the pullback sense. Pullback convergence involves starting at earlier initial times with a fixed end time. It is, generally, not the same as forward convergence in the sense usually understood in dynamical systems, but in this case it is the same due to the uniform boundedness of the contractive rate w.r.t. ω.

Corollary 8.8. *For any norm $\|\cdot\|$ on $\mathbb{R}^N$, $\mathbf{p}^{(0)} \in \Sigma_N$ and $\omega \in \Omega$*

$$\left\|\mathbf{p}^{(n)}(\omega) - a_{\theta_n\omega}\right\| \to 0 \quad as \quad n \to \infty.$$

The random attractor is, in fact, asymptotically Lyapunov stable in the conventional forward sense.

8.4 Approximating Invariant Measures

Consider now a compact metric space (X, d). A random difference equation (56) on X driven by the noise process θ generates a random dynamical system (θ, φ). It can be reformulated as a difference equation with a triangular or skew-product structure

$$(\omega, x) \mapsto F(\omega, x) := \begin{pmatrix} \theta(\omega) \\ f(\omega, x) \end{pmatrix}$$

An invariant measure μ of $F = (\theta, \varphi)$ on $\Omega \times X$ defined by $\mu = F^*\mu$ (which is shorthand for an integral expression) can be decomposed as

$$\mu(\omega, B) = \mu_\omega(B) \mathbb{P}(d\omega) \quad \text{for all } B \in \mathcal{B}(X),$$

where the measures μ_ω on X are θ-invariant w.r.t. f, i.e.,

$$\mu_{\theta(\omega)}(B) = \mu_\omega\left(f^{-1}(\omega, B)\right) \quad \text{for all } B \in \mathcal{B}(X), \omega \in \Omega.$$

This decomposition is very important since only the state space X, but not the sample space Ω, can be discretized.

To compute a given invariant measure μ consider a sequence of finite subsets X_N of X given by

$$X_N = \{x_1^{(N)}, \cdots, x_N^{(N)}\} \subset X,$$

for $N \in \mathbb{N}$ with maximal step size

$$h_N = \sup_{x \in X} \text{dist}(x, X_N)$$

such that $h_N \to 0$ as $N \to \infty$.

Then the invariant measure μ will be approximated by a sequence of invariant stochastic vectors associated with *random Markov chains* describing transitions between the states of the discretized state spaces X_N. These involve random $N \times N$ matrices, i.e., measurable mappings

$$P_N : \Omega \to \mathcal{S}_N,$$

where $\mathcal{S}_N$ denotes the set of $N \times N$ (nonrandom) stochastic matrices, satisfying the property

$$P_N^n(\theta^m(\omega)) P_N^m(\omega) = P_N^{m+n}(\omega) \quad \text{for all } m, n \in \mathbb{Z}_+. \tag{62}$$

Recall that a stochastic matrix has non-negative entries with the columns summing to 1.

Consider a random Markov chain $\{P_N(\omega), \omega \in \Omega\}$ and a random probability vector $\{p_N(\omega), \omega \in \Omega\}$ on the deterministic grid X_N. Then

$$p_{N,n+1}(\theta^{n+1}(\omega)) = p_{N,n}(\theta^n(\omega)) P_N(\theta^n(\omega))$$

and an equilibrium probability vector is defined by

$$\bar{p}_N(\theta(\omega)) = \bar{p}_N(\omega) P_N(\omega) \quad \text{for all } \omega \in \Omega.$$

It can be represented trivially as a random measure $\mu_{N,\omega}$ on X.

The distance between random probability measures will be given with the *Prokhorov metric* ρ and the distance of a random Markov chain $P : \Omega \to \mathcal{S}_N$ and the generating mapping f of the random dynamical system is defined by

$$D(P(\omega), f) = \sum_{i,j=1}^{N} \left(p_{i,j}(\omega) \operatorname{dist}_{X \times X}((x_i^{(N)}, x_j^{(N)}), \operatorname{Gr} f(\omega, \cdot)) \right), \tag{63}$$

where the distance to the random graph is given by

$$\operatorname{dist}_{X \times X}((x, y), \operatorname{Gr} f(\omega, \cdot)) = \inf_{z \in X} \max\{d(x, z), d(y, f(\omega, z))\} \quad \text{for all } x, y \in X.$$

The following necessary and sufficient result holds if θ-semi-invariant rather than θ-invariant families of decomposed probability measures are used.

Definition 8.9. A family of probability measures μ_ω on X is called θ-semi-invariant w.r.t. f, if

$$\mu_{\theta(\omega)}(B) \leq \mu_\omega\left(f^{-1}(\omega, B)\right) \quad \text{for all } B \in \mathcal{B}(X), \omega \in \Omega.$$

Such θ-semi-invariant families are, in fact, θ-invariant when the mappings $x \mapsto f(\omega, x)$ are continuous.

Theorem 8.10. *A random probability measure* $\{\mu_\omega, \omega \in \Omega\}$ *is* θ-*semi-invariant w.r.t.* f *on* X *if and only if it is* <u>randomly stochastically</u> <u>approachable</u>, *i.e., for each* N *there exist*

- *(i) A grid* X_N *with fineness* $h_N \to 0$ *as* $N \to \infty$
- *(ii) A random Markov chain* $\{P_N(\omega), \omega \in \Omega\}$ *on* X_N
- *(iii) Random probability measure* $\{\mu_{N,\omega}, \omega \in \Omega\}$ *on* X *corresponding to a random equilibrium probability vector* $\{\bar{p}_N(\omega), \omega \in \Omega\}$ *of* $\{P_N(\omega), \omega \in \Omega\}$ *on* X_N

with the expected convergences

$$\mathbb{E}D\left(P_N(\omega), f(\omega, \cdot)\right) \to 0, \qquad \mathbb{E}\rho\left(\mu_{N,\omega}, \mu_\omega\right) \to 0 \quad as\ n \to \infty.$$

Proof. See Imkeller and Kloeden [16]. □

The double terminology "random stochastic" seems to be an overkill, but just think of a Markov chain for which the transition probabilities are not fixed, but can vary randomly in time.

Acknowledgements This work was partially supported by the DFG grant KL 1203/7-1, the Ministerio de Ciencia e Innovación project MTM2011-22411, the Consejería de Innovación, Ciencia y Empresa (Junta de Andalucía) under the Ayuda 2009/FQM314 and the Proyecto de Excelencia P07-FQM-02468 (Peter E. Kloeden). Martin Rasmussen was supported by an EPSRC Career Acceleration Fellowship.

The authors thank Dr. Thomas Lorenz for carefully reading parts of the article.

References

1. R.P. Agarwal, in *Difference Equations and Inequalities*. Monographs and Textbooks in Pure and Applied Mathematics, vol. 228 (Marcel Dekker, New York, 2000)
2. L. Arnold, in *Random Dynamical Systems*. Monographs in Mathematics (Springer, Berlin, 1998)
3. L. Arnold, B. Schmalfuss, Lyapunov's second method for random dynamical systems. J. Differ. Equat. **177**, 235–265 (2001)
4. J.-P. Aubin, H. Frankowska, in *Set-Valued Analysis*. Systems and Control: Foundations and Applications, vol. 2 (Birkhäuser, Boston, 1990)
5. B. Aulbach, The fundamental existence theorem on invariant fiber bundles. J. Differ. Equat. Appl. **3**, 501–537 (1998)
6. B. Aulbach, S. Siegmund, The dichotomy spectrum for noninvertible systems of linear difference equations. J. Differ. Equat. Appl. **7**(6), 895–913 (2001)
7. B. Aulbach, T. Wanner, Invariant foliations and decoupling of non-autonomous difference equations. J. Differ. Equat. Appl. **9**(5), 459–472 (2003)
8. B. Aulbach, T. Wanner, Topological simplification of nonautonomous difference equations. J. Differ. Equat. Appl. **12**(3–4), 283–296 (2006)
9. W.-J. Beyn, J.-M. Kleinkauf, The numerical computation of homoclinic orbits for maps. SIAM J. Numer. Anal. **34**(3), 1209–1236 (1997)
10. N.P. Bhatia, G.P. Szegö, *Stability Theory of Dynamical Systems* (Springer, Berlin, 2002)

11. D. Cheban, P.E. Kloeden, B. Schmalfuß, The relationship between pullback, forwards and global attractors of nonautonomous dynamical systems. Nonlinear Dynam. Syst. Theor. **2**(2), 9–28 (2002)
12. C. Dafermos, An invariance principle for compact processes. J. Differ. Equat. **9**, 239–252 (1971)
13. P. Diamond, P.E. Kloeden, Spatial discretization of mappings. J. Comp. Math. Appl. **26**, 85–94 (1993)
14. J.K. Hale, in *Asymptotic Behavior of Dissipative Systems*. Mathematical Surveys and Monographs, vol. 25 (AMS, Providence, 1988)
15. T. Hüls, A model function for non-autonomous bifurcations of maps. Discrete Contin. Dynam. Syst. Ser. B **7**(2), 351–363 (2007)
16. P. Imkeller, P.E. Kloeden, On the computation of invariant measures in random dynamical systems. Stochast. Dynam. **3**, 247–265 (2003)
17. G. Iooss, in *Bifurcation of Maps and Applications*. Mathematics Studies, vol. 36 (North-Holland, Amsterdam, 1979)
18. J. Kato, A.A. Martynyuk, A.A. Shestakov, *Stability of Motion of Nonautonomous Systems (Method of Limiting Equations)* (Gordon and Breach, Amsterdam, 1996)
19. P.E. Kloeden, Asymptotic invariance and limit sets of general control systems. J. Differ. Equat. **19**, 91–105 (1975)
20. P.E. Kloeden, Eventual stability in general control systems. J. Differ. Equat. **19**, 106–124 (1975)
21. P.E. Kloeden, Lyapunov functions for cocycle attractors in nonautonomous difference equations. Izvetsiya Akad Nauk Rep Moldovia Mathematika **26**, 32–42 (1998)
22. P.E. Kloeden, A Lyapunov function for pullback attractors of nonautonomous differential equations. Electron. J. Differ. Equat. Conf. **05**, 91–102 (2000)
23. P.E. Kloeden, Pullback attractors in nonautonomous difference equations. J. Differ. Equat. Appl. **6**, 33–52 (2000)
24. P.E. Kloeden, Pullback attractors of nonautonomous semidynamical systems. Stochast. Dynam. **3**(1), 101–112 (2003)
25. P.E. Kloeden, M. Rasmussen, in *Nonautonomous Dynamical Systems*. Mathematical Surveys and Monographs, vol. 176 (American Mathematical Society, Providence, 2011)
26. P.E. Kloeden, B. Schmalfuß, Lyapunov functions and attractors under variable time–step discretization. Discrete Contin. Dynam. Syst. **2**, 163–172 (1996)
27. P.E. Kloeden, H. Keller, B. Schmalfuß, Towards a theory of random numerical dynamics, in *Stochastic Dynamics*, ed. by H. Crauel, V.M. Gundlach (Springer, Berllin, 1999), pp. 259–282
28. P.E. Kloeden, V. Kozyakin, Asymptotic behaviour of random tridiagonal Markov chains in biological applications. Discrete Contin. Dynam. Syst. Series B (2013, to appear)
29. P.E. Kloeden, C. Pötzsche, M. Rasmussen, Limitations of pullback attractors of processes. J. Differ. Equat. Appl. **18**, 693–701 (2012)
30. Y.A. Kuznetsow, in *Elements of Applied Bifurcation Theory*. Applied Mathematical Sciences, vol. 112, 3rd edn. (Springer, Berlin, 2004)
31. S. Lang, in *Real and Functional Analysis*, Graduate Texts in Mathematics, vol. 42 (Springer, Berlin, 1993)
32. J.P. LaSalle, *The Stability of Dynamical Systems* (SIAM, Philadelphia, 1976)
33. C. Pötzsche, Stability of center fiber bundles for nonautonomous difference equations, in *Difference and Differential Equations*, ed. by S. Elaydi et al. Fields Institute Communications, vol. 42 (AMS, Providence, 2004), pp. 295–304
34. C. Pötzsche, Discrete inertial manifolds. Math. Nachr. **281**(6), 847–878 (2008)
35. C. Pötzsche, A note on the dichotomy spectrum. J. Differ. Equat. Appl. **15**(10), 1021–1025 (2009), see also the Corrigendum (2011)
36. C. Pötzsche, Nonautonomous bifurcation of bounded solutions II: A shovel bifurcation pattern. Discrete Contin. Dynam. Syst. Ser. A **31**(3), 941–973 (2011)
37. C. Pötzsche, Robustness of hyperbolic solutions under parametric perturbations. J. Differ. Equat. Appl. **15**(8–9), 803–819 (2009)

38. C. Pötzsche, *Geometric Theory of Discrete Nonautonomous Dynamical Systems*. Lecture Notes in Mathematics, vol. 2002 (Springer, Berlin, 2010)
39. C. Pötzsche, Nonautonomous bifurcation of bounded solutions I: A Lyapunov-Schmidt approach. Discrete Contin. Dynam. Syst. Ser. B **14**(2), 739–776 (2010)
40. C. Pötzsche, Nonautonomous continuation of bounded solutions. Comm. Pure Appl. Anal. **10**(3), 937–961 (2011)
41. C. Pötzsche, M. Rasmussen, Taylor approximation of invariant fiber bundles for non-autonomous difference equations. Nonlinear Anal. (TMA) **60**(7), 1303–1330 (2005)
42. C. Pötzsche, S. Siegmund, C^m-smoothness of invariant fiber bundles. Topological Meth. Nonlinear Anal. **24**(1), 107–146 (2004)
43. M. Rasmussen, Towards a bifurcation theory for nonautonomous difference equation. J. Differ. Equat. Appl. **12**(3–4), 297–312 (2006)
44. M. Rasmussen, in *Attractivity and Bifurcation for Nonautonomous Dynamical Systems*. Lecture Notes in Mathematica, vol. 1907 (Springer, Berlin, 2007)
45. M. Rasmussen, Nonautonomous bifurcation patterns for one-dimensional differential equations. J. Differ. Equat. **234**, 267–288 (2007)
46. M. Rasmussen, An alternative approach to Sacker-Sell spectral theory. J. Differ. Equat. Appl. **16**(2–3), 227–242 (2010)
47. R. Sacker, G. Sell, A spectral theory for linear differential systems. J. Differ. Equat. **27**, 320–358 (1978)
48. G.R. Sell, Y. You, in *Dynamics of Evolutionary Equations*. Applied Mathematical Sciences, vol. 143 (Springer, Berlin, 2002)
49. K.S. Sibirsky, *Introduction to Topological Dynamics* (Noordhoff International Publishing, Leiden, 1975)
50. S. Siegmund, Normal forms for nonautonomous difference equations. Comput. Math. Appl. **45**(6–9), 1059–1073 (2003)
51. A. Stuart, A. Humphries, in *Dynamical Systems and Numerical Analysis*. Monographs on Applied and Computational Mathematics (Cambridge University Press, Cambridge, 1998)
52. V.I. Tkachenko, On the exponential dichotomy of linear difference equations. Ukr. Math. J. **48**(10), 1600–1608 (1996)
53. T. Yoshizawa, *Stability Theory by Liapunov's Second Method* (The Mathematical Society of Japan, Tokyo, 1966)

Resonance Problems for Some Non-autonomous Ordinary Differential Equations

Jean Mawhin

Abstract Recent years have seen a lot of activity in the study of quasilinear non-autonomous ordinary differential equations or systems of the form $(\phi(y'))' = f(t, y, y')$, where $\phi : A \subset \mathbb{R}^n \to B \subset \mathbb{R}^n$ is some homeomorphism such that $\phi(0) = 0$ between the open sets A and B. The situation generalizes the classical case where $A = B = \mathbb{R}^n$ and ϕ is the identity, and the well-studied case of the p-Laplacian ($p > 1$) where $\phi(s) = \|s\|^{p-2}s$. Contemporary researches concern less standard situations where $\phi : B(a) \to \mathbb{R}^n$ (singular homeomorphism) and $\phi : \mathbb{R}^n \to B(a)$ (bounded homeomorphism), where $B(a)$ is the open ball of centre 0 and radius a. For $n = 1$, a model for the first case, namely $\phi(s) = \frac{s}{\sqrt{1-s^2}}$, corresponds to acceleration in special relativity, and a model for the second situation, namely $\phi(s) = \frac{s}{\sqrt{1+s^2}}$, corresponds to problem with curvature satisfying various conditions. In those case, both topological and variational methods, and sometimes combination of them give new complementary existence and multiplicity results. We will describe some of them. Some attention will be given to the generalized forced pendulum equation $(\phi(y'))' + A \sin y = h(t)$ when ϕ is singular or bounded.

1 Introduction, Notations and Preliminary Results

1.1 Introduction

Recent years have seen a lot of activity in the study of quasilinear non-autonomous ordinary differential equations of the form

J. Mawhin (✉)
Institut de Recherche en Mathématique et Physique, Université Catholique de Louvain, chemin du cyclotron, 2, 1348 Louvain-la-Neuve, Belgium
e-mail: jean.mawhin@uclouvain.be

A. Capietto et al., *Stability and Bifurcation Theory for Non-Autonomous Differential Equations*, Lecture Notes in Mathematics 2065, DOI 10.1007/978-3-642-32906-7_3, © Springer-Verlag Berlin Heidelberg 2013

$$(\phi(u'))' = f(t, u, u')$$

where $\phi : (-a, a) \to (-b, b)$ is an increasing homeomorphism such that $\phi(0) = 0$ between the open intervals $(-a, a)$ and $(-b, b)$, with $0 < a, b \leq +\infty$. The situation generalizes the classical case where $a = b = +\infty$ and ϕ is the identity, and the well-studied case of the p-Laplacian ($p > 1$) where $a = b = +\infty$ and $\phi(s) = |s|^{p-2}s$. In this last case, the Fredholm alternative for the solvability of

$$(|u'|^{p-2}u')' - \lambda|u|^{p-2}u = h(t)$$

with classical Dirichlet, Neumann or periodic boundary conditions on $[0, T]$

$$u(0) = 0 = u(T), \ u'(0) = 0 = u'(T), \ u(0) = u(T), \ u'(0) = u'(T) \qquad (1)$$

is far to be fully understood, despite of recent interesting partial results [6, 27, 29, 40, 43, 44, 70, 128, 129].

Contemporary researches concern less standard situations where $\phi : (-a, a) \to \mathbb{R}$ (singular homeomorphism) and $\phi : \mathbb{R} \to (-a, a)$ (bounded homeomorphism). A model for the first case, namely $\phi(s) = \frac{s}{\sqrt{1-s^2}}$, corresponds to acceleration in special relativity. A model for the second situation, namely $\phi(s) = \frac{s}{\sqrt{1+s^2}}$, corresponds to curvature or capillarity problem. In those cases, both topological and variational methods give new complementary existence and multiplicity results. The topological approach essentially makes use of Brouwer and Leray–Schauder's degrees, as described for example in [39, 73, 75, 76]. The variational approach uses various tools from the direct method of calculus of variations and critical point theory, as described for example in the monographs [85, 109] and is some papers mentioned when they are used. Some attention will be given to the generalized forced pendulum equation

$$(\phi(u'))' + \mu \sin u = h(t).$$

The case of differential systems

$$(\phi(u'))' = f(t, u, u')$$

where now $\phi : B(a) \to B(b)$ is a homeomorphism of the open ball $B(a) \subset \mathbb{R}^n$ onto the open ball $B(b) \subset \mathbb{R}^n$ ($0 < a, b \leq +\infty$) will be considered as well, when $\phi(0) = 0$ and $\phi = \nabla\Phi$, with $\Phi : \overline{B}(a) \to \mathbb{R}$ a strictly convex continuous function of class C^1 on $B(a)$ and such that $\Phi(0) = 0$. In the case of variational systems of the form

$$(\phi(u'))' = \nabla_u F(t, u) + h(t)$$

with $F : [0, T] \times \mathbb{R}^n \to \mathbb{R}$ such that $\nabla_u F$ is a Carathéodory function, some results will be obtained using the Lagrangian formulation associated to the corresponding action

$$I(u) = \int_0^T [\Phi(u'(t)) + F(t, u(t)) + \langle h(t), u(t) \rangle] \, dt$$

defined on a suitable set of T-periodic functions. Other results will be based upon a Hamiltonian approach, consisting in writing the system above in the equivalent form

$$u' = \phi^{-1}(v), \quad v' = \nabla_u F(t, u) + h(t),$$

which has the Hamiltonian structure

$$u' = \nabla_v H(t, u, v), \quad v' = -\nabla_u H(t, u, v)$$

for the Hamiltonian H defined by

$$H(t, u, v) = \Phi^*(v) - F(t, u) - \langle h(t), u \rangle$$

with Φ^* the Legendre–Fenchel transform of Φ.

The study of radial solutions on a ball or an annulus of some partial differential equations or systems with Dirichlet or Neumann boundary conditions have led to differential equations or systems of the form

$$(r^{N-1} \phi(v'))' = r^{N-1} f(r, v, v')$$

for which existence and multiplicity results have been recently obtained in [17–19], [21–25] by using methods similar to the ones described in this work. Those questions will not be considered here.

1.2   Notations

We will consider quasilinear second order differential systems of the form

$$(\phi(u'))' = f(t, u, u'), \tag{2}$$

where $f : [0, T] \times \mathbb{R}^{2n} \to \mathbb{R}^n$ is continuous, $\phi : B(a) \to B(b)$ belong to a suitable class of homeomorphism with $B(\rho) \subset \mathbb{R}^n$ the open ball of center 0 and radius ρ, $B(+\infty) = \mathbb{R}^n$, $0 < a \leq +\infty$, $0 < b \leq +\infty$ and $a + b = +\infty$. We assume moreover that $\phi(0) = 0$. A *solution* of (2) on $[0, T]$ is a function $u \in C^1([0, T], \mathbb{R}^n)$ such that $u'(t) \in B(a)$ for all $t \in [0, T]$, $\phi \circ u' \in C^1([0, T], \mathbb{R}^n)$ and (2) holds.

We have assumed that f is continuous for simplicity. The case of Carathéodory f can be treated as well.

In $\mathbb{R}^n$, we denote the usual inner product by $\langle \cdot, \cdot \rangle$ and the corresponding Euclidian norm by $|\cdot|$. We denote the usual norm in $L^p := L^p(0, T; \mathbb{R}^n)$ $(1 \le p \le \infty)$ by $|\cdot|_p$. We set $C := C([0, T], \mathbb{R}^n)$, $C^1 = C^1([0, T], \mathbb{R}^n)$, and $W^{1,\infty} := W^{1,\infty}([0, T], \mathbb{R}^n)$. The usual norm $|\cdot|_\infty$ is considered on C. The spaces $W^{1,\infty}$ and C^1 are endowed with the norm

$$|v|_{1,\infty} = |v|_\infty + |v'|_\infty.$$

Each $v \in L^1$ can be written $v(t) = \bar{v} + \tilde{v}(t)$, with

$$\bar{v} := T^{-1} \int_0^T v(t)\, dt, \qquad \int_0^T \tilde{v}(t)\, dt = 0,$$

and each $v \in C$ can be written $v(t) = v_0 + \hat{v}(t)$, with $v_0 = v(0)$ and $\hat{v}(0) = 0$. For $u \in C^1$, we have the estimate

$$|\hat{u}|_\infty = \max_{t \in [0,T]} \left| \int_0^t u'(s)\, ds \right| \le T|u'|_\infty, \tag{3}$$

and (3) is optimal as shown by taking $n = 1$ and $\hat{u}(t) = t$. If we assume in addition that $u(0) = u(T)$, then

$$\hat{u}(t) = \int_0^t u'(s)\, ds = -\int_t^T u'(s)\, ds,$$

and $\max_{[0,T]} |\hat{u}|$ being reached either in $[0, T/2]$ or in $[T/2, T]$, we deduce from the above inequality that

$$|\hat{u}|_\infty \le \frac{T}{2} |u'|_\infty \tag{4}$$

when $u(0) = u(T)$. Again it is easily shown that the constant $T/2$ is optimal. Now, any $u \in C^1$ such that $u(0) = u(T)$ can also be decomposed as $u(t) = \bar{u} + \tilde{u}(t)$. Using a result of Northcott [88], one has

$$|\tilde{u}|_\infty \le \frac{2T}{\pi^2} \left[\sum_{n=0}^\infty \frac{1}{(2n+1)^2} \right] |u'|_\infty = \frac{2T}{\pi^2} \cdot \frac{\pi^2}{8} |u'|_\infty = \frac{T}{4} |u'|_\infty. \tag{5}$$

Comparing (5) with (4) shows that the last decomposition will give better results. Define, for $k \ge 0$ any integer,

$$C_T^k := \{ u \in C^k([0, T], \mathbb{R}^n) : u(0) = u(T), \ldots, u^{(k)}(0) = u^{(k)}(T) \},$$

and

$$\widetilde{C}_T^k := \{u \in C_T^k : \bar{u} = 0\}.$$

It is well know that $u \in C^k$ has an antiderivative in C_T^{k+1} if and only if $u \in \widetilde{C}_T^k$, and, if it is the case, it has a unique antiderivative belonging to $\widetilde{C}_T^{k+1}$. We denote by $\widetilde{H}u$ this antiderivative so that

$$\widetilde{H} : \widetilde{C}_T^k \rightarrow \widetilde{C}_T^{k+1}.$$

Let us define the mapping $N_f : C^1 \rightarrow C^1$ by

$$N_f(u)(t) = \int_0^t f[s, u(s), u'(s)] \, ds \quad (t \in [0, T]),$$

the integration operator $H : C \rightarrow C^1$ by

$$Hu(t) = \int_0^t u(s) \, ds \quad (t \in [0, T]),$$

and the linear projectors $P : C \rightarrow \mathbb{R}^n \subset C$ and $Q : C \rightarrow \mathbb{R}^n \subset C$ by

$$Pu = u(0) = u_0, \quad Qu = T^{-1} \int_0^T u(t) \, dt = \bar{u},$$

where $\mathbb{R}^n$ is identified with the subspace of C of constant mappings. In any vector space E, we denote by $B(\rho)$ the open ball of center 0 and radius $\rho > 0$. We write $B(+\infty) = E$.

1.3 Classes of Homeomorphisms

A homeomorphism $\phi : \mathbb{R}^n \rightarrow \mathbb{R}^n$ is called *classical*, a homeomorphism $\phi : \mathbb{R}^n \rightarrow B(b)$ ($b < +\infty$) *bounded*, and a homeomorphism $\phi : B(a) \rightarrow \mathbb{R}^n$ ($a < +\infty$) *singular*. All those types were already considered in [53] in the scalar case and for periodic or Neumann problems with a nonlinearity depending only on the derivative.

Standard examples of classical homeomorphisms correspond to $\phi(s) = s$, for which (2) is the semilinear system

$$u'' = f(t, u, u'),$$

or to

$$\phi(s) = \phi_p(s) := |s|^{p-2}s \quad (p > 1),$$

($|\cdot|$ the Euclidian norm in $\mathbb{R}^n$), for which (2) is the quasilinear system associated to the *p-Laplacian*

$$(|u'|^{p-2}u')' = f(t, u, u').$$

An example of bounded homeomorphism corresponds to

$$\phi(s) = \phi_C(s) := \frac{s}{\sqrt{1 + |s|^2}},$$

for which (2) reduces for $n = 1$ to quasilinear equations associated to *curvature* or *capillarity* problems

$$\left(\frac{u'}{\sqrt{1 + u'^2}}\right)' = f(t, u, u'),$$

considered for example in [7,8,30,31,37,46,53–55,58,61,89–93,96,97,108] and in papers of the author and Bereanu listed in the bibliography. An example of singular homeomorphism corresponds to

$$\phi(s) = \phi_R(s) := \frac{s}{\sqrt{1 - |s|^2}},$$

for which (2) reduces to quasilinear equations associated to *relativistic* acceleration

$$\left(\frac{u'}{\sqrt{1 - |u'|^2}}\right)' = f(t, u, u'),$$

considered for example in [35, 36, 46, 53, 121, 122], and in papers of the author and Bereanu listed in the bibliography. Notice that if ϕ is classical, the same is true for ϕ^{-1}, if ϕ is bounded, ϕ^{-1} is singular, and if ϕ is singular, ϕ^{-1} is bounded. In particular $\phi_p^{-1} = \phi_{p/(p-1)}$ and $\phi_C^{-1} = \phi_R$.

The class of homeomorphisms ϕ occuring in (2) is characterized by the following condition.

(H_Φ) ϕ *is a homeomorphism from* $B(a) \subset \mathbb{R}^n$ *onto* $\mathbb{R}^n$ *such that* $\phi(0) = 0$, $\phi = \nabla\Phi$, *with* $\Phi : \overline{B(a)} \to \mathbb{R}$ *of class* C^1 *on* $\overline{B(a)}$, *continuous strictly convex on* $\overline{B(a)}$, *and such that* $\Phi(0) = 0$.

So, ϕ is strictly monotone on $B(a)$, in the sense that

$$\langle\phi(u) - \phi(v), u - v\rangle > 0 \quad \text{for} \quad u \neq v,$$

and Φ reaches its minimum 0 at 0.

If $\Phi^* : \mathbb{R}^n \to \mathbb{R}$ is the Legendre–Fenchel transform of Φ defined by

$$\Phi^*(v) = \langle\phi^{-1}(v), v\rangle - \Phi[\phi^{-1}(v)] = \sup_{u \in \overline{B(a)}} \{\langle u, v\rangle - \Phi(u)\},$$

then Φ^* is also strictly convex, and, if $d := \max_{u \in \overline{B(a)}} \Phi(u)$,

$$\langle u, v \rangle - d \leq \langle u, v \rangle - \Phi(u) \leq \langle u, v \rangle$$

for all $u \in \overline{B(a)}$ and $v \in \mathbb{R}^n$. Consequently,

$$a|v| - d \leq \Phi^*(v) \leq a|v| \quad (v \in \mathbb{R}^n), \tag{6}$$

so that Φ^* is coercive on $\mathbb{R}^n$. Adapting the reasoning of Proposition 2.4 in [85], we obtain that Φ^* is of class C^1. Hence $\phi^{-1} = \nabla \Phi^*$, so that

$$v = \nabla \Phi(u) = \phi(u), \quad u \in B(a) \quad \Leftrightarrow \quad u = \phi^{-1}(v) = \nabla \Phi^*(v), \quad v \in \mathbb{R}^n.$$

1.4 A Nonlinear Projector

A technical result is needed for the construction of the equivalent fixed point problems in the Dirichlet and periodic cases. For simplicity, we only consider the cases where ϕ is singular, so that $\phi^{-1} : \mathbb{R}^n \to B(a)$ ($0 < a \leq +\infty$). The classical case has been considered elsewhere (see e.g. [68,69]), and the bounded case requires some restrictions.

Given $h \in C$ and $b \in \mathbb{R}^n$, let us define $\Gamma(b; h)$ by

$$\Gamma(b; h) = \int_0^T \phi^{-1}[h(t) - b] \, dt = \int_0^T \nabla_b \Phi^*[h(t) - b] \, dt$$

$$= \nabla_b \int_0^T \Phi^*[h(t) - b] \, dt = \nabla_b \gamma(b; h),$$

where $\gamma(b; h)$ is defined by

$$\gamma(b; h) = \int_0^T \Phi^*[h(t) - b] \, dt.$$

The following Lemma is taken from [12].

Lemma 1. If $\phi = \nabla \Phi$, with Φ verifying Assumption (H_Φ), then, for each $h \in C$, the system

$$\int_0^T \phi^{-1}[h(t) - b] \, dt = 0$$

has a unique solution $b := Q_\phi(h)$. Moreover, $Q_\phi : C \to \mathbb{R}^n$ is continuous, and Q_ϕ takes bounded sets of C into bounded sets of $\mathbb{R}^n$.

Proof. For each b, $c \in \mathbb{R}^n$, $b \neq c$, and any $\lambda \in]0, 1[$, we have

$$\gamma[(1 - \lambda)b + \lambda c] = \int_0^T \Phi^*[(1 - \lambda)(b - h(t)) + \lambda(c - h(t))]\, dt$$

$$< \int_0^T \{(1 - \lambda)\Phi^*(b - h(t)) + \lambda\Phi^*(c - h(t))\}\, dt$$

$$\leq (1 - \lambda)\gamma(b; h) + \lambda\gamma(c; h),$$

so that $\gamma(\cdot; h)$ is strictly convex on $\mathbb{R}^n$ for each $h \in C$. Hence, $\Gamma(\cdot; h) = \nabla_b \gamma(\cdot; h)$ is strictly monotone on $\mathbb{R}^n$ for each $h \in C$. On the other hand, using (6), we get

$$Ta|b| - Td - |h|_1 \leq \gamma(b; h) \leq Ta|b| + |h|_1, \tag{7}$$

so that, for each $h \in C$, $\gamma(b; h)$ is coercive. Consequently, for each $h \in C$, $\gamma(\cdot; h)$ admits a unique minimum $b := Q_\phi(h)$, which corresponds to the unique critical point of $\gamma(\cdot; h)$. This implies that, for each $h \in C$, the system $\Gamma(b; h) = 0$ has a unique solution $b := Q_\phi(h)$.

Let us now show that Q_ϕ is continuous. Let (h_n) be a sequence converging in C to $h \in C$. Then (h_n) is bounded. Let $b_n = Q_\phi(h_n)$. Then, by (7),

$$\gamma(0; h_n) \geq \gamma(b_n; h_n) \geq Ta|b_n| - Td - |h_n|_1, \tag{8}$$

so that

$$|b_n| \leq (Ta)^{-1}[|h_n|_1 + Td + \gamma(0; h_n)]$$

which shows that (b_n) is bounded. Going if necessary to a subsequence, we can assume that (b_n) converges to β. From the relations

$$\int_0^T \phi^{-1}[h_n(t) - b_n]\, dt = 0 \quad (n \in \mathbb{N}),$$

and the dominated convergence theorem, we deduce that

$$\int_0^T \phi^{-1}[h(t) - \beta]\, dt = 0,$$

i.e. by the uniquess of the solutions, $\beta = Q_\phi(h)$, a limit independent of the subsequence. Hence

$$Q_\phi(h) = \lim_{n \to \infty} Q_\phi(h_n),$$

and Q_ϕ is continuous. Notice also that $Q_\phi(0) = 0$.

Finally, to show that Q_ϕ takes bounded sets of C into bounded sets of $\mathbb{R}^n$, we use again (8) to obtain

$$\gamma(0; h) \geq Ta|Q_\phi(h)| - Td - |h|_1,$$

and hence, using (7),

$$|Q_\phi(h)| \leq (Ta)^{-1}[(2|h|_1 + Td)].$$

$\square$

Remark 1. Lemma 1 shows that the mapping Q_ϕ verifies the identity

$$Q \circ \phi^{-1} \circ (I - Q_\phi) \circ u = 0 \quad \text{for all} \quad u \in C. \tag{9}$$

In other terms, $\phi^{-1} \circ (I - Q_\phi) : C \to \widetilde{C}$.

Furthermore, from the homeomorphic character of ϕ and $\phi(0) = 0$, we have

$$Q_\phi(0) = 0.$$

Remark 2. In the (classical) case where $\phi = I$, one has $Q_I = Qh = \overline{h}$.

Remark 3. It is easy to see that, for $n = 1$, Lemma 1 holds, with an elementary proof, for all increasing homeomorphisms $\phi : (-a, a) \to \mathbb{R}$ such that $\phi(0) = 0$. So, for $n = 1$, there is no need of assumption of the existence of the primitive of ϕ on $[-a, a]$.

Example 1. Let us consider the C^∞-mapping $\Phi \cdot \overline{B(1)} \subset \mathbb{R}^n \to \mathbb{R}$, given by

$$\Phi(u) = 1 - \sqrt{1 - |u|^2} \quad (u \in \overline{B(1)}), \tag{10}$$

so that $0 \leq \Phi(u) \leq 1$ for all $u \in \overline{B(1)}$, and

$$\phi(u) = \nabla\Phi(u) = \frac{u}{\sqrt{1 - |u|^2}} \quad (u \in B(1)).$$

As $|\cdot|^2$ is strictly convex on $\mathbb{R}^n$, it follows that Φ is strictly convex on $\overline{B(1)}$. Furthermore, $\phi : B(1) \to \mathbb{R}^n$ is a homeomorphism such that, for any $v \in \mathbb{R}^n$.

$$\phi^{-1}(v) = \frac{v}{\sqrt{1 + |v|^2}} = \nabla\Phi^*(v),$$

where $\Phi^*(v) = \sqrt{1 + |v|^2} - 1$ is strictly convex and of class C^∞ on $\mathbb{R}^n$. Hence, Assumption (H_Φ) with $a = 1$ holds for Φ given by (10).

2   Topological Approach

2.1   Introduction

The topological approach consists, as expected, in writing the boundary value problem as a fixed point problem in a suitable function space, and using Leray–Schauder degree to study the existence and multiplicity of the fixed points. A somewhat suprising fact is that, in this approach, the case of singular ϕ, which may look the most difficult one when seeing the differential equation, is finally even simpler than the classical case where ϕ is identity. This is especially examplified by the study of the corresponding Dirichlet problem. A reason may be seen in the fact that problems with singular ϕ have automatically an a priori bound for the derivative of solution, which is an important step in using Leray–Schauder continuation method.

2.2   Dirichlet Problem

Let $f : [0, T] \times \mathbb{R}^{2n} \to \mathbb{R}^n$ be continuous and consider for simplicity the homogeneous Dirichlet problem

$$(\phi(u'))' = f(t, u, u'), \quad u(0) = 0 = u(T). \tag{11}$$

The non-homogeneous case can be treated in a similar way, at the expense of a generalization of Lemma 1 (see [15] for details).

2.2.1   Equivalent Fixed Point Problem

The following fixed point operator was introduced for scalar classical ϕ in [51,52], for bounded ϕ in [10] and for singular ϕ in [11].

Theorem 1. *u is a solution of problem (11) if and only if $u \in C^1$ is a fixed point of the operator S defined on C^1 by*

$$S(u) = H \circ \phi^{-1} \circ (I - Q_\phi) \circ N_f(u).$$

Proof. If u is a solution of (11), then

$$\phi(u') = c + N_f(u),$$

where $c = \phi(u'(0))$, and hence

$$u' = \phi^{-1} \circ [c + N_f(u)], \tag{12}$$

so that $u(T) = 0$ if

$$\int_0^T \phi^{-1}[c + N_f(u)(t)]\, dt = 0,$$

i.e., using Lemma 1, if

$$c = -Q_\phi[N_f(u)].$$

Hence, (12) and the boundary condition at 0 give

$$u = H \circ \phi^{-1} \circ (I - Q_\phi) \circ N_f(u) = S(u).$$

Conversely, if $u \in C^1$ is a fixed point of S, then taking respectively $t = 0$ and $t = T$ in $u = S(u)$, we get $u(0) = 0$ and

$$u(T) = \int_0^T \phi^{-1}[(I - Q_\phi)(N_f(u)(t))]\, dt = 0$$

by (9). On the other hand, differentiating the fixed point equation gives

$$u' = \phi^{-1} \circ (I - Q_\phi) \circ N_f(u)$$

i.e.

$$\phi(u') = (I - Q_\phi) \circ N_f(u),$$

and hence, $\phi(u') \in C^1$ and, differentiating again,

$$(\phi(u'))' = f(t, u, u').$$

$\square$

It is standard to prove that $S : C^1 \to C^1$ is completely continuous.

Remark 4. The following diagram summarizes the construction of the fixed point operator S :

$$C^1 \xrightarrow{\ N_f\ } C^1 \xrightarrow{\ I - Q_\phi\ } C^1 \xrightarrow{\ \phi^{-1}\ } C \xrightarrow{\ H\ } C^1$$

The operator S maps C^1 into itself, but its fixed points satisfy the Dirichlet boundary conditions.

2.2.2   Existence Result for Singular ϕ

With this reduction to a fixed point problem, the existence of a solution to (11) for any continuous f follows from Schauder's fixed point theorem [105], as first proved in [11].

Theorem 2. *For any continuous $f : [0, T] \times \mathbb{R}^{2n} \to \mathbb{R}^n$, problem (11) has at least one solution.*

Proof. For any $u \in C^1$, we have

$$(S(u))' = \phi^{-1} \circ (I - Q_\phi) \circ N_f(u),$$

and hence

$$|(S(u))'|_\infty < a. \tag{13}$$

The second part of the proof of Theorem 1 also shows that, for any $u \in C^1$,

$$S(u)(0) = 0 = S(u)(T),$$

which, together with (13), gives

$$|S(u)|_1 = |S(u)|_\infty + |(S(u))'|_\infty \leq (T + 1)a$$

for all $u \in C^1$. Thus S maps C^1 into the closed ball $\overline{B}((T + 1)a) \subset C^1$ and has a fixed point using Schauder's fixed point theorem.   □

A direct consequence of (2) if the following

Corollary 1. *Given any $h \in C$ and any continuous $g : \mathbb{R}^n \to \mathbb{R}^n$, the problem*

$$(\phi(u'))' + g(u) = h(t), \quad u(0) = 0 = u(T) \tag{14}$$

has at least one solution.

If we recall that the classical linear Dirichlet problem

$$u'' + g(u) = h(t), \quad u(0) = 0 = u(T)$$

is called *non-resonant* if it is solvable for any $h \in C$, one can conclude that *the Dirichlet problem (14) is always non-resonant.*

2.3 Periodic Problem

2.3.1 Equivalent Fixed Point Problem for Classical or Singular ϕ

Let us consider now the periodic problem

$$(\phi(u'))' = f(t, u, u'), \quad u(0) = u(T), \quad u'(0) = u'(T). \tag{15}$$

Notice first that the problem

$$(\phi(u'))' = 1, \quad u(0) = u(T), \quad u'(0) = u'(T)$$

has no solution, because the existence of a solution would imply, by integration of both members of the differential equation and use of the boundary conditions, that $0 = T$. Hence we cannot expect an existence result like (2). This can be interpreted also by saying that the periodic problem

$$(\phi(u'))' = h(t), \quad u(0) = u(T), \quad u'(0) = u'(T)$$

is *resonant*.

The following result was essentially proved for classical ϕ in [68], for bounded ϕ in [10] and for singular ϕ in [11], in a slightly different setting.

Theorem 3. *u is a solution of problem (15) if and only if $u \in C^1$ is a fixed point of the operator $M : C^1 \to C^1$ defined by*

$$M(u) = Pu - N_f(u)(T) + H \circ \phi^{-1} \circ (I - Q_\phi) \circ N_f(u). \qquad (16)$$

Furthermore, $|(M(u))'|_\infty < a$ for all $u \in C^1$ and M is completely continuous on C^1.

Proof. If u is a solution of problem (15), then $u \in C^1$, $\phi(u') \in C^1$ and, integrating both members of the differential equation over $[0, T]$ gives

$$N_f(u)(T) = 0. \qquad (17)$$

The differential equation in (15) is equivalent to

$$\phi(u') = c + N_f(u)$$

and hence to

$$u' = \phi^{-1}[c + N_f(u)] \qquad (18)$$

where $c = \phi(u'(0))$. The first boundary condition implies that c must be such that

$$\int_0^T \phi^{-1}[c + N_f(u)(t)] \, dt = 0$$

which, using Lemma 1, gives

$$c = -Q_\phi[N_f(u)].$$

Thus, (18) becomes

$$u' = \phi^{-1} \circ (I - Q_\phi) \circ N_f(u) \tag{19}$$

which is equivalent to the integrated form

$$u - Pu = H \circ \phi^{-1} \circ (I - Q_\phi) \circ N_f(u). \tag{20}$$

Finally, as (17) and (20) take values in supplementary subspaces of C, they can be written as the unique fixed point problem

$$u = Pu - N_f(u)(T) + H \circ \phi^{-1} \circ (I - Q_\phi) \circ N_f(u). \tag{21}$$

Conversely, if u is a fixed point of M, i.e. a solution of (21), then taking $t = 0$ in (21) gives (17). Differentiating both members of (21) gives (19) which, integrated over $[0, T]$ gives

$$u(T) - u(0) = TQ \circ \phi^{-1} \circ (I - Q_\phi) \circ N_f(u) = 0$$

using (9), so that the first boundary condition is satisfied. Now (19) is equivalent to

$$\phi(u') = (I - Q_\phi) \circ N_f(u),$$

which gives, using (17)

$$\phi(u'(T)) - \phi(u'(0)) = N_f(u)(T) = 0, \quad \text{i.e.} \quad u'(T) = u'(0),$$

and, by differentiating both members

$$(\phi(u'))' = f(t, u, u').$$

$\square$

Remark 5. The following diagram summarizes the construction of the non-constant part of the fixed point operator M :

$$C^1 \xrightarrow{N_f} C^1 \xrightarrow{I - Q_\phi} C^1 \xrightarrow{\phi^{-1}} C \xrightarrow{H} C^1$$

The operator M maps C^1 into itself, but its fixed points satisfy the periodic boundary conditions.

2.3.2   Existence Theorem for Singular ϕ

Again we concentrate on the case of a singular $\phi : B(a) \to \mathbb{R}^n$. In order to find conditions for the existence of fixed points of M using Leray–Schauder degree, we introduce the homotopy $\mathscr{M}$ defined on $C \times [0, 1]$ by

$$\mathscr{M}(u) = Pu - N_f(u)(T) + H \circ \phi^{-1} \circ (I - Q_\phi) \circ \lambda N_f(u).$$

Notice that, for $\lambda \in (0, 1]$, an argument entirely similar to that of Theorem 3 shows that the fixed points of $\mathscr{M}(\cdot, \lambda)$ are the solutions of the problem

$$(\phi(u')' = \lambda f(t, u, u'), \quad u(0) = u(T), \quad u'(0) = u'(T).$$

For $\lambda = 0$, the fixed points of $\mathscr{M}(\cdot, 0)$ are the solutions of

$$u = Pu - N_f(u)(T),$$

so that they are constant $u \in \mathbb{R}^n$, solutions of the n-dimensional system

$$N_f(u)(T) = 0.$$

We can now prove the following Leray–Schauder type existence result for problem (15). We denote by d_B the Brouwer degree for continuous mappings in $\mathbb{R}^n$, and by d_{LS} the Leray–Schauder degree [60] for compact perturbations of identity in a Banach space (see e.g. [39]).

Theorem 4. *Assume that there exists an open bounded set $\Omega \subset C$ such that the following conditions hold :*

1. For each $\lambda \in (0, 1]$, there is no solution of problem

$$(\phi(u'))' = \lambda f(t, u, u'), \quad u(0) = u(T), \quad u'(0) = u'(T) \tag{22}$$

such that $u \in \partial\Omega$.
2. There is no solution $u \in \partial\Omega \cap \mathbb{R}^n$ of equation

$$\overline{f}(u) := N_f(u)(T) = 0, \tag{23}$$

where $\mathbb{R}^n$ denotes the subspace of constant functions in C.
3. $d_B[\overline{f}, \Omega \cap \mathbb{R}^n, 0] \neq 0$.

Then problem (15) has at least one solution such that

$$u \in \Omega_a := \{u \in C^1 : u \in \Omega, \ |u'|_\infty < a\},$$

and, for the associated fixed point operator M, one has

$$d_{LS}[I - M, \Omega_\rho, 0] = d_B[\overline{f}, \Omega \cap \mathbb{R}, 0],$$

for any $\rho \geq a$.

Proof. Let u be a possible fixed point of $\mathcal{M}(\cdot, \lambda)$. If $\lambda \in (0, 1]$, then, by the reasoning above, u is a solution of problem (22), and $u \notin \partial\Omega$ by Assumption 1. Furthermore,

$$|u'|_\infty = |(\mathcal{M}(\cdot, \lambda))'|_\infty < a.$$

If $\lambda = 0$, then, by the reasoning above, u is a constant solution of (23). By Assumption 2, $u \notin \partial\Omega$. Consequently, for $\lambda \in [0, 1]$, and any $\rho \geq a$, $\mathcal{M}(\cdot, \lambda)$ has no fixed point on $\partial\Omega_\rho$. The homotopy invariance of Leray–Schauder degree implies that

$$d_{LS}[I - M, \Omega_\rho, 0] = d_{LS}[I - \mathcal{M}(\cdot, 0), \Omega_\rho, 0]. \tag{24}$$

Now,

$$\mathcal{M}(\cdot, 0) : C^1 \to \mathbb{R}^n$$

with $\mathbb{R}^n$ identified to the the subset of constant functions in C, and hence the reduction formula for Leray–Schauder degree gives

$$d_{LS}[I - \mathcal{M}(\cdot, 0), \Omega_\rho, 0] = d_B[(I - \mathcal{M}(\cdot, 0))|_{\mathbb{R}^n}, \Omega_\rho \cap \mathbb{R}^n, 0]$$
$$= d_B[\overline{f}, \Omega \cap \mathbb{R}^n, 0] \neq 0, \tag{25}$$

by Assumption 3. The result follows from relations (24), (25) and the existence property of Leray–Schauder degree. □

2.3.3 Forced Planar Polynomial Systems with Singular ϕ

In this section, let us provide $\mathbb{R}^2$ with the multiplication structure of the complex plane $\mathbb{C}$, and consider the planar periodic problem

$$(\phi(z'))' = p(z) + h(t), \quad z(0) = z(T), \quad z'(0) = z'(T), \tag{26}$$

where $h \in C$ and $p : \mathbb{C} \to \mathbb{C}$ is a polynomial of effective degree $N \geq 1$, namely

$$p(z) = \sum_{k=0}^{N} a_k z^k \quad (a_k \in \mathbb{C} \quad (k = 0, \ldots, N), \quad a_N \neq 0).$$

Theorem 5. *Problem (26) has at least one solution for every* $h \in C$.

Proof. In order to use Theorem 4, consider the family of problems

$$(\phi(z'))' = \lambda[p(z) + h(t)], \quad z(0) = z(T), \quad z'(0) = z'(T), \quad (\lambda \in (0, 1]), \ (27)$$

and let z be a possible solution of (27). We know that

$$|z'|_\infty < a,$$

so that letting

$$z(t) = z_0 + \widehat{z}(t)$$

with $z_0 = z(0)$, we have, by (4),

$$|\widehat{z}|_\infty < \frac{Ta}{2}.$$

Integrating both members of (27) over $[0, T]$ and using the boundary conditions, we get

$$0 = \int_0^T [p(z_0 + \widehat{z}(t)) + h(t)] \, dt = 0,$$

i.e., explicitly,

$$0 - \int_0^T \left[\sum_{k=0}^N a_k \left(\sum_{j_k=0}^k \frac{k!}{j_k!(k - j_k)!} z_0^{j_k} \widehat{z}(t)^{k-j_k} \right) + h(t) \right] dt,$$

This equation has the form

$$0 = \int_0^T [a_N z_0^N + \sum_{j=0}^{N-1} p_j[\widehat{z}(t)] z_0^j + h(t)] \, dt,$$

where the p_j are polynomials on $\mathbb{C}$ whose coefficients only depend upon those of p, and hence

$$|a_N| |z_0|^N \le \sum_{j=0}^{N-1} b_j |z_0|^j + T^{-1} |h|_1,$$

where

$$|b_j| = \max_{|u| \le Ta/2} |p_j(u)| \quad (j = 0, 1, \dots, N).$$

This implies the existence of $R_1 > 0$ such that $|z_0| < R$ and hence such that

$$|z|_1 < R_1 + a \left(\frac{T}{2} + 1 \right).$$

On the other hand, for any $\lambda \in [0, 1]$ and any possible zero u of

$$\overline{F}(u, \lambda) := T \left[a_N u^N + \sum_{k=0}^{N-1} a_k u^k + \overline{h} \right]$$

we have

$$|a_N||u|^N \leq \sum_{k=0}^{N-1} |a_k||u|^k + |\overline{h}|$$

and hence there exists $R_2 > 0$ such that $|u| < R_2$. Consequently, for any $R \geq R_2$,

$$d_B[\overline{f}, B(R), 0] = d_B[a_N z^N, B(R), 0] = N.$$

Taking

$$\Omega = \{z \in C^1 : |z|_\infty < B(R), \ |z'|_\infty < a\}$$

with $R \geq \max\{R_1 + a[(T/2) + 1], R_2\}$, all the assumptions of Theorem 4 are satisfied.   □

Remark 6. Such a result does not holds in classical case, as shown by the example

$$z'' = -z + \sin t, \quad z(0) = z(2\pi), \quad z'(0) = z'(2\pi),$$

whose first term in right-hand side is a polynomial of degree one and which has no solution, as shown by multiplying each member by $\sin t$ and integrating the result over $[0, 2\pi]$.

2.3.4 Asymptotic Sign Conditions for Scalar Problems with Singular ϕ

We now restrict ourself to *scalar* equations ($n = 1$). For $u \in C_T^1$, we write

$$u_L := \min_{[0,T]} u, \ u_M := \max_{[0,T]} u.$$

The following results can be found in [11]. Conditions of type (28) were first introduced by Villari [123] in classical periodic problems.

Theorem 6. *Assume that there exist $R > 0$ and $\epsilon \in \{-1, 1\}$ such that*

$$\epsilon \int_0^T f[t, u(t), u'(t)] \, dt > 0 \quad if \quad u \in C_T^1, \quad u_L \geq R, \quad |u'|_\infty < a, \qquad (28)$$

$$\epsilon \int_0^T f(t, u(t), u'(t)) \, dt < 0 \quad if \quad u \in C_T^1, \quad u_M \leq -R, \quad |u'|_\infty < a.$$

Then problem (15) with $n = 1$ has at least one solution and, for the associated fixed point operator M, one has, for any $\rho_1 \geq R + Ta$ and $\rho_2 \geq a$.

$$d_{LS}[I - M, \Omega_{\rho_1, \rho_2}, 0] = \text{sgn } \varepsilon,$$

where

$$\Omega_{\rho_1, \rho_2} = \{u \in C^1 : |u|_\infty < \rho_1, \ |u'|_\infty < \rho_2\}.$$

Proof. We construct an open bounded set $\Omega \subset C$ having the properties requested by Theorem 4. Let $\lambda \in (0, 1]$ and u be a possible solution of (22). Then, $|u'|_\infty < a$. Integrating both members of (22) on $[0, T]$ gives

$$\int_0^T f[s, u(s), u'(s)] \, ds = 0 \qquad (29)$$

If $u_M \leq -R$ (resp. $u_L \geq R$) then it follows from (28) that

$$\epsilon \int_0^T f[t, u(t), u'(t)] \, dt < 0 \quad (\text{resp.} \quad \epsilon \int_0^T f[t, u(t), u'(t)] \, dt > 0).$$

Using (29) we deduce that

$$u_M > -R \quad \text{and} \quad u_L < R.$$

From

$$u_M \leq u_L + \int_0^T |u'(t)| \, dt,$$

we obtain

$$-(R + Ta) < u_L \leq u_M < R + Ta.$$

and hence

$$|u|_\infty < R + Ta.$$

Hence if we take, for any $\rho_1 \geq R + Ta$ and $\rho_2 \geq a$,

$$\Omega_{\rho_1,\rho_2} = \{u \in C^1 : |u|_\infty < \rho_1, \ |u'|_\infty < \rho_2\},$$

in Theorem 4, its Assumption (1) and (2) hold. Using (24) and (25) and elementary results on the one-dimensional Brouwer degree, we get, for $\rho_1 \geq R + Ta$ and $\rho_2 \geq a$,

$$d_{LS}[I - M, \Omega_{\rho_1,\rho_2}, 0] = d_B[\overline{f}, (-\rho_1, \rho_1), 0] = \text{sign } \epsilon.$$

The result follows from Theorem 4.   □

Corollary 2. *Let $h : [0, T] \times \mathbb{R}^2 \to \mathbb{R}$, $k : \mathbb{R} \to \mathbb{R}$, and $g : [0, T] \times \mathbb{R} \to \mathbb{R}$ be continuous, with h is bounded on $[0, T] \times \mathbb{R} \times (-a, a)$ and g satisfies condition*

$$\lim_{u \to -\infty} g(t, u) = +\infty, \qquad \lim_{u \to +\infty} g(t, u) = -\infty$$

$$(resp. \quad \lim_{u \to -\infty} g(t, u) = -\infty, \qquad \lim_{u \to +\infty} g(t, u) = +\infty)$$

uniformly in $t \in [0, T]$. Then the problem

$$(\phi(u'))' + k(u)u' + g(t, u) = h(t, u, u'), \quad u(0) = u(T), \quad u'(0) = u'(T)$$

has at least one solution.

Example 2. If $c \in \mathbb{R} \setminus \{0\}$, $d \in \mathbb{R}$, $q \geq 0$ and $p > 1$, the problem

$$\left(\frac{u'}{\sqrt{1 - u'^2}}\right)' + d|u'|^q + c|u|^{p-1}u = h(t),$$

$$u(0) = u(T), \quad u'(0) = u'(T),$$

has at least one solution for all $h \in C$.

Corollary 3. *Let $k : \mathbb{R} \to \mathbb{R}$ and $h : [0, T] \times \mathbb{R}^2 \to \mathbb{R}$ be continuous, with h bounded on $[0, T] \times \mathbb{R} \times (-a, a)$. Then, for each $\mu \neq 0$, the problem*

$$(\phi(u'))' + k(u)u' + \mu u = h(t, u, u'), \quad u(0) = u(T), \quad u'(0) = u'(T)$$

has at least one solution, and, for the associated fixed point operator M we have, for all sufficiently large $\rho_1 > 0$, $\rho_2 \geq a$ and

$$\Omega_{\rho_1,\rho_2} = \{u \in C^1 : |u|_\infty < \rho_1, \ |u'|_\infty < \rho_2\},$$

$$d_{LS}[I - M, \Omega_{\rho_1,\rho_2}, 0] = -\text{sgn } \mu.$$

When $h(t, u, w) = h(t)$ only depends upon t, Corollary 3 shows that, for $\mu \neq 0$, problem

$$(\phi(u'))' + \mu u = h(t), \quad u(0) = u(T), \quad u'(0) = u'(T) \tag{30}$$

has at least one solution for any $h \in C$. In this sense one can say that problem (30) with $\mu \neq 0$ is *non-resonant*. Consequently, $\mu = 0$ is the only value for which *resonance* occurs in (30).

Another easy consequence is a *Landesman-Lazer-type existence condition* for the forced Liénard equation with singular ϕ.

Corollary 4. *Let $k, g : \mathbb{R} \to \mathbb{R}$ be continuous. Then problem*

$$(\phi(u'))' + k(u)u' + g(u) = h(t), \quad u(0) = u(T), \quad u'(0) = u'(T)$$

has at least one solution for each $h \in C$ such that

$$\limsup_{u \to -\infty} g(u) < \overline{h} < \liminf_{u \to +\infty} g(u)$$

or such that

$$\limsup_{u \to +\infty} g(u) < \overline{h} < \liminf_{u \to -\infty} g(u).$$

Proof. Let us consider, say, the first case, the proof of the other one being similar. By Assumptions, there exists $R > 0$ such that

$$\overline{h} \quad g(u) < 0 \quad \text{for} \quad u \geq R, \quad \overline{h} - g(u) < 0 \quad \text{for} \quad u \leq -R.$$

Consequently, for all $u \in C_T^1$ with $u_L \geq R$ we have

$$\int_0^T [h(s) - k(u(s))u'(s) - g(u(s))]\, ds < 0$$

and, for all $u \in C_T^1$ with $u_M \leq -R$ we have

$$\int_0^T [h(s) - k(u(s))u'(s) - g(u(s))]\, ds > 0.$$

$\square$

Example 3. Assume that $k : \mathbb{R} \to \mathbb{R}$ is continuous and $c \neq 0$. The problem

$$\left(\frac{u'}{\sqrt{1 - u'^2}}\right)' + k(u)u' + c \arctan u = h(t), \quad u(0) = u(T), \quad u'(0) = u'(T)$$

has at least one solution if and only if

$$-\frac{|c|\pi}{2} < \overline{h} < \frac{|c|\pi}{2}.$$

The problem

$$\left(\frac{u'}{\sqrt{1-u'^2}}\right)' + k(u)u' + c\exp u = h(t), \quad u(0) = u(T), \quad u'(0) = u'(T)$$

has at least one solution if and only if

$$a\overline{h} > 0.$$

If $p > 1$, the problem

$$\left(\frac{u'}{\sqrt{1-u'^2}}\right)' + k(u)u' + c|u|^{p-2}u = h(t), \quad u(0) = u(T), \quad u'(0) = u'(T)$$

has at least one solution for all $h \in C$.

2.3.5 Another Equivalent Fixed Point Problem

To obtain sharper existence results under conditions which are not of asymptotic type, it is useful to introduce another fixed point problem equivalent to problem (15), involving an operator mapping C_T^1 into itself associated to another decomposition of a function into a constant and a non-constant part.

Define the operator $M_\#$ on C_T^1 by

$$M_\#(u) = Qu - N_f(u)(T) + \widetilde{H} \circ \phi^{-1} \circ (I - Q_\phi) \circ N_{(I-Q)f}(u). \qquad (31)$$

The sequence of relations

$$C_T^1 \xrightarrow{N_{(I-Q)f}} C_T^1 \xrightarrow{I-Q_\phi} C_T^1 \xrightarrow{\phi^{-1}} \widetilde{C}_T \xrightarrow{\widetilde{H}} \widetilde{C}_T^1,$$

where use has been made of (1), shows that $M_\# : C_T^1 \to C_T^1$ and it is standard to show that $M_\#$ is completely continuous. In a way very similar to Theorem 3, on can prove the following result.

Theorem 7. *u is a solution of problem (15) if and only if $u \in C_T^1$ is a fixed point of the operator $M_\# : C_T^1 \to C_T^1$ defined by (31). Furthermore, $|(M_\#(u))'|_\infty < a$ for all $u \in C_T^1$ and $M_\#$ is completely continuous on C_T^1.*

2.3.6   A Continuum Containing the Solution Set for Singular ϕ

When ϕ is singular, a variant of Theorem 7 is useful in scalar problems, first proved in [11] in a slightly different form. We assume that $n = 1$.

Lemma 2. *The set $\mathscr{S}$ of solutions $[\bar{u}, \tilde{u}] \in \mathbb{R} \times \widetilde{C}_T^1$ of the modified problem*

$$(\phi(\tilde{u}'))' = (I - Q)f(\cdot, \bar{u} + \tilde{u}, \tilde{u}'), \quad \tilde{u}(0) = \tilde{u}(T), \quad \tilde{u}'(0) = \tilde{u}'(T) \quad (32)$$

contains a continuum $\mathscr{C}$ whose projection on $\mathbb{R}$ is $\mathbb{R}$ and projection on $\widetilde{C}_T^1$ is contained in the ball $B(a[(T/4) + 1])$.

Proof. Using an argument similar to the one of Theorem 7, on can show that, for each fixed $\bar{u} \in \mathbb{R}$, problem (32) is equivalent to the fixed point problem in $\widehat{C}_T^1$

$$\tilde{u} = \widetilde{H} \circ \phi^{-1} \circ (I - Q_\phi) \circ N_{(I-Q)f}(\bar{u} + \tilde{u}) := \widetilde{M}_{\#}(\bar{u}, \tilde{u}).$$

Again, $\widetilde{M}_{\#}$ is completely continuous on $\mathbb{R} \times \widehat{C}_T^1$, and, for each $[\bar{u}, \tilde{u}] \in \mathbb{R} \times \widetilde{C}_T^1$, we have, using (5),

$$|(\widetilde{M}_{\#}(\bar{u}, \tilde{u}))'|_\infty < a, \quad |\widetilde{M}_{\#}(\bar{u}, \tilde{u})|_\infty < Ta/4. \quad (33)$$

It follows from (33) that, for each $\bar{u} \in \mathbb{R}$, any possible fixed point $\tilde{u}$ of $\widetilde{M}_{\#}(\bar{u}, \cdot)$ is such that

$$|\tilde{u}|_1 < a[(T/4) + 1]. \quad (34)$$

Furthermore, for each $\lambda \in [0, 1]$, and each $\bar{u} \subset \mathbb{R}$, any possible fixed point $\hat{u}$ of

$$\widetilde{\mathscr{M}}(\bar{u}, \cdot, \lambda) := \widetilde{H} \circ \phi^{-1} \circ (I - Q_\phi) \circ [\lambda N_{(I-Q)f}(\bar{u} + \cdot)]$$

satisfies, for the same reasons, inequality (34), which implies that

$$\begin{aligned}
& d_{LS}[I - \widetilde{M}_{\#}(0, \cdot), B(a[(T/4) + 1]), 0] \\
& = d_{LS}[I - \widetilde{\mathscr{M}}(0, \cdot, 1), B(a[(T/4) + 1]), 0] \quad (35) \\
& = d_{LS}[I - \widetilde{\mathscr{M}}(0, \cdot, 0), B(a[(T/4) + 1]), 0] = d_{LS}[I, B(a[(T/4) + 1]), 0] = 1.
\end{aligned}$$

Conditions (34), (35) and Leray–Schauder theory [50, 60, 72] then imply the existence of $\mathscr{C}$.   □

2.3.7   Weak Asymptotic Sign Conditions for Singular ϕ

The existence part of Theorem 6 can be obtained under non-strict asymptotic sign conditions upon f.

Theorem 8. *Assume that there exist $R > 0$ and $\epsilon \in \{-1, 1\}$ such that*

$$\epsilon \int_0^T f(t, u(t), u'(t))\, dt \geq 0 \quad \text{if} \quad u \in C_T^1, \quad u_L \geq R, \quad |u'|_\infty < a, \qquad (36)$$

$$\epsilon \int_0^T f(t, u(t), u'(t))\, dt \leq 0 \quad \text{if} \quad u \in C_T^1, \quad u_M \leq -R, \quad |u'|_\infty < a,$$

Then problem (15) has at least one solution.

Proof. Consider the continuum $\mathscr{C}$ given by Lemma 2. If $[R + (Ta/4), \widetilde{u}] \in \mathscr{C}$, then, for each $t \in [0, T]$,

$$R + \frac{Ta}{4} + \widetilde{u}(t) > R$$

and hence, using (36)

$$\epsilon \int_0^T f(t, R + Ta/4 + \widetilde{u}(t), \widetilde{u}'(t))\, dt \geq 0.$$

Similarly, if $[-R - (Ta/4), \widetilde{u}] \in \mathscr{C}$, then

$$\epsilon \int_0^T f(t, -R - Ta/4 + \widetilde{u}(t), \widetilde{u}'(t))\, dt \leq 0.$$

The existence of $[\bar{u}, \widetilde{u}] \in \mathscr{C}$ such that

$$\epsilon \int_0^T f(t, \bar{u} + \widetilde{u}(t), \widetilde{u}'(t))\, dt = 0,$$

and hence such that $u = \bar{u} + \widetilde{u}$ is a solution of (15) follows from the intermediate value theorem for a continuous function on a connected set. $\qquad \square$

In the special case where $f(t, u, w) = h(t)$ only depends upon t, condition (36) reduces to $\bar{h} = 0$, which is easily seen to be also necessary for the existence of a solution to the problem

$$(\phi(u'))' = h(t), \quad u(0) = u(T), \quad u'(0) = u'(T). \qquad (37)$$

So we have the result.

Corollary 5. *Problem (37) has at least one solution u if and only if $\bar{h} = 0$, in which case it has the one parameter family of solutions $c + u$ ($c \in \mathbb{R}$).*

Corollary 6. *Let $h \in C$, $k : \mathbb{R} \to \mathbb{R}$, $g : \mathbb{R} \to \mathbb{R}$ be continuous. If $\bar{h} = 0$, and if there exists $R > 0$ and $\epsilon \in \{-1, 1\}$ such that*

$$\epsilon g(u)u \le 0 \quad whenever \quad |u| \ge R,$$

then the periodic problem for the Liénard equation with singular ϕ

$$(\phi(u'))' + k(u)u' + g(u) = h(t), \quad u(0) = u(T), \quad u'(0) = u'(T)$$

has at least one solution.

Proof. Notice that, if $u \in C_T^1$, $u_L \ge R$ and $|u'|_\infty < a$, using the boundary conditions,

$$\epsilon \int_0^T \{h(t) - k[u(t)]u'(t) - g[u(t)]\} \, dt = -\epsilon \int_0^T g[u(t)] \, dt \ge 0,$$

and if $u \in C_T^1$, $u_M \le R$ and $|u'|_\infty < a$,

$$\epsilon \int_0^T \{h(t) - k[u(t)]u'(t) - g[u(t)]\} \, dt = -\epsilon \int_0^T g[u(t)] \, dt \le 0.$$

$\square$

2.3.8 A Localized Sign Condition and Periodic Nonlinearities for Singular ϕ

We first prove, generalizing [20], an existence theorem for (15) which does not involve asymptotic conditions upon the nonlinearity. Let $g : [0, T] \times \mathbb{R}^2 \to \mathbb{R}$ be continuous and $h \in C$.

Theorem 9. *Assume that there exists $r < s$ and $A \le B$ such that*

$$T^{-1} \int_0^T g[t, r + \widetilde{u}(t), \widetilde{u}'(t)] \, dt \le A$$

$$and \tag{38}$$

$$T^{-1} \int_0^T [g(t, s + \widetilde{u}(t), \widetilde{u}'(t)] \, dt \ge B$$

or

$$T^{-1} \int_0^T g[t, r + \widetilde{u}(t), \widetilde{u}'(t)] \, dt \ge B$$

$$and \tag{39}$$

$$T^{-1} \int_0^T g[t, s + \widetilde{u}(t), \widetilde{u}'(t)] \, dt \le A$$

for any $\widetilde{u} \in \widetilde{C}_T^1$ *satisfying* $\lceil \widetilde{u} \rceil_\infty < \frac{Ta}{4}$. *If* $A < B$, *then for any* $h \in C$ *satisfying*

$$A < \bar{h} < B, \qquad (40)$$

the problem

$$(\phi(u'))' + g(t, u, u') = h(t), \quad u(0) = u(T), \quad u'(0) = u'(T) \qquad (41)$$

has at least one solution u *such that* $r < \bar{u} < s$. *If* $A \leq B$, *then for any* $h \in C$ *satisfying*

$$A \leq \bar{h} \leq B, \qquad (42)$$

problem (41) has at least one solution u *such that* $r \leq \bar{u} \leq s$.

Proof. Assume, say, that (38) holds, and let $\mathscr{C}$ be the continuum in $\mathbb{R} \times \widetilde{C}_T^1$ associated to (41) by Lemma 2. If $A < B$ and condition (40) holds, then,

$$\bar{h} - T^{-1} \int_0^T g[t, r + \widetilde{u}(t), \widetilde{u}'(t)] \, dt \geq \bar{h} - A > 0$$

$$\bar{h} - T^{-1} \int_0^T [g(t, s + \widetilde{u}(t), \widetilde{u}'(t)] \, dt \leq \bar{h} - B < 0$$

for all $\widetilde{u} \in \widetilde{C}_T^1$ such that $\lceil \widetilde{u} \rceil_\infty < \frac{Ta}{4}$. Hence, there is some $[r, \widetilde{v}]$ on $\mathscr{C}$ such that

$$\bar{h} - T^{-1} \int_0^T g[t, r + \widetilde{v}(t), \widetilde{v}'(t)] \, dt \geq \bar{h} - A > 0,$$

and some $[s, \widetilde{w}]$ on $\mathscr{C}$ such that

$$\bar{h} - T^{-1} \int_0^T [g(t, s + \widetilde{w}(t), \widetilde{w}'(t)] \, dt \leq \bar{h} - B < 0.$$

Consequently, there is some $[\bar{u}, \widetilde{u}] \in \mathscr{C}$ with $r < \bar{u} < s$ such that

$$\bar{h} - T^{-1} \int_0^T [g(t, \bar{u} + \widetilde{u}(t), \widetilde{u}'(t)] \, dt = 0,$$

i.e. such that $u = \bar{u} + \widetilde{u}$ is a solution of (41). The other cases are entirely similar.
□

An immediate consequence of Lemma 9 is the following result of [20], improving an earlier one of Torres [121] (see also [122]).

Theorem 10. *Let $k : \mathbb{R} \to \mathbb{R}$ be continuous, $h \in C$ and $\mu > 0$. If*

$$Ta < 2\pi, \quad |\bar{h}| < \mu \cos\left(\frac{Ta}{4}\right),$$

then the problem

$$(\phi(u'))' + k(u)u' + \mu \sin u = h(t), \quad u(0) = u(T), \quad u'(0) = u'(T) \quad (43)$$

has at least two solution u_1, u_2 such that $-\frac{\pi}{2} < \bar{u}_1 < \frac{\pi}{2} < \bar{u}_2 < \frac{3\pi}{2}$. If $Ta = 2\pi$, then problem (43) has at least one solution for any $h \in C$ with $\bar{h} = 0$.

Proof. A simple computation shows that we can apply Theorem 9 *(i)* with

$$r = -\frac{\pi}{2}, \ s = \frac{\pi}{2} \quad \text{and} \quad A = \mu \sin\left(-\frac{\pi}{2} + \frac{Ta}{4}\right) = -B;$$

$$r = \frac{\pi}{2}, \ s = \frac{3\pi}{2} \quad \text{and} \quad A = \mu \sin\left(-\frac{\pi}{2} + \frac{Ta}{4}\right) = \mu \sin\left(\frac{3\pi}{2} + \frac{Ta}{4}\right) = -B.$$

$\square$

Example 4. Let $k : \mathbb{R} \to \mathbb{R}$ be continuous, $h \in C$ and $\mu > 0$. If

$$T < 2\pi, \quad |\bar{h}| < \mu \cos\left(\frac{T}{4}\right),$$

then the problem

$$\left(\frac{u'}{\sqrt{1 - u'^2}}\right)' + k(u)u' + \mu \sin u = h(t), \quad u(0) = u(T), \quad u'(0) = u'(T) \quad (44)$$

has at least two solution u_1, u_2 such that $-\frac{\pi}{2} < \bar{u}_1 < \frac{\pi}{2} < \bar{u}_2 < \frac{3\pi}{2}$. If

$$Ta = 2\pi,$$

then problem (44) has at least one solution for any $h \in C$ with $\bar{h} = 0$.

2.3.9 Lower and Upper Solutions for Singular ϕ

In this subsection, we extend, following [11], the method of upper and lower solutions [38, 100, 101] to the periodic boundary value problem (15) with $n = 1$ and ϕ singular.

Definition 1. A *lower solution* α (resp. *upper solution* β) of (15) is a function $\alpha \in C^1$ such that $|\alpha'|_\infty < a$, $\phi(\alpha') \in C^1$, $\alpha(0) = \alpha(T)$, $\alpha'(0) \geq \alpha'(T)$ (resp. $\beta \in C^1$, $|\beta'|_\infty < a$, $\phi(\beta') \in C^1$, $\beta(0) = \beta(T)$, $\beta'(0) \leq \beta'(T)$) and

$$(\phi(\alpha'(t)))' \geq f(t, \alpha(t), \alpha'(t)) \quad (resp. \quad (\phi(\beta'(t)))' \leq f(t, \beta(t), \beta'(t))) \quad (45)$$

for all $t \in [0, T]$. Such a lower or upper solution is called *strict* if the inequality (45) is strict for all $t \in [0, T]$.

Theorem 11. *If (15) has a lower solution α and a upper solution β such that $\alpha(t) \leq \beta(t)$ for all $t \in [0, T]$, then problem (15) has a solution u such that $\alpha(t) \leq u(t) \leq \beta(t)$ for all $t \in [0, T]$. Moreover, if α and β are strict, then $\alpha(t) < u(t) < \beta(t)$ for all $t \in [0, T]$, and $d_{LS}[I - M, \Omega_{\alpha,\beta,a}, 0] = 1$, where*

$$\Omega_{\alpha,\beta,a} = \{u \in C^1 : \alpha(t) < u(t) < \beta(t) \quad \text{for all} \quad t \in [0, T], \quad |u'|_\infty < a\},$$

and M is the fixed point operator associated to (15).

Proof. Let $\gamma : [0, T] \times \mathbb{R} \to \mathbb{R}$ be the continuous function defined by

$$\gamma(t, u) = \begin{cases} \beta(t) & \text{if } u > \beta(t) \\ u & \text{if } \alpha(t) \leq u \leq \beta(t) \\ \alpha(t) & \text{if } u < \alpha(t), \end{cases}$$

and define $F : [0, T] \times \mathbb{R}^2 \to \mathbb{R}$ by $F(t, u, v) = f(t, \gamma(t, u), v)$. We consider the modified problem

$$(\phi(u'))' = F(t, u, u') + u - \gamma(t, u), \quad u(0) = u(T), \quad u'(0) = u'(T), \quad (46)$$

and first show that if u is a solution of (46) then $\alpha(t) \leq u(t) \leq \beta(t)$ for all $t \in [0, T]$, so that u is a solution of (15). Suppose by contradiction that there is some $t_0 \in [0, T]$ such that $[\alpha - u]_M = \alpha(t_0) - u(t_0) > 0$. If $t_0 \in (0, T)$; then $\alpha'(t_0) = u'(t_0)$ and there are sequences (t_k) in $[t_0 - \varepsilon, t_0[$ and (t'_k) in $(t_0, t_0 + \varepsilon]$ converging to t_0 such that $\alpha'(t_k) - u'(t_k) \geq 0$ and $\alpha'(t'_k) - u'(t'_k) \leq 0$. As ϕ is an increasing homeomorphism, this implies $(\phi(\alpha'(t_0)))' \leq (\phi(u'(t_0)))'$. Hence, because α is a lower solution of (15) we obtain

$$(\phi(\alpha'(t_0)))' \leq (\phi(u'(t_0)))' = f(t_0, \alpha(t_0), \alpha'(t_0)) + u(t_0) - \alpha(t_0)]$$
$$< f(t_0, \alpha(t_0), \alpha'(t_0)) \leq (\phi(\alpha'(t_0)))',$$

a contradiction. If $[\alpha - u]_M = \alpha(0) - u(0) = \alpha(T) - u(T)$, then $\alpha'(0) - u'(0) \leq 0$, $\alpha'(T) - u'(T) \geq 0$. Using that $\alpha'(0) \geq \alpha'(T)$, we deduce that $\alpha'(0) - u'(0) = 0 = \alpha'(T) - u'(T)$. This implies that $\phi(\alpha'(0)) = \phi(u'(0))$. On the other hand, $[\alpha - u]_M = \alpha(0) - u(0)$ implies, reasoning in a similar way as for $t_0 \in (0, T)$, that

$$(\phi(\alpha'(0)))' \leq (\phi(u'(0)))'.$$

Using the inequality above and $\alpha'(0) = u'(0)$, we can proceed as in the case $t_0 \in (0, T)$ to obtain again a contradiction. In consequence we have that $\alpha(t) \leq u(t)$ for all $t \in [0, T]$. Analogously, using the fact that β is a upper solution of (15), we can show that $u(t) \leq \beta(t)$ for all $t \in [0, T]$. We remark that if α, β are strict, then $\alpha(t) < u(t) < \beta(t)$ for all $t \in [0, T]$.

We now apply Corollary 3 to the modified problem (46) to obtain the existence of a solution, and the relation

$$d_{LS}[I - \widetilde{M}, B(\rho), 0] = 1 \tag{47}$$

for the equivalent fixed point operator $\widetilde{M}$ and all sufficiently large $\rho > 0$.

Moreover, if α and β are strict, then $\alpha(t) < u(t) < \beta(t)$ for all $t \in [0, T]$. If ρ is large enough, then, using (47) and the additivity-excision property of the Leray–Schauder degree, we have

$$d_{LS}[I - \widetilde{M}, \Omega_{\alpha,\beta,a}, 0] = d_{LS}[I - \widetilde{M}, B(\rho), 0] = 1.$$

On the other hand, as the completely continuous operator M associated to (15) is equal to $\widetilde{M}$ on $\overline{\Omega_{\alpha,\beta,a}}$, we deduce that $d_{LS}[I - M, \Omega_{\alpha,\beta,a}, 0] = 1$. $\qquad\square$

Remark 7. In contrast to the case of a classical ϕ, no Nagumo-type condition is required upon f in Theorem 11.

Remark 8. A careful analysis of the above proof implies that Theorem 11 holds also if $f : [0, T] \times (0, +\infty) \times \mathbb{R} \to \mathbb{R}$ is continuous.

We now show, using an argument introduced by Amann–Ambrosetti–Mancini [2] for semilinear Dirichlet problems with bounded nonlinearity, that the existence conclusion in Theorem 11 also holds when the lower and upper solutions are not ordered.

Theorem 12. *If (15) has a lower solution α and an upper solution β, then problem (15) has at least one solution.*

Proof. Let $\mathscr{C}$ be given by Lemma 2. If there is some $[\overline{u}, \widetilde{u}] \in \mathscr{C}$ such that

$$\int_0^T f(t, \overline{u} + \widetilde{u}(t), \widetilde{u}'(t)) \, dt = 0,$$

then $\overline{u} + \widetilde{u}$ solves (15). If

$$\int_0^T f(t, \overline{u} + \widetilde{u}(t), \widetilde{u}'(t)) \, dt > 0$$

for all $[\overline{u}, \widetilde{u}] \in \mathscr{C}$, then, using (32), $\overline{u} + \widetilde{u}$ is an upper solution for (15) for each $[\overline{u}, \widetilde{u}] \in \mathscr{C}$. Then, for $[\alpha_M + (Ta/4), \widetilde{u}] \in \mathscr{C}$, $\alpha_M + (Ta/4) + \widetilde{u}(t) \geq \alpha(t)$ for all $t \in [0, T]$ is an upper solution and the existence of a solution to (15) follows from Theorem 11. Similarly, if

$$\int_0^T f(t, \overline{u} + \widetilde{u}(t), \widetilde{u}'(t)) \, dt < 0$$

for all $[\bar{u},\tilde{u}] \in \mathscr{C}$, then $[\beta_L -(Ta/4),\tilde{u}] \in \mathscr{C}$ gives the lower solution $\beta_L -(Ta/4)+ \tilde{u}(t) \le \beta(t)$ for all $t \in [0, T]$ and the existence of a solution.   □

The choice of constant lower and upper solutions in Theorems 11 and 12 leads to the following simple existence condition.

Corollary 7. *Problem (15) has at least one solution if there exist constants a and b such that*

$$f(t,a,0) \cdot f(t,b,0) \le 0$$

for all $t \in [0, T]$.

Another application of Theorems 11 and 12 gives necessary and sufficient conditions for the existence of a solution of problem

$$(\phi(u'))' = g(t,u), \quad u(0) = u(T), \quad u'(0) = u'(T) \tag{48}$$

when $g : [0, T] \times \mathbb{R} \to \mathbb{R}$ is continuous and $g(t, \cdot)$ monotone for each fixed $t \in [0, T]$.

Corollary 8. *If $g : [0, T] \times \mathbb{R} \to \mathbb{R}$ is continuous and $g(t, \cdot)$ is either non decreasing for all $t \in [0, T]$ or non increasing for all $t \in [0, T]$, then problem (48) is solvable if and only if there exists $c \in \mathbb{R}$ such that*

$$\int_0^T g(s,c) \, ds = 0. \tag{49}$$

Proof. Necessity. If problem (48) has a solution u, then, integrating both members of the differential equation in (48) and using the boundary condition, it follows that

$$\int_0^T g(s, u(s)) \, ds = 0. \tag{50}$$

Assuming for example that $g(s, \cdot)$ is non decreasing for every $s \in [0, T]$, we deduce from (50) that

$$\int_0^T g(s, u_L) \, ds \le 0 \le \int_0^T g(s, u_M) \, ds,$$

so that, by the intermediate value theorem, there exists some $c \in [u_L, u_M]$ satisfying (49). The reasoning is similar when $g(t, \cdot)$ is non decreasing for each $t \in [0, T]$. *Sufficiency.* If $c \in \mathbb{R}$ satisfies (49), then, by Corollary 5, the problem

$$(\phi(u'))' = g(t,c), \quad u(0 = u(T), \quad u'(0) = u'(T)$$

has a one-parameter family of solutions of the form $d + \tilde{u}(t)$ with $\tilde{u} \in \tilde{C}_T^1$. There exists $d_1 \le d_2$ such that, for all $t \in [0, T]$,

$$\alpha(t) := d_1 + \tilde{u}(t) \le c \le d_2 + \tilde{u}(t) =: \beta(t).$$

Hence, if $g(t, \cdot)$ is non decreasing for each $t \in [0, T]$,

$$(\phi(\alpha'(t)))' = (\phi(\widetilde{u}'(t)))' = g(t, c) \geq g(t, \alpha(t))$$

and α is a lower solution for (48). Similarly β is an upper solution for (48). A similar argument shows that, if $g(t, \cdot)$ is non increasing for every $t \in [0, T]$, α is an upper solution and β a lower solution for (48). So the result follows from Theorem 12.

□

Example 5. For $h \in C$, $p > 0$ and $c \neq 0$, the problem

$$\left(\frac{u'}{\sqrt{1 - u'^2}}\right)' + c(u^+)^p = h(t), \quad u(0) = u(T), \quad u'(0) = u'(T)$$

has a solution if and only if

$$c \int_0^T h(t)\, dt \geq 0.$$

2.3.10 Periodic Nonlinearity for Singular or Bounded ϕ

We describe in this subsection some existence results of [20] on the forced pendulum problem

$$(\phi(u'))' + k(u)u' + \mu \sin u = h(t), \quad u(0) = u(T), \quad u'(0) = u'(T), \quad (51)$$

where $k : \mathbb{R} \to \mathbb{R}$ is continuous, $\mu > 0$ and $h \in C$.

We first assume that $\phi : (-a, a) \to \mathbb{R}$ is singular.

Theorem 13. *Let $\mu > 0$ and assume that $h \in C$ satisfies*

$$|h|_\infty \leq \mu.$$

Then problem (51) has at least one solution. Moreover, if

$$|h|_\infty < \mu,$$

then problem (51) has at least two solutions not differing by a multiple of 2π.

Proof. Assume that $|h|_\infty \leq \mu$. Then $\alpha = -\frac{3\pi}{2}$ is a constant lower solution for (51) and $\beta = -\frac{\pi}{2}$ is a constant upper solution for (51) such that $\alpha < \beta$. Hence, using Theorem 11, it follows that (51) has a solution u_1 such that $\alpha \leq u_1 \leq \beta$. Note that if $|h|_\infty < \mu$, then α, β are strict and $\alpha < u_1 < \beta$. In this case, let M_μ be the fixed point operator associated to (51) and let

$$\Omega = \Omega_{-\frac{3\pi}{2}, \frac{3\pi}{2}, a} \setminus \left(\overline{\Omega}_{-\frac{3\pi}{2}, -\frac{\pi}{2}, a} \cup \overline{\Omega}_{\frac{\pi}{2}, \frac{3\pi}{2}, a}\right).$$

Then using the additivity property of the Leray–Schauder degree and Theorem 11, we deduce that

$$d_{LS}[I - M_\mu, \Omega, 0] = -1.$$

Hence, the existence property of the Leray–Schauder degree yields the existence of a solution $u_2 \in \Omega$ of (51). If we assume that $u_2 = u_1 + 2j\pi$ for some $j \in \mathbb{Z}$ then, as $-3\pi/2 < u_1 < -\pi/2$, one has

$$-\frac{3\pi}{2} + 2j\pi < u_2 = u_1 + 2j\pi < -\frac{\pi}{2} + 2j\pi.$$

This leads to one of the contradictions : $u_2 \in \Omega_{\frac{\pi}{2}, \frac{3\pi}{2}, a}$ if $j = 1$ or $u_2 = u_1 \in \Omega_{-\frac{3\pi}{2}, -\frac{\pi}{2}, a}$ for $j = 0$. $\square$

Example 6. If $\mu > 0$, problem (44) has at least two solutions not differing by a multiple of 2π for all $h \in C$ such that $|h|_\infty < \mu$, and at least one solution for all $h \in C$ such that $|h|_\infty = \mu$.

To obtain the same type of result for the case of bounded ϕ-Laplacians, i.e. when $\phi : \mathbb{R} \to (-b, b)$ is an increasing homeomorphism such that $\phi(0) = 0$, we use the following a priori estimate result [14].

Lemma 3. *Let* $0 < b, c \le \infty$, $\psi : (-c, c) \to (-b, b)$ *be a homeomorphism such that* $\psi(0) = 0$ *and* $f : [0, T] \times \mathbb{R}^2 \to \mathbb{R}$ *be a continuous function. Assume that there exists* $e \in C$ *such that* $2|e^-|_1 < b$ *and*

$$f(t, u, v) \ge e(t) \quad \text{for all} \quad (t, u, v) \in [0, T] \times \mathbb{R}^2. \tag{52}$$

If u is a possible solution of the problem

$$(\psi(u'))' = f(t, u, u'), \quad u(0) = u(T), \quad u'(0) = u'(T), \tag{53}$$

then $|u'|_\infty \le R_\psi$, *where* $R_\psi = \max(|\psi^{-1}(\pm 2|e^-|_1)|)$.

Proof. Let u be a solution of (53). This implies that

$$\int_0^T f(t, u(t), u'(t)) \, dt = 0. \tag{54}$$

Using (52), we deduce the inequality

$$|f(t, u, v)| \le f(t, u, v) + 2e^-(t) \quad \text{for all} \quad (t, u, v) \in [0, T] \times \mathbb{R}^2. \tag{55}$$

From (53), (54) and (55) it follows that

$$|(\psi(u'))'|_1 = |f(\cdot, u(\cdot), u'(\cdot))|_1 \le \int_0^T f(t, u(t), u'(t)) \, dt + 2|e^-|_1 = 2|e^-|_1.$$

$$\tag{56}$$

Because $u \in C_T^1$, there exists $\xi \in [0, T]$ such that $u'(\xi) = 0$, hence $\psi(u'(\xi)) = 0$ and

$$\psi(u'(t)) = \int_{\xi}^{t} (\psi(u'(s)))' \, ds \quad (t \in [0, T]).$$

Using the equality above and (56) we have that

$$|\psi(u'(t))| \leq 2|e^-|_1 \quad (t \in [0, T]),$$

and hence $|u'|_\infty \leq R_\psi$. $\qquad\qquad\square$

We can use this result to obtain a sufficient condition for the existence of multiple solution for a forced pendulum equation with bounded ϕ.

Corollary 9. *Assume that* $\psi : \mathbb{R} \to (-b, b)$ $(0 < b \leq \infty)$ *is an increasing homeomorphism such that* $\psi(0) = 0$ *and*

$$|[h - \mu]^-|_1 < \frac{b}{2}.$$

If $|h|_\infty = \mu$, *the problem*

$$(\psi(u'))' + \mu \sin u = h(t), \quad u(0) = u(T), \quad u'(0) = u'(T) \tag{57}$$

has at least one solution. Moreover, if $|h|_\infty < \mu$, *problem (57) has at least two solutions not differing by a multiple of* 2π.

Proof. Let R_ψ be the constant given in Lemma 3 with $e = h - \mu$. Let $c = R_\psi + 1$ and consider an increasing homeomorphism $\phi : (-c, c) \to \mathbb{R}$ such that $\phi = \psi$ on $[-R_\psi, R_\psi]$. It follows that $R_\psi = R_\phi$ and applying Lemma 3, we deduce that u is a solution of (51) with $k \equiv 0$ if and only if u is a solution of (57). Now the result follows from Theorem 13. $\qquad\qquad\square$

Example 7. If $h \in C$ is such that $\overline{h} = 0$ and $|h|_\infty < \mu < \frac{1}{2T}$, the problem

$$\left(\frac{u'}{\sqrt{1 + u'^2}}\right)' + \mu \sin u = h(t), \quad u(0) = u(T), \quad u'(0) = u'(T),$$

has at least two classical solutions not differing by a multiple of 2π.

2.3.11 Ambrosetti–Prodi Problem: Coercive Restoring Force and Singular ϕ

In this subsection, we consider, following [11], periodic problems of the type

$$(\phi(u'))' + k(u)u' + g(t, u, u') = s, \quad u(0) = u(T), \quad u'(0) - u'(T), \tag{58}$$

where $s \in \mathbb{R}$, $k : \mathbb{R} \to \mathbb{R}$ and $g : [0, T] \times \mathbb{R}^2 \to \mathbb{R}$ are continuous and g satisfies the *coercivity condition*

$$(H_f) \quad g(t, u, v) \to +\infty \quad \text{if} \quad |u| \to \infty \quad \text{uniformly in} \quad [0, T] \times (-a, a). \quad (59)$$

We are interested in studying the existence and multiplicity of the solutions of (58) in terms of the value of the parameter s *(Ambrosetti–Prodi problem* [3, 45, 78]).

We first obtain an a priori bound for the set of possible solutions.

Lemma 4. *For each $b \in \mathbb{R}$, there exists $\rho = \rho(b) > 0$ such that any possible solution u of (58) with $s \leq b$ belongs to the open ball $B(\rho)$.*

Proof. Let $s \leq b$ and u be a solution of (58). This implies that u satisfies $|u'|_\infty < a$ and

$$T^{-1} \int_0^T g(t, u(t), u'(t)) \, dt = s. \quad (60)$$

Using (59) we can find $R > 0$ such that

$$g(t, u, v) > b \quad \text{if} \quad |u| \geq R, \quad (t, v) \in [0, T] \times (-a, a). \quad (61)$$

If $u_L \geq R$, then using (61), we deduce that

$$T^{-1} \int_0^T g(t, u(t), u'(t)) \, dt > b,$$

which, together with (60) gives $s > b$, a contradiction. So we have $u_L < R$. Analogously we can show that $u_M > -R$. Then using the inequality

$$u_M \leq u_L + \int_0^T |u'(\tau)| \, d\tau,$$

we obtain $|u|_\infty < R + Ta$. We can take any $\rho \geq R + (T + 1)a$. $\square$

Theorem 14. *If g satisfies condition (59), there exists $s_1 \in \mathbb{R}$ such that problem (58) has zero, at least one or at least two solutions according to $s < s_1, s = s_1$ or $s > s_1$.*

Proof. For $j \geq 1$ an integer, let

$$S_j = \{s \in \mathbb{R} : (58) \text{ has at least } j \text{ solutions}\}.$$

(a) $S_1 \neq \varnothing$.

Take $s^* > \max_{t \in [0, T]} g(t, 0, 0)$ and use (59) to find $R_+^* > 0$ such that

$$\max_{t \in [0, T]} g(t, R_+^*, 0) > s^*.$$

Then $\beta \equiv R_+^* > 0$ is a strict upper solution and $\alpha \equiv 0$ is a strict lower solution for (58) with $s = s^*$. Hence, using Theorem 11, $s^* \in S_1$.

(b) If $\widetilde{s} \in S_1$ and $s > \widetilde{s}$ then $s \in S_1$.

Let $\widetilde{u}$ be a solution of (58) with $s = \widetilde{s}$, and let $s > \widetilde{s}$. Then $\widetilde{u}$ is a strict upper solution for (58). Take now $R_- < \widetilde{u}_L$ such that $\min_{t \in [0,T]} g(t, R_-, 0) > s$: $\alpha \equiv R_-$ is a strict lower solution for (58). From Theorem 11, $s \in S_1$.

(c) $s_1 = \inf S_1$ is finite and $S_1 \supset (s_1, \infty)$.

Let $s \in \mathbb{R}$ and suppose that (58) has a solution u. Then (60) hold, implying that $s \geq c$, with $c = \inf_{[0,T] \times \mathbb{R} \times]-a,a[} g$. To obtain the second part of claim (c), we apply (b).

(d) $S_2 \supset (s_1, \infty)$.

Let $s_3 < s_1 < s_2$. For each $s \in \mathbb{R}$, let $\mathscr{M}(s, \cdot)$ be the fixed point operator in C^1 associated to problem (58). Using Lemma 4 we find ρ such that each possible zero of $I - \mathscr{M}(s, \cdot)$ with $s \in [s_3, s_2]$ is such that $u \in B(\rho)$. Consequently, the Leray–Schauder degree $d_{LS}[I - \mathscr{M}(s, \cdot), B(\rho), 0]$ is well defined and does not depend upon $s \in [s_3, s_2]$. However, using (c), we see that $u - \mathscr{M}(s_3, u) \neq 0$ for all $u \in C^1$. This implies that $d_{LS}[I - \mathscr{M}(s_3, \cdot), B(\rho), 0] = 0$, so that $d_{LS}[I - \mathscr{M}(s_2, \cdot), B(\rho), 0] = 0$ and, by excision property of Leray–Schauder degree, $d_{LS}[I - \mathscr{M}(s_2, \cdot), B(\rho'), 0] = 0$ if $\rho' > \rho$. Let $s \in (s_1, s_2)$ and $\widehat{u}$ be a solution of (58) (using (c)). Then $\widehat{u}$ is a strict upper solution of (58) with $s = s_2$. Let $R < \widehat{u}_L$ be such that $\min_{t \in [0,T]} g(t, R, 0) > s_2$. Then R is a strict lower solution of (58) with $s = s_2$. Consequently, using Theorem 11, problem (58) with $s = s_2$ has a solution in $\Omega_{R,\widehat{u}}$ and $d_{LS}[I - \mathscr{M}(s_2, \cdot), \Omega_{R,\widehat{u}}, 0] = -1$. Taking ρ' sufficiently large, we deduce from the additivity property of Leray–Schauder degree that

$$d_{LS}[I - \mathscr{M}(s_2, \cdot), B(\rho') \backslash \Omega_{R,\widehat{u}}, 0] = d_{LS}[I - \mathscr{M}(s_2, \cdot), B(\rho'), 0]$$
$$-d_{LS}[I - \mathscr{M}(s_2, \cdot), \Omega_{R,\widehat{u}}, 0] = -d_{LS}[I - \mathscr{M}(s_2, \cdot), \Omega_{R,\widehat{u}}, 0] = -1,$$

and (58) with $s = s_2$ has a second solution in $B(\rho') \backslash \overline{\Omega}_{R,\widehat{u}}$.

(e) $s_1 \in S_1$.

Let (τ_k) be a sequence in $(s_1, +\infty)$ converging to s_1, and let u_k be a solution of (58) with $s = \tau_k$ given by (c). Using Theorem 3 we deduce that

$$u_k = \mathscr{M}(\tau_k, u_k). \tag{62}$$

From Lemma 4, there exists $\rho > 0$ such that $|u_k|_1 < \rho$ for all $k \geq 1$. The complete continuity of $\mathscr{M}$ implies that, up to a subsequence, the right-hand member of (62) converges in C^1, and hence (u_k) converges to some $u \in C^1$ such that $u = \mathscr{M}(s_1, u)$, i.e. to a solution of (58) with $s = s_1$. $\square$

A similar proof provides the following dual Ambrosetti–Prodi condition.

Theorem 15. *If g satisfies the anticoercivity condition*

$$g(t, u, v) \to -\infty \quad \text{if} \quad |u| \to \infty \quad \text{uniformly in} \quad [0, T] \times (-a, a).$$

there exists $s_1 \in \mathbb{R}$ such that problem (58) has zero, at least one or at least two solutions according to $s > s_1, s = s_1$ or $s < s_1$.

Corollary 10. *Let $e \in C, k, g : \mathbb{R} \to \mathbb{R}$ be continuous and such that*

$$g(u) \to +\infty \quad (resp. -\infty) \quad \text{if} \quad |u| \to \infty.$$

Then, there exists $s_1 \in \mathbb{R}$ such that the problem

$$(\phi(u'))' + k(u)u' + g(u) = s + e(t), \quad u(0) = u(T), \quad u'(0) = u'(T)$$

has no solution if $s < s_1$ (resp. $s > s_1$), at least one solution if $s = s_1$ and at least two solutions if $s > s_1$ (resp. $s < s_1$).

Example 8. For each $e \in C, k : \mathbb{R} \to \mathbb{R}$ continuous, $p > 0$ and $c > 0$ (resp. $c < 0$), there exists $s_1 \in \mathbb{R}$ such that the problem

$$\left(\frac{u'}{\sqrt{1 - u'^2}} \right)' + f(u)u' + c|u|^p = s + e(t), \quad u(0) = u(T), \quad u'(0) = u'(T)$$

has no solution if $s < s_1$ (resp. $s > s_1$), at least one solution if $s = s_1$ and at least two solutions if $s > s_1$ (resp. $s < s_1$).

2.3.12 Ambrosetti–Prodi Problem: Bounded Restoring Force and Singular ϕ

The coercivity condition upon g can be replaced by a boundedness condition without loosing the Ambrosetti–Prodi type conclusion. Consider the periodic boundary value problem

$$(\phi(u'))' + k(u)u' + g(u) = e(t) + s, \quad u(0) = u(T), \quad u'(0) = u'(T), \quad (63)$$

where $s \in \mathbb{R}$ is a parameter, $e \in C, k : \mathbb{R} \to \mathbb{R}, g : \mathbb{R} \to \mathbb{R}$ are continuous and the following conditions hold :

(H1) $\bar{e} = 0$.
(H2) $g(u) > 0$ *for all $u \in \mathbb{R}$.*
(H3) $\lim_{u \to \pm\infty} g(u) = 0$.

We write $g_M := \max_{\mathbb{R}} g$. The classical problem

$$u'' + g(u) = e(t) + s$$

was considered by Ward in [125] using a variational method and under the supplementary condition that the indefinite integral G is g is bounded. The case of singular ϕ presented here was given in [13].

Consider, for any integer $j \geq 1$

$$S_j = \{s \in \mathbb{R} : (63) \text{ has at least } j \text{ solutions}\}.$$

Lemma 5. *If $s \in S_1$, then $0 < s \leq g_M$.*

Proof. Assumptions (H2) and (H3) imply that g is bounded and $0 < g(u) \leq g_M$ for all $u \in \mathbb{R}$. Hence, if u is a solution of (63) then, using (H1), it follows that

$$T^{-1} \int_0^T g(u(t))dt = s, \tag{64}$$

and $0 < s \leq g_M$. $\qquad\square$

For $s \in \mathbb{R}$, we define the continuous operator $N_f : C^1 \times \mathbb{R} \to C^1$ by

$$N_f(u,s)(t) = \int_0^t [e(\tau) + s - k(u(\tau))u'(\tau) - g(u(\tau))] \, d\tau \quad (t \in [0,T]).$$

Using Theorem 3, it follows that $u \in C^1$ is a solution of (63) if and only if

$$u = Pu - N_f(u,s)(T) + H \circ \phi^{-1} \circ (I - Q_\phi) \circ N_f(u,s) := \mathscr{G}(u,s),$$

and the nonlinear operator $\mathscr{G}(\cdot, s) : C^1 \to C^1$ is completely continuous.

Let $\widetilde{M} : \mathbb{R} \times \widetilde{C}_T^1 \to \widetilde{C}_T^1$ be the completely continuous operator defined by

$$\widetilde{M}(\overline{u}, \widetilde{u}) = \widetilde{H} \circ \phi^{-1} \circ (I - Q_\phi) \circ N_f(\overline{u} + \widetilde{u}, 0).$$

If u is a solution of (63), then (64) holds and $\widetilde{u} = \widetilde{M}(\overline{u}, \widetilde{u})$. Reciprocally, if $[\overline{u}, \widetilde{u}] \in \mathbb{R} \times \widetilde{C}_T^1$ is such that $\widetilde{u} = \widetilde{M}(\overline{u}, \widetilde{u})$, then $u = \overline{u} + \widetilde{u}$ is a solution of (63) with $s = T^{-1} \int_0^T g(u(t))dt$.

Using Lemma 2, we deduce the following useful result.

Lemma 6. *The set of the solutions $[\overline{u}, \widetilde{u}] \in \mathbb{R} \times \widetilde{C}_T^1$ of the problem*

$$(\phi(\widetilde{u}'))' + k(\overline{u} + \widetilde{u})\widetilde{u}' + g(\overline{u} + \widetilde{u}) = e(t) + T^{-1} \int_0^T g(\overline{u} + \widetilde{u}(t), \widetilde{u}'(t)) \, dt$$

contains a continuum $\mathscr{C}$ whose projection on $\mathbb{R}$ is $\mathbb{R}$ and projection on $\widehat{C}_T^1$ is contained in the open ball $B(a[(T/4) + 1])$.

Let $\gamma : \mathbb{R} \times \widehat{C}_T^1 \to \mathbb{R}$ be the continuous function defined by

$$\gamma(\overline{u}, \widetilde{u}) = T^{-1} \int_0^T g(\overline{u} + \widetilde{u}(t)) \, dt.$$

Lemma 7. $S_1 \neq \varnothing$.

Proof. Let $[\overline{u}, \widetilde{u}] \in \mathscr{C}$. Then $u = \overline{u} + \widetilde{u}$ is a solution of problem (63) with $s = \gamma(\overline{u}, \widetilde{u})$. □

Let us consider $s^*(e) := \sup S_1$.

Lemma 8. *We have that* $0 < s^*(e) \leq g_M$ *and* $s^*(e) \in S_1$.

Proof. The first assertion follows from Lemma 5. Let $\{s_n\}$ be a sequence belonging to S_1 which converges to $s^*(e)$. Let $u_n = \overline{u}_n + \widetilde{u}_n$ be a solution of (63) with $s = s_n = \gamma(\overline{u}_n, \widetilde{u}_n)$. It follows that $\widetilde{u}_n = \widetilde{M}(\overline{u}_n, \widetilde{u}_n)$ and $\{\widetilde{u}_n\}$ belongs to $B(a[(T/4)+1])$. Hence, if up to a subsequence $u_n \to \pm\infty$, then using (H3) it follows that $\gamma(\overline{u}_n, \widetilde{u}_n) \to 0$, which means that $s^*(e) = 0$, a contradiction. We have proved that $\{(\overline{u}_n, \widetilde{u}_n)\}$ is a bounded sequence in $\mathbb{R} \times \widetilde{C}_T^1$. Because $\widetilde{M}$ is completely continuous, we can assume, passing to a subsequence, that $\widetilde{M}(\overline{u}_n, \widetilde{u}_n) \to \widetilde{u}$ and $\overline{u}_n \to \overline{u}$. We deduce that $\widetilde{u} = \widetilde{M}(\overline{u}, \widetilde{u})$, $\gamma(\overline{u}, \widetilde{u}) = s^*(e)$ and $u = \overline{u} + \widetilde{u}$ is a solution of (63) with $s = s^*(e)$. □

Arguing as in the proof of Lemma 8 we deduce the following a priori estimate.

Lemma 9. *Let* $0 < s_1 < s^*(e)$. *Then, there is* $\rho' > 0$ *such that any possible solution* u *of problem (63) with* $s \in [s_1, s^*(e)]$ *belongs to* $B(\rho')$.

Lemma 10. *We have* $(0, s^*(e)) \subset S_2$.

Proof. Let $s_1, s_2 \in \mathbb{R}$ such that $0 < s_1 < s^*(e) < s_2$. Using Lemma 5, Lemma 9 and the invariance property of Leray–Schauder degree, it follows that there is $\rho' > 0$ sufficiently large such that $d_{LS}[I - \mathscr{G}(s, \cdot), B(\rho'), 0]$ is well defined and independent of $s \in [s_1, s_2]$. However, using Lemma 5 we deduce that $u - \mathscr{G}(s_2, u) \neq 0$ for all $u \in C^1$. This implies that $d_{LS}[I - \mathscr{G}(s_2, \cdot), B(\rho'), 0] = 0$, so that $d_{LS}[I - \mathscr{G}(s_1, \cdot), B(\rho'), 0] = 0$ and, by excision property of Leray–Schauder degree,

$$d_{LS}[I - \mathscr{G}(s_1, \cdot), B_{\rho''}, 0] = 0 \quad \text{if} \quad \rho'' \geq \rho'. \tag{65}$$

Let u_* be a solution of (63) with $s = s^*(e)$ (using Lemma 8). Then, u_* is a strict lower solution of (63) with $s = s_1$. Using Lemma 6 and (H3), there is $[\overline{u^*}, \widetilde{u^*}] \in \mathscr{C}$ such that $u^* = \overline{u^*} + \widetilde{u^*} > u_*$ on $[0, T]$ and $\gamma(\overline{u^*}, \widetilde{u^*}) < s_1$. It follows that u^* is an upper solution of (63) with $s = s_1$. So, using Theorem 11, we have that

$$d_{LS}[I - \mathscr{G}(s_1, \cdot), \Omega_{u_*, u^*, a}, 0] = 1, \tag{66}$$

and (63) has a solution in $\Omega_{u_*, u^*, a}$. Taking ρ'' sufficiently large and using (65) and (66), we deduce from the additivity property of Leray–Schauder degree that

$$d_{LS}[I - \mathscr{G}(s_1, \cdot), B_{\rho''} \backslash \overline{\Omega}_{u_*, u^*, a}, 0] = d_{LS}[I - \mathscr{G}(s_1, \cdot), B_{\rho''}, 0]$$
$$-d_{LS}[I - \mathscr{G}(s_1, \cdot), \Omega_{u_*, u^*, a}, 0] = -d_{LS}[I - \mathscr{G}(s_1, \cdot), \Omega_{u_*, u^*, a}, 0] = 1,$$

and (63) with $s = s_1$ has a second solution in $B_{\rho''} \backslash \overline{\Omega}_{u_*, u^*, a}$. □

Theorem 16. *If conditions (H1)–(H3) hold, there exists $s^*(e) \in (0, g_M]$ such that problem (63) has zero, at least one or at least two solutions according to $s \notin (0, s^*(e)]$, $s = s^*(e)$ or $s \in (0, s^*(e))$.*

Proof. The conclusion of Theorem 16 follows from Lemmas 5, 8 and 10.   □

Example 9. Let $e \in C$ be such that $\bar{e} = 0$. If $b > 0$ and $c \geq 0$ then there is $s^* > 0$ such that the problem

$$\left(\frac{u'}{\sqrt{1 - u'^2}}\right)' + \frac{cu'^4 + b}{1 + |u|} = e(t) + s, \quad u(0) = u(T), \quad u'(0) - u'(T)$$

has zero, at least one or at least two solutions according to $s \notin (0, s^*]$, $s = s^*$ or $s \in (0, s^*)$.

2.3.13   Singular Restoring Forces and Singular ϕ

In this subsection we prove, following [11], the existence of *positive* solutions for the following periodic problems with *singular attractive* restoring force

$$(\phi(u'))' + g(u) = h(t), \quad u(0) = u(T), \quad u'(0) = u'(T) \tag{67}$$

or with *singular repulsive* restoring force

$$(\phi(u'))' - g(u) = h(t), \quad u(0) = u(T), \quad u'(0) = u'(T) \tag{68}$$

where $h \in C$ and $g : (0, +\infty) \to (0, |\infty)$ is continuous and such that

$$g(u) \to +\infty \quad \text{as} \quad u \to 0+, \tag{69}$$

$$g(u) \to 0 \quad \text{as} \quad u \to +\infty. \tag{70}$$

The classical case where $\phi(s) \equiv s$ was first considered by Lazer and Solimini in [59] (see also [74] for another approach) and the case of a p-Laplacian in [56].

Theorem 17. *Suppose that g satisfies conditions (69) and (70). Then problem (67) has at least one solution if and only if $\bar{h} > 0$.*

Proof. If u is a solution of (67), then $\bar{h} = Qg(u) > 0$ because g is positive. Conversely, suppose that $\bar{h} > 0$. Using (69), there exists $\epsilon > 0$ such that $g(\epsilon) > h(t)$ for all $t \in [0, T]$. Hence, $\alpha \equiv \epsilon$ is a strict lower solution for (67). On the other hand, using Corollary 5, there exists $w \in C_T^1$ such that $(\phi(w'))' = \tilde{h}(t)$. Using (70), there exists some $\delta > 0$ such that $\beta(t) = \delta + w(t) > \alpha(t)$ and $g(\beta(t)) < \bar{h}$ for all $t \in [0, T]$. Then, β is a strict upper solution for (67) and Theorem 11 implies the result.   □

Example 10. If $\mu > 0$ and $h \in C$, the problem

$$\left(\frac{u'}{\sqrt{1-u'^2}}\right)' + \frac{1}{u^\mu} = h(t), \quad u(0) = u(T), \quad u'(0) = u'(T)$$

has at least one solution if and only if $\bar{h} > 0$.

To solve (68) we need the following supplementary condition

$$\int_0^1 g(u)\, du = +\infty. \tag{71}$$

Lemma 11. *Suppose that g satisfies conditions (69), (70), and (71). There exists $\epsilon > 0$ such that if $\lambda \in [0, 1]$ and u is any positive solution of problem*

$$(\phi(u'))' = (1 - \lambda)[Qg(u) + \bar{h}] + \lambda g(u) + \lambda h(t),$$
$$u(0) = u(T), \quad u'(0) = u'(T), \tag{72}$$

then $u(t) > \epsilon$ for all $t \in [0, T]$.

Proof. Let $\lambda \in [0, 1]$ and u be a possible positive solution of (72). Then

$$Qg(u) + \bar{h} = 0 \tag{73}$$

and hence, if $\lambda \in (0, 1]$, (72) is equivalent to

$$(\phi(u'))' = \lambda g(u) + \lambda h(t), \quad u(0) = u(T), \quad u'(0) = u'(T). \tag{74}$$

Using the positivity of g, we deduce that

$$|g(u) + h(t)| \le g(u) + |h(t)| = g(u) + h(t) + 2h^-(t) \tag{75}$$

for all $(t, u) \in [0, T] \times \mathbb{R}$. From (74), (73) and (75) it follows that

$$|(\phi(u'))'|_1 = \lambda|g(u) + h|_1 \le 2\lambda|h^-|_1. \tag{76}$$

Because $u \in C^1$ is such that $u(0) = u(T)$, there exists $\eta \in [0, T]$ such that $u'(\eta) = 0$, which implies $\phi(u'(\eta)) = 0$ and

$$\phi(u'(t)) = \int_\eta^t (\phi(u'(s)))'\, ds \quad (t \in [0, T]).$$

Using the equality above and (76) we have that

$$|\phi(u'(t))| \le 2\lambda|h^-|_1 \quad (t \in [0, T]), \tag{77}$$

Using (69), there exists $\xi > 0$ such that

$$g(u) > -\overline{h} \quad \text{for all} \quad 0 < u \leq \xi. \tag{78}$$

and therefore, by (78) and (73), there exists $t_1 \in [0, T]$ such that $u(t_1) > \xi$. Now, let

$$v(t) = \phi(u'(t)) \tag{79}$$

which implies

$$u'(t) = \phi^{-1}(v(t)) \tag{80}$$

for all $t \in [0, T]$. Introducing (79) in (74) we obtain

$$v'(t) - \lambda g(u(t)) = \lambda h(t) \tag{81}$$

for all $t \in [0, T]$. Multiplying (80) by $v'(t)$ and (81) by $u'(t)$ and subtracting we get

$$v'(t)\phi^{-1}(v(t)) - \lambda g(u(t))u'(t) = \lambda h(t)u'(t)$$

i.e.

$$\left(\int_0^{v(t)} \phi^{-1}(s)\, ds \right)' - \lambda g(u(t))u'(t) = \lambda h(t)u'(t)$$

for all $t \in [0, T]$. This implies that

$$\int_0^{v(t)} \phi^{-1}(s)\, ds - \int_0^{v(t_1)} \phi^{-1}(s)\, ds - \lambda \int_{u(t_1)}^{u(t)} g(s)ds = \lambda \int_{t_1}^{t} h(s)u'(s)ds$$

for all $t \in [0, T]$. Using the fact that $\int_0^v \phi^{-1}(s)\, ds \geq 0$ for all $v \in \mathbb{R}$, we deduce that

$$\lambda \int_{u(t)}^{u(t_1)} g(s)ds \leq \int_0^{v(t_1)} \phi^{-1}(s)\, ds + \lambda \int_{t_1}^{t} h(s)u'(s)ds \tag{82}$$

for all $t \in [0, T]$. Using (77), (79) and (82), we obtain

$$\int_{u(t)}^{u(t_1)} g(s)ds \leq \frac{1}{\lambda} \left(\int_0^{2\lambda|h^-|_1} \phi^{-1}(s)\, ds + \int_0^{-2\lambda|h^-|_1} \phi^{-1}(s)\, ds \right) + a|h|_1$$

$$\leq \max_{[-2|h^-|_1,\, 2|h^-|_1]} 2|\phi^{-1}||h^-|_1 + a|h|_1 := c \tag{83}$$

for all $t \in [0, T]$. Using (71) we can find $0 < \epsilon < \xi$ such that

$$\int_\epsilon^\xi g(t)dt > c. \qquad (84)$$

Since $u(t_1) > \xi$ and g is positive, from (83) and (84) one gets $u(t) > \epsilon$ for all $t \in [0, T]$. Now, for $\lambda = 0$, the solutions of (72) are the constant functions u solutions of

$$g(u) + \overline{h} = 0$$

and they satisfy $u > \xi > \epsilon$. $\qquad\qquad\qquad\qquad\qquad\qquad\qquad\qquad\qquad\qquad\square$

Theorem 18. *Suppose that $e \in C$ and g satisfies conditions (69), (70), and (71). Then problem (68) has at least one positive solution if and only if $\overline{h} < 0$.*

Proof. If u is a solution, then $\overline{h} = -Qg(u) < 0$. For sufficiency, we use the homotopy (72) and the corresponding homotopy for the associated family of fixed point operators $\mathscr{M}(\lambda, \cdot)$ defined in (16) with $f = g + h$. Let $\lambda \in [0, 1]$ and u be a possible positive solution of (72). We already know from Lemma 11 that $u(t) > \epsilon$ for some $\epsilon > 0$ and all $t \in [0, T]$. From assumption (70) follows easily the existence of $R > 0$ such that $g(u) < -\overline{h}$ if $u \geq R$. Hence, because of (73), there exists $t_2 \in [0, T]$ such that $u(t_2) < R$, which implies $u(t) < R + Ta$ ($t \in [0, T]$). Hence, all the possible positive solutions of problem (72) are contained in the open bounded set

$$\Omega := \{u \in C^1 : \epsilon < u(t) < R + Ta \ (0 \leq t \leq T), \ |u'|_\infty < a\}.$$

From the homotopy invariance of Leray–Schauder degree, we obtain

$$
\begin{aligned}
d_{LS}[I - \mathscr{M}(1, \cdot), \Omega, 0] &= d_{LS}[I - \mathscr{M}(0, \cdot), \Omega, 0] \\
&= d_B[g(\cdot) + \overline{h}, \Omega \cap \mathbb{R}, 0] \\
&= d_B[g(\cdot) + \overline{h}, (\epsilon, R), 0] = -1,
\end{aligned}
$$

so that that $\mathscr{M}(1, \cdot)$ has a fixed point. $\qquad\qquad\qquad\qquad\qquad\qquad\qquad\square$

Example 11. If $\mu \geq 1$ and $h \in C$, the problem

$$\left(\frac{u'}{\sqrt{1 - u'^2}}\right)' - \frac{1}{u^\mu} = h(t), \quad u(0) = u(T), \quad u'(0) - u'(T)$$

has at least one positive solution if and only if $\overline{h} < 0$.

2.4 Neumann Problem for Singular ϕ

2.4.1 Equivalent Fixed Point Problem

Let us consider the homogeneous Neumann problem for continuous $f : [0, T] \times \mathbb{R}^{2n} \to \mathbb{R}^n$

$$(\phi(u'))' = f(t, u, u'), \quad u'(0) = 0 = u'(T). \tag{85}$$

The proofs of the following results are similar to the corresponding ones for the periodic problem and left to the reader as an exercise. The approach already appeared in [9] together with an application to sign conditions for existence.

Theorem 19. *u is a solution of the problem (85) if and only if $u \in C^1$ is a fixed point of the operator K defined on C^1 by*

$$K(u) = Pu - N_f(u)(T) + H \circ \phi^{-1} \circ N_f(u).$$

Furthermore, $|(K(u))'|_\infty < a$ for all $u \in C^1$ and K is completely continuous on C^1.

Theorem 20. *Assume that there exists an open bounded set $\Omega \subset C$ such that the following conditions hold :*

1. For each $\lambda \in (0, 1]$, there is no solution of problem

$$(\phi(u')' = \lambda f(t, u, u'), \quad u'(0) = 0 = u'(T)$$

such that $u \in \partial\Omega$.
2. There is no solution on $\partial\Omega \cap \mathbb{R}^n$ of equation

$$\overline{f}(u) := N_f(u)(T) = 0,$$

where $\mathbb{R}^n$ denotes the subspace of constant functions in C.
3. $d_B[\overline{f}, \Omega \cap \mathbb{R}^n, 0] \neq 0$.

Then problem (85) has at least one solution such that $u \in \Omega$, and, for the associated fixed point operator K, one has

$$d_{LS}[I - K, \Omega_\rho, 0] = d_B[\overline{f}, \Omega \cap \mathbb{R}, 0],$$

where $\rho \geq a$ and $\Omega_\rho \subset C^1$ is the open bounded set defined by

$$\Omega_\rho = \{u \in C^1 : u \in \Omega, \ |u'|_\infty < \rho\}.$$

Let C_N^1 be the space

$$C_N^1 := \{u \in C^1 : \quad u'(0) = 0 = u'(T)\}.$$

For $u \in C_N^1$, we write

$$u_L := \min_{[0,T]} u, \quad u_M := \max_{[0,T]} u, \quad \bar{u} = T^{-1} \int_0^T u(s)\,ds, \quad \widetilde{u}(t) = u(t) - \bar{u}.$$

Theorem 21. *The set $\mathscr{S}$ of the solutions $[\bar{u}, \widetilde{u}] \in \mathbb{R} \times \widetilde{C}^1$ of problem*

$$(\phi(\widetilde{u}'))' = (I - Q)f(\cdot, \bar{u} + \widetilde{u}, \widetilde{u}'), \quad u'(0) = 0 = u'(T)$$

contains a continuum $\mathscr{C}$ whose projection on $\mathbb{R}$ is $\mathbb{R}$ and projection on $\widetilde{C}^1$ is contained in the ball $B(a(T + 1))$.

2.4.2   Existence Theorems

One can prove, in a similar way as in the periodic case, the Neumann version of Theorem 8 with C_T^1 replaced by C_N^1, so that condition $\bar{h} = 0$ is also necessary and sufficient for the existence of a solution to the problem

$$(\phi(u'))' = h(t), \quad u'(0) = 0 = u'(T).$$

Similarly, the conclusion of Corollary 2 holds, with the same assumptions, for the problem

$$(\phi(u'))' + g(t, u) = h(t, u, u'), \quad u'(0) = 0 = u'(T)$$

and the conclusion of Corollary 3 holds, with the same assumptions, for the problem

$$(\phi(u'))' + \mu u = h(t, u, u'), \quad u(0) = u(T), \quad u'(0) = u'(T)$$

In particular, the problem

$$(\phi(u'))' + \mu u = h(t), \quad u(0) = u(T), \quad u'(0) = u'(T)$$

has at least one solution for any $h \in C$, and one can say again that the problem is *non-resonant*. Consequently, $\mu = 0$ is the only value for which *resonance* occurs for Neumann boundary conditions.

One can easily check also that the conclusions of Corollaries 6 and 4 hold, with the same assumptions, but for "relativistic Duffing equations" only

$$(\phi(u'))' + g(u) = h(t), \quad u'(0) = 0 = u'(T),$$

instead of the more general class of Liénard equations.

The results about lower and upper solutions, Ambrosetti–Prodi-type results, singular nonlinearities and periodic nonlinearities have their counterpart, with similar proofs, for the Neumann problem.

3   Lagrangian Variational Approach for Periodic Solutions

3.1   Introduction

In this section, we consider the existence of solutions of the periodic problem

$$(\phi(u'))' = \nabla_u F(t, u) + h(t), \quad u(0) = u(T), \quad u'(0) = u'(T), \qquad (86)$$

where $\phi : B(a) \subset \mathbb{R}^n \to \mathbb{R}^n$ is a homeomorphism satisfying Assumption (H_ϕ), $F : [0, T] \times \mathbb{R}^n \to \mathbb{R}$ is such that $\nabla_u F : [0, T] \times \mathbb{R}^n \to \mathbb{R}^n$ exists and satisfies Carathéodory conditions, $F(t, 0) = 0$ for a.e. $t \in [0, T]$ (without loss of generality) and $h \in L^1$. For the classical case

$$u'' = \nabla_u F(t, u) + h(t), \quad u(0) = u(T), \quad u'(0) = u'(T), \qquad (87)$$

existence results were proved through the direct method of the calculus of variations by Berger and Schechter [26,28], when $F(t, u) - \langle h(t), u \rangle$ is coercive in u uniformly in $t \in [0, T]$, and in [85] when $\int_0^T [F(t, u) - \langle h(t), u \rangle] \, dt$ is coercive in u and, either $\nabla_u F(t, u)$ is bounded or $F(t, \cdot)$ is convex for a.e. $t \in [0, T]$. Many extensions have been given for wider classes of potentials F and for u'' replaced by the p-Laplacian $(|u'|^{p-2}u')'$ (see [64,106–120,124 130] and their references). The case of a bounded ϕ has been considered in [33], whose results are described in this section using a variant of the method of [33] introduced in [23].

An existence theorem was proved for (87) in [84,85], when F is periodic in each variable u_i for a.e. $t \in [0, T]$ and h has mean value zero, easily extended to the case of the p-Laplacian as shown in [79]. When u'' is replaced by a "relativistic" differential operator $(\phi(u'))'$ like above, the scalar case was considered in [32] and the case of system (86), under conditions upon F of the type covered in [28] and [85] was studied in [33]. No proof based upon topological methods of the results given here is known by now.

3.2   The Functional Framework

3.2.1   The Functional

The following variational setting for dealing with equations or systems of type (86) was first introduced in [23].

Putting

$$K := \{v \in W^{1,\infty} : |v'|_\infty \le a, \quad v(0) = v(T)\},$$

it is clear that K is a convex subset of $W^{1,\infty}$.

Let $\Psi : C \to (-\infty, +\infty]$ be defined by

$$\Psi(v) = \begin{cases} \varphi(v), & \text{if } v \in K, \\ \\ +\infty, & \text{otherwise,} \end{cases} \tag{88}$$

with $\varphi : K \to \mathbb{R}$ given by

$$\varphi(v) = \int_0^T \Phi(v'(t)) \, dt, \quad v \in K.$$

Obviously, Ψ is proper and convex. The following lemma allows to prove the lower semicontinuity of Ψ.

Lemma 12. *If $\{u_n\} \subset K$ and $u \in C$ are such that $u_n(t) \to u(t)$ for all $t \in [0, T]$, then*

(i) $u \in K$;
(ii) $u'_n \to u'$ in the w^*-topology $\sigma(L^\infty, L^1)$.

Proof. From the relation

$$|u_n(t_1) - u_n(t_2)| = \left| \int_{t_2}^{t_1} u'_n(t) \, dt \right| \le a|t_1 - t_2|,$$

letting $n \to \infty$, we get

$$|u(t_1) - u(t_2)| \le a|t_1 - t_2| \quad (t_1, t_2 \in [0, T]),$$

which yields $u \in K$.

Next, we show that that if $\{u'_k\}$ is a subsequence of $\{u'_n\}$ with $u'_k \to v \in L^\infty$ in the w^*-topology $\sigma(L^\infty, L^1)$ then

$$v = u' \quad \text{a.e. on } [0, T]. \tag{89}$$

Indeed, as

$$\int_0^T u'_k(t) f(t) \, dt \to \int_0^T v(t) f(t) \, dt \quad \text{for all} \quad f \in L^1(0, T; \mathbb{R}),$$

taking $f \equiv \chi_{t_1,t_2}$, the characteristic function of the interval having the endpoints $t_1, t_2 \in [0, T]$, it follows

$$\int_{t_1}^{t_2} u_k'(t)\, dt \to \int_{t_1}^{t_2} v(t)\, dt \quad (t_1, t_2 \in [0, T]).$$

Then, letting $k \to \infty$ in

$$u_k(t_2) - u_k(t_1) = \int_{t_1}^{t_2} u_k'(t)\, dt$$

we obtain

$$u(t_2) - u(t_1) = \int_{t_1}^{t_2} v(t)\, dt \quad (t_1, t_2 \in [0, T])$$

which, clearly implies (89).

Now, to prove *(ii)* it suffices to show that if $\{u_j'\}$ is an arbitrary subsequence of $\{u_n'\}$, then it contains itself a subsequence $\{u_k'\}$ such that $u_k' \to u'$ in the w^*–topology $\sigma(L^\infty, L^1)$. Since L^1 is separable and $\{u_j'\}$ is bounded in $L^\infty = (L^1)^*$, we know that it has a subsequence $\{u_k'\}$ convergent to some $v \in L^\infty$ in the w^*–topology $\sigma(L^\infty, L^1)$. Then, as shown before (see (89)), we have $v = u'$.   □

Consequently, if $\{u_n\} \subset K$ and $u \in C$ are such that $u_n(t) \to u(t)$ for all $t \in [0, T]$, then $u \in K$ and

$$\varphi(u) \le \liminf_{n \to \infty} \varphi(u_n). \tag{90}$$

This implies that Ψ is lower semicontinuous on C. Also, note that K is closed in C.

Next, let $\mathscr{G} : C \to \mathbb{R}$ be defined by

$$\mathscr{G}(u) = \int_0^T [F(t, u(t)) + \langle h(t), u(t) \rangle]\, dt, \quad u \in C.$$

A standard reasoning shows that $\mathscr{G}$ is of class C^1 on C and its derivative is given by

$$\mathscr{G}'(u)(v) = \int_0^T \langle \nabla F(t, u(t)) + h(t), v(t) \rangle\, dt, \quad u, v \in C.$$

3.2.2 Critical Points and Solutions of Differential Systems

The functional $I : C \to (-\infty, +\infty]$ defined by

$$I = \Psi + \mathscr{G} \tag{91}$$

has the structure required by Szulkin's critical point theory [111], namely the sum of a proper convex lower semicontinuous function and of a function of class C^1.

Accordingly, a function $u \in C$ is *a critical point* of I if $u \in K$ and satisfies the inequality

$$\Psi(v) - \Psi(u) + \mathscr{G}'(u)(v - u) \geq 0 \quad \text{for all } v \in C, \tag{92}$$

or, equivalently

$$\int_0^T [\Phi(v'(t)) - \Phi(u'(t))] \, dt + \int_0^T \langle \nabla F(t, u(t)) + h(t), v(t) - u(t) \rangle \, dt \geq 0$$

$$\text{for all } v \in K.$$

The following simple result is given in [111].

Lemma 13. *Each local minimum of I is a critical point of I.*

Proof. Let u be a local minimum of I. By the convexity of Ψ, given $v \in C$, we have, for all $t \in (0, 1]$ sufficiently small,

$$0 \leq \lambda^{-1}\{I[(1 - \lambda)u + \lambda v] - I(u)\}$$
$$= \lambda^{-1}\{\Psi[(1 - \lambda)u + \lambda v] - \Psi(u) + \mathscr{G}[u + \lambda(v - u)] - \mathscr{G}(u)\}$$
$$\leq \Psi(v) - \Psi(u) + \lambda^{-1}\{\mathscr{G}[u + \lambda(v - u)] - \mathscr{G}(u)\}$$

which gives (92) by letting $\lambda \to 0$. $\square$

Now, we consider the periodic boundary value problem (86) under the basic hypothesis (H_Φ). Recall that by a *solution* of (86) we mean a function $u \in C^1$, such that $|u'|_\infty < a$, $\phi(u')$ is differentiable a.e. and (86) is satisfied a.e. The following elementary lemma will be useful is relating the critical points of I to the solutions of (86).

Lemma 14. *For every $f \in L^1$, the problem*

$$(\phi(u'))' = \bar{u} + f, \quad u(0) = u(T), \quad u'(0) = u'(T) \tag{93}$$

has a unique solution u_f, which is also the unique solution of the variational inequality

$$\int_0^T [\Phi(v'(t)) - \Phi(u'(t)) + \langle \bar{u}(t), \bar{v}(t) - \bar{u}(t) \rangle + \langle f(t), v(t) - u(t) \rangle] \, dt \geq 0$$

$$\text{for all } v \in K, \tag{94}$$

and the unique minimum over K of the strictly convex functional J defined by

$$J(u) = \int_0^T \left[\Phi(u'(t)) + \frac{|\bar{u}|^2}{2} + \langle f(t), u(t) \rangle \right] dt. \tag{95}$$

Proof. Problem (93) is equivalent to finding $u = \bar{u} + \tilde{u}$ with $\bar{u}$ and $\tilde{u}$ solutions of

$$
\begin{cases}
(\phi(\tilde{u}'))' = \tilde{f}, & \tilde{u}(0) = \tilde{u}(T), \quad \tilde{u}'(0) = \tilde{u}'(T), \\
\bar{u} = -\overline{f}, & \displaystyle\int_0^T \tilde{u}(t)\, dt = 0.
\end{cases}
$$

Now the first equation gives

$$
\phi(\tilde{u}'(t)) - \overline{\phi(\tilde{u}')} = \widetilde{H}\tilde{f}
$$

so that, if we let $c = \overline{\phi(\tilde{u}')}$,

$$
\tilde{u}'(t) = \phi^{-1}[c + \widetilde{H}\tilde{f}(t)].
$$

Now u will satisfy the first boundary condition if c is such that

$$
\int_0^T \phi^{-1}[c + \widetilde{H}\tilde{f}(t)]\, dt = 0.
$$

Lemma 1 implies the existence and uniqueness of $c = -Q_\phi(\widetilde{H}\tilde{f})$, and hence the unique solvability of problem (93).

Now, if u is a solution of (93), then, taking $v \in K$, taking the inner product of each member of the differential system by $v - u$, integrating over $[0, T]$, and using integration by parts and the boundary conditions, we get

$$
\int_0^T [\langle \phi(u'(t)), v'(t) - u'(t)\rangle + \langle \bar{u}, \bar{v} - \bar{u}\rangle + \langle f(t), v(t) - u(t)\rangle]\, dt = 0,
$$

which gives (94) if we use the convexity inequality for Φ

$$
\Phi(v') - \Phi(u') \geq \langle \phi(u'), v' - u'\rangle.
$$

The convexity inequality

$$
\frac{|\bar{v}|^2}{2} - \frac{|\bar{u}|^2}{2} \geq \langle \bar{u}, \bar{v} - \bar{u}\rangle
$$

introduced in (94) implies that

$$
\int_0^T \left[\Phi(v'(t)) - \Phi(u'(t)) + \frac{|\bar{v}|^2}{2} + \langle f(t), v(t)\rangle - \frac{|\bar{u}|^2}{2} - \langle f(t), u(t)\rangle \right] dt \geq 0
$$

$$
\text{for all } v \in K,
$$

which shows that J has a minimum on K at u. Conversely if it is the case, then, for all $\lambda \in (0, 1]$ and all $v \in K$, we get

$$\int_0^T \{\Phi[(1 - \lambda)u'(t) + \lambda v'(t)] + \frac{|(1 - \lambda)\bar{u} + \lambda\bar{v}|^2}{2}$$
$$+ \langle f(t), (1 - \lambda)u(t) + \lambda v(t)\rangle\} \, dt$$
$$\geq \int_0^T [\Phi(u'(t)) + \frac{|\bar{u}|^2}{2} + \langle f(t), u(t)\rangle] \, dt,$$

which, using the convexity of Φ, simplifying, dividing both members by λ and letting $\lambda \to 0$, gives the variational inequality (94). Thus solving (94) is equivalent to minimizing (95) over K. Now, it is straightforward to check that J is strictly convex over K and therefore has a unique minimum there, which gives the required uniqueness conclusions of Lemma 32. $\qquad\square$

The idea of proof of the result below first occured in [32].

Proposition 1. *If u is a critical point of I, then u is a solution of problem (86).*

Proof. For u a critical point of I, we set

$$f_u := \nabla F(\cdot, u) + h - \bar{u} \in L^1$$

and consider the problem

$$(\phi(w'))' = \bar{w} + f_u(t), \quad w(0) = w(T), \quad w'(0) = w'(T). \tag{96}$$

By virtue of Lemma 32, problem (96) has an unique solution $u^\#$ and it is also the unique solution of the variational inequality

$$\int_0^T [\Phi(v'(t)) - \Phi(u^{\#'}(t)) + \langle \bar{u^\#}, (\bar{v} - \bar{u^\#})\rangle + \langle f_u(t), v(t) - u^\#(t)\rangle] \, dt \geq 0$$

$$\text{for all } v \in K.$$

Since u is a critical point of I, we infer that

$$\int_0^T [\Phi(v'(t)) - \Phi(u'(t)) + \langle \bar{u}, \bar{v} - \bar{u}\rangle + \langle f_u(t), v(t) - u(t)\rangle] \, dt \geq 0$$

$$\text{for all } v \in K.$$

It follows by uniqueness that $u = u^\#$. Hence, u solves problem (86). $\qquad\square$

3.3   Ground State Solutions

The following results come from [33] and [17].

3.3.1   A Sufficient Condition for Minimization

We begin by a lemma which is the main tool for the minimization problems in this subsection. With this aim, for any $\rho > 0$, set

$$\widehat{K}_\rho := \{u \in K \ : \ |\bar{u}| \le \rho\}.$$

Lemma 15. *Assume that there is some $\rho > 0$ such that*

$$\inf_{\widehat{K}_\rho} I = \inf_K I. \tag{97}$$

Then I is bounded from below on C and attains its infimum at some $u \in \widehat{K}_\rho$, which solves problem (86).

Proof. By virtue of (97) and $\inf_C I = \inf_K I$, it suffices to prove that there is some $u \in \widehat{K}_\rho$ such that

$$I(u) = \inf_{\widehat{K}_\rho} I. \tag{98}$$

Then, we get that u is a minimum point of I on C, so, on account of Lemma 13, u is a critical point of I, and by virtue of Proposition 1, a solution of (86).
   If $v \in \widehat{K}_\rho$ then, using (5) we obtain

$$|v|_\infty \le |\bar{v}| + |\widetilde{v}|_\infty \le \rho + \frac{Ta}{4}.$$

This, together with $|v'|_\infty \le a$ show that $\widehat{K}_\rho$ is bounded in $W^{1,\infty}$ and, by the compactness of the embedding $W^{1,\infty} \subset C$, the set $\widehat{K}_\rho$ is relatively compact in C. Let $\{u_n\} \subset \widehat{K}_\rho$ be a minimizing sequence for I. Passing to a subsequence if necessary and using Lemma 12, we may assume that $\{u_n\}$ converges uniformly to some $u \in K$. It is easily seen that actually $u \in \widehat{K}_\rho$. From (90) and the continuity of $\mathscr{F}$ on C, we obtain

$$I(u) \le \liminf_{n \to \infty} I(u_n) = \lim_{n \to \infty} I(u_n) = \inf_{\widehat{K}_\rho} I,$$

showing that (98) holds true. □

3.3.2   Periodic Nonlinearities

The following result deals with problem (86) with periodic nonlinearities. It was first proved in the scalar case in [32] and in the vector case in [33] using slightly different arguments.

Theorem 22. *If there are some $\omega_1 > 0, \ldots, \omega_n > 0$ such that*

$$F(t, u) = F(t, u_1 + \omega_1, \ldots, u_n + \omega_n)$$

for all $(t, u) \in [0, T] \times \mathbb{R}^n$, then, for any $h \in L^1$ with $\overline{h} = 0$, problem (86) has at least one solution which minimizes I on C (or K).

Proof. Let $\omega = n^{1/2} \max_{1 \leq j \leq n} \omega_j$, so that

$$[0, \omega_1] \times \ldots \times [0, \omega_n] \subset \overline{B}(\omega).$$

Due to the periodicity of $F(t, \cdot)$ and because of $\overline{h} = 0$, it holds

$$I(v + j_1 \omega_1 e_1 + \ldots + j_n \omega_n e_n) = I(v)$$

for all $v \in K$ and $(j_1, \ldots, j_n) \in \mathbb{Z}^n$. Then, the conclusion follows from the equality

$$\{I(v) \ : \ v \in K\} = \{I(v) \ : \ v \in \widehat{K}_\omega\}$$

and Lemma 15.   □

For the use in examples, let us introduce the continuous mapping $S : \mathbb{R}^n \to \mathbb{R}^n$ by

$$S(u) := (\mu_1 \sin u_1, \mu_2 \sin u_2, \ldots, \mu_n \sin u_n) \quad (\mu_j \in \mathbb{R}, \ j = 1, \ldots, n)$$

so that

$$S(u) = \nabla c(u) \text{ with } c(u) := \sum_{j=1}^{n} \mu_j (1 - \cos u_j). \tag{99}$$

Example 12. For any $\mu \in \mathbb{R}$ and any $h \in L^1$ such that $\overline{h} = 0$, the problem

$$\left(\frac{u'}{\sqrt{1 - |u'|^2}} \right)' + S(u) = h(t), \quad u(0) = u(T), \quad u'(0) = u'(T)$$

has at least one solution.

This corresponds to $F(t, u) = c(u)$ such that $F(t, u_1 + 2\pi, \ldots, u_n + 2\pi) = F(t, u)$ for all $t \in [0, T]$ and all $u \in \mathbb{R}^n$.

In particular, in the scalar case, *the forced relativistic pendulum problem*

$$\left(\frac{u'}{\sqrt{1-u'^2}}\right)' + \mu \sin u = h(t), \quad u(0) = u(T), \quad u'(0) = u'(T)$$

has at least one solution whenever $\overline{h} = 0$.

With respect to Example 4, we suppress the restriction $4T < 1$, but the existence only holds for $k \equiv 0$ and $\overline{h} = 0$. With respect to Example 6, we have $\overline{h} = 0$ instead of $|h|_\infty < \mu$. The multiplicity conclusion will be proved in Sect. 4.

3.3.3 Asymptotically Positive Potential

Theorem 23. *If*

$$\liminf_{|u|\to\infty}[F(t,u) + \langle h(t), u\rangle] > 0 \quad uniformly\ a.e.\ in \quad t \in [0,T], \qquad (100)$$

then problem (86) has at least one solution which minimizes I *on* C.

Proof. Using (5) and (100) it follows that there exists $\rho > 0$ such that

$$F(t, u(t)) + \langle h(t), u(t)\rangle > 0$$

for any $u \in K$ such that $|\overline{u}| > \rho$. It follows that $I(u) > 0$ provided that $u \in K$ and $|\overline{u}| > \rho$. The proof follows from Lemma 15, as $I(0) = 0$. $\qquad\square$

Example 13. The problem

$$\left(\frac{u'}{\sqrt{1-u'^2}}\right)' = \frac{u+h(t)}{1+[u+h(t)]^2} + \cos u, \quad u(0) = u(T), \quad u'(0) = u'(T)$$

has at least one solution for all $h \in C$.

Example 14. If $b \in L^1$ and essinf $b > 0$, the problem

$$\left(\frac{u'}{\sqrt{1-u'^2}}\right)' = b(t)\frac{u}{\sqrt{1+|u|^2}} + S(u) + h(t), \qquad (101)$$

$$u(0) = u(T), \ u'(0) = u'(T)$$

has at least one solution for all $h \in L^\infty$ such that $|h|_\infty < $ essinf b.

Indeed,

$$F(t,u) = b(t)(\sqrt{1+|u|^2} - 1) + c(u)$$

and

$$\liminf_{|u|\to\infty}[b(t)(\sqrt{1+|u|^2}-1)+c(u)-\langle h(t),u\rangle]$$

$$\geq \liminf_{|u|\to\infty}[|u|(\text{essinf } b-|h|_\infty)-\sum_{j=1}^{n}|\mu_j|-\text{essinf } b]>0.$$

Example 15. For any $b \in L^1$ such that $b(t) \geq 0$ for a.e. $t \in [0,T]$ and $\overline{b} > 0$, the problem

$$\left(\frac{u'}{\sqrt{1-|u'|^2}}\right)' = b(t)e^{|u|^2}u+S(u)+h(t), \quad u(0)=u(T), \ u'(0)=u'(T)$$

has at least one solution for all $h \in L^\infty$.

Indeed,

$$F(t,u) = \frac{1}{2}b(t)[e^{|u|^2}-1]+c(u)+\langle h(t),u\rangle$$

and

$$\liminf_{|u|\to\infty}[(1/2)b(t)[e^{|u|^2}-1]+c(u)+\langle h(t),u\rangle$$

$$\geq \liminf_{|u|\to\infty}[(1/2)(\text{essinf } b)e^{|u|^2}-\sum_{j=1}^{n}|\mu_j|-(1/2)\text{essinf } b-|h|_\infty|u|]>0.$$

3.3.4 Nonlinearities with Power Growth Restriction

Let us consider the case of problem (86) with a nonlinearity $\nabla_u F$ having an arbitrary power growth in u and a potential F satisfying a semi-coercivity condition of the Ahmad–Lazer–Paul type [1]. The growth condition on $\nabla_u F$ is :

(H_P) *There exists $\alpha \geq 0$, $g \in L^1$ nonnegative and $k \in L^1$ nonnegative such that, for a.e. $t \in [0,T]$ and all $u \in \mathbb{R}^n$, one has*

$$|\nabla_u F(t,u)| \leq g(t)|u|^\alpha + k(t).$$

For the classical problem with $(\phi(u'))'$ replaced by u'', the case where $\alpha = 0$ was considered in [85], and the case where $\alpha \in [0,1)$ in [116]. Define the mapping $\overline{F} : \mathbb{R}^n \to \mathbb{R}$ by

$$\overline{F}(u) = T^{-1}\int_0^T F(t,u)\,dt.$$

Theorem 24. *Assume that Assumptions (H_ϕ), (H_F) and (H_P) hold. Then, for all $h \in L_1$ such that*

$$\lim_{|u|\to\infty} |u|^{-\alpha} \left[\overline{F}(u) - \langle \overline{h}, u \rangle \right] = +\infty, \tag{102}$$

problem (86) has at least one solution which minimizes I over C (or K).

Proof. Using the elementary inequality in $\mathbb{R}^n$

$$|y + z|^\alpha \le 2^\alpha (|y|^\alpha + |z|^\alpha),$$

we have, for all $u \in K$,

$$
\begin{aligned}
I(u) &= \int_0^T [\Phi(u'(t)) + F(t, \overline{u}) + F(t, u(t)) - F(t, \overline{u}) + \langle h(t), u(t) \rangle] \, dt \\
&\ge \int_0^T [F(t, \overline{u}) + \langle h(t), \overline{u} \rangle] \, dt \\
&\quad + \int_0^T \langle \int_0^1 \nabla_u F(t, \overline{u} + s\widetilde{u}(t)) \, ds + \widetilde{h}(t), \widetilde{u}(t) \rangle \, dt \\
&\ge T[\overline{F}(\overline{u}) + \langle \overline{h}, \overline{u} \rangle] \\
&\quad - \int_0^T \int_0^1 [g(t)|\overline{u} + s\widetilde{u}(t)|^\alpha + k(t)] |\widetilde{u}| \, ds \, dt - |h|_1 |\widetilde{u}|_\infty \\
&\ge T[\overline{F}(\overline{u}) + \langle \overline{h}, \overline{u} \rangle] \\
&\quad - |g|_1 2^\alpha [|\overline{u}|^\alpha + (Ta/4)^\alpha](Ta/4) - (|k|_1 + |h|_1)(Ta/4),
\end{aligned}
$$

where we have used inequality (5). Hence

$$
\begin{aligned}
I(u) &\ge |\overline{u}|^\alpha \left\{ T|\overline{u}|^{-\alpha} [\overline{F}(\overline{u}) + \langle \overline{h}, \overline{u} \rangle] - |g|_1 2(\alpha - 2)Ta \right\} \\
&\quad - [(Ta/2)^\alpha |g|_1 + |k|_1 + |h|_1](Ta/4). \tag{103}
\end{aligned}
$$

Assumption (102) implies the existence of some $\rho > 0$ such that the second term in the right-hand member of (103) is positive for $|\overline{u}| \ge \rho$. As $I(0) = 0$, the result follows from Lemma 15. $\qquad\square$

Remark 9. The classical case where $(\phi(u'))'$ is replaced by u'' requires, as shown in [116], $\alpha \in [0, 1)$ and Assumption (102) replaced by the stronger one

$$\lim_{|u|\to\infty} |u|^{-2\alpha} \left[\overline{F}(u) - \langle \overline{h}, u \rangle \right] = +\infty.$$

Example 16. For any $b \in L^1$ such that $\overline{b} > 0$, problem (101) has at least one solution for all $h \in L^1$ such that $|\overline{h}| < \overline{b}$.

Indeed, we have in this case $F(t, u) = b(t)[\sqrt{1 + |u|^2} - 1] + c(u)$, $\alpha = 0$, and, for any $v \in \mathbb{R}^n \setminus \{0\}$,

$$\overline{b}[\sqrt{1 + |v|^2} - 1] + c(v) + \langle \overline{h}, v \rangle \geq |v|[\overline{b} - |\overline{h}|] - \sum_{j=1}^{n} |\mu_j| - \overline{b},$$

with the right-hand member tending to $+\infty$ when $|v| \to \infty$.

With respect to Example 14, the use of the boundedness condition allows weakening the conditions upon b and h from essinf $b > 0$ and $|h|_\infty <$ essinf b to $\overline{b} > 0$ and $|\overline{h}| < \overline{b}$.

In particular, in the scalar case, *for every $h \in L^1$ such that $-\frac{\pi}{2} < \overline{h} < \frac{\pi}{2}$, the problem*

$$\left(\frac{u'}{\sqrt{1 - u'^2}} \right)' - \arctan u - \cos u = h(t), \quad u(0) = u(T), \quad u'(0) = u'(T)$$

has at least one solution. Corollary 4 does not apply to this example.

Remark 10. When $\alpha = 0$, condition (102) is of the type introduced by Ahmad–Lazer–Paul [1] for the Laplacian with Dirichlet conditions. The reader will observe that for $\alpha = 0$, the conclusion of Theorem 24 still remains true if (102) is replaced by the weaker but more technical condition

$$\liminf_{|u| \to \infty} [\overline{F}(u) + \langle \overline{h}, u \rangle] \, dt > Ta \int_0^T [g(t) + k(t) + |\widetilde{h}(t)|] \, dt.$$

3.3.5 Convex Potentials

Theorem 25. *If $F(t, \cdot)$ is convex for all $t \in [0, T]$, then, problem (86) has at least one solution which minimizes I on C (or K) for all $h \in L^1$ such that condition (102) with $\alpha = 0$ holds.*

Proof. By Assumption (102) with $\alpha = 0$, the real function $\overline{F} + \langle \overline{h}, \cdot \rangle$ achieves a minimum at some $\overline{v} \in \mathbb{R}^n$, for which

$$\nabla \overline{F}(\overline{v}) + \overline{h} = 0.$$

Now, by the convexity of $F(t, \cdot)$,

$$I(u) \geq T[\overline{F}(\overline{v}) + \langle \overline{h}, \overline{v} \rangle] + \int_0^T \langle \nabla_u F(t, \overline{v}) + h(t), u(t) - \overline{v} \rangle \, dt$$

$$= T[\overline{F}(\overline{v}) + \langle \overline{h}, \overline{v} \rangle] + \int_0^T \langle \nabla_u F(t, \overline{v}) + h(t), \widetilde{u}(t) \rangle \, dt$$

$$\geq T[\overline{F}(\overline{v}) + \langle \overline{h}, \overline{v} \rangle] - (Ta/4)|\nabla_u F(\cdot, \overline{v}) + h|_1.$$

From (102) we can find $\rho > 0$ such that $I(u) > 0$ provided that $|\bar{u}| \geq \rho$. Furthermore $I(0) = 0$. Therefore (97) is fulfilled and the result follows from Lemma 15.   □

Example 17. If $b \in L^1$ is such that $b(t) \geq 0$ for a.e. $t \in [0, T]$ and $\bar{b} > 0$, the problem

$$\left(\frac{u'}{\sqrt{1 - |u'|^2}}\right)' = b(t)e^{|u|^2}u + h(t), \quad u(0) = u(T), \quad u'(0) = u'(T)$$

has at least one solution for any $h \in L^1$.

Indeed, $F(t, u) = \frac{b(t)}{2}[e^{|u|^2} - 1]$ is convex in u for a.e. $t \in [0, T]$,

$$\overline{F}(u) + \langle \bar{h}, u \rangle \geq \frac{\bar{b}}{2}[e^{|u|^2} - 1] - |\bar{h}||u|,$$

and the right-hand member tends to $+\infty$ as $|u| \to \infty$.

With respect to Example 15, the use of the convexity of F allows to replace the assumption essinf $b > 0$ by the weaker one $b(t) \geq 0$ and $\bar{b} > 0$. But the oscillatory term $S(u)$ has to be dropped.

3.4   Saddle Point Solutions for Bounded Nonlinearities

The following results are inspired from [17].

3.4.1   Palais–Smale Condition

Towards the application of the minimax results obtained by Szulkin in [111] to the functional I defined by (91) we have to know when I satisfies the compactness *Palais–Smale* (in short (PS)) *condition*.

We say that a sequence $\{u_n\} \subset K$ is a *(PS)-sequence* if $I(u_n) \to c \in \mathbb{R}$ and

$$\int_0^T [\Phi(v'(t)) - \Phi(u'_n(t)) + \langle \nabla_u F(t, u_n(t)) + h(t), v(t) - u_n(t) \rangle] \, dt$$

$$\geq -\varepsilon_n |v - u_n|_\infty \quad \text{for all} \quad v \in K, \quad (104)$$

where $\varepsilon_n \to 0+$. According to [111], the functional I is said to satisfy the *(PS) condition* if any (PS)-sequence has a convergent subsequence in C.

The lemma below provides useful properties of the *(PS)*-sequences.

Lemma 16. *Let* $\{u_n\}$ *be a (PS)-sequence. Then the following hold true:*

(i) *The sequence* $\left\{\int_0^T [F(t, u_n(t)) + \langle h(t), u_n(t) \rangle] \, dt\right\}$ *is bounded;*

(ii) *If* $\{\bar{u}_n\}$ *is bounded, then* $\{u_n\}$ *has a convergent subsequence in* C;

(iii) One has that

$$\left| \int_0^T \langle \nabla_u F(t, u_n(t)) + h(t), u_n(t) \rangle \, dt \right| \leq \varepsilon_n \quad \text{for all} \quad n \in \mathbb{N}. \quad (105)$$

Proof. (i) This is immediate from the fact that $\{I(u_n)\}$ and Φ are bounded.

(ii) From (5) and $u_n \in K$, the sequence $\{\widetilde{u}_n\}$ is bounded in $W^{1,\infty}$. By the compactness of the embedding $W^{1,\infty} \subset C$, we deduce that $\{\widetilde{u}_n\}$ has a convergent subsequence in C. Using then the boundedness of $\{\overline{u}_n\} \subset \mathbb{R}$ it follows that $\{u_n\}$ has a convergent subsequence in C.

(iii) Taking $v = u_n + w$ with $w \in \mathbb{R}^n$ such that $|w| = 1$ in (104) one obtains

$$\left\langle \int_0^T [\nabla_u F(t, u_n(t)) + h(t)] \, dt, w \right\rangle \geq -\varepsilon_n$$

for all $w \in \mathbb{R}^n$ with $|w| = 1$, and hence (105).

$\square$

3.4.2 Bounded Nonlinearities with Anti-coercive Potential

Theorem 26. *Let $F : [0, T] \times \mathbb{R}^n \to \mathbb{R}$ and $g, k \in L^1$ nonnegative be such that condition (H_P) with $\alpha = 0$ is satisfied. If $h \in L^1$ is such that*

$$\lim_{|x| \to \infty} [\overline{F}(x) + \langle \overline{h}, x \rangle] = -\infty, \quad (106)$$

then problem (86) has at least one solution.

Proof. We apply the Saddle Point theorem for functionals of Szulkin's type [111, Theorem 3.5]. From (106) the functional I is not bounded from below. Indeed, if $v = c \in \mathbb{R}^n$ is a constant function then

$$I(c) = \int_0^T [F(t, c) + \langle h(t), c \rangle] \, dt \to -\infty \quad \text{as} \quad |c| \to \infty. \quad (107)$$

We split $C = \mathbb{R}^n \oplus \widetilde{C}$, where $\widetilde{C} = \{v \in C \; : \; \overline{v} = 0\}$. Note that

$$I(v) \geq \int_0^T [F(t, \widetilde{v}(t)) + \langle h(t), \widetilde{v}(t) \rangle] \, dt \quad \text{for all } v \in K \cap \widetilde{C},$$

which together with (5) imply that there is a constant $\alpha \in \mathbb{R}$ such that

$$I(v) \geq \alpha \quad \text{for all } v \in X. \quad (108)$$

Using (107) and (108) we can find some $R > 0$ so that

$$\sup_{S_R} I < \inf_C I,$$

where $S_R = \{c \in \mathbb{R}^n : |c| = R\}$.

It remains to show that I satisfies the (PS) condition. Let $\{u_n\} \subset K$ be a (PS)-sequence. $\{I(u_n)\}$, $\{\varphi(u_n)\}$ are bounded and, by (H_P) with $\alpha = 0$, we have, letting $l(t) = g(t) + k(t)$,

$$\left| \int_0^T [F(t, u_n(t)) + \langle h(t), u_n(t) \rangle - F(t, \bar{u}_n) - \langle h(t), \bar{u}_n \rangle] \, dt \right|$$

$$\leq \int_0^T \int_0^1 |\langle \nabla_u F(t, \bar{u}_n(t) + s\widetilde{u}_n(t)) + h(t), \widetilde{u}_n(t) \rangle| \, ds \, dt$$

$$\leq (Ta/4) \int_0^T [g(t) + k(t)] \, dt.$$

From

$$I(u_n) = \varphi(u_n) + \int_0^T [F(t, \bar{u}_n) + \langle h(t), \bar{u}_n \rangle] \, dt$$

$$+ \int_0^T [F(t, u_n(t)) - F(t, \bar{u}_n) + \langle h(t), \widetilde{u}_n(t) \rangle] \, dt$$

it follows that there exists a constant $\beta \in \mathbb{R}$ such that

$$\int_0^T [F(t, \bar{u}_n) + \langle h(t), \bar{u}_n \rangle] \, dt \geq \beta.$$

Then by (106) the sequence $\{\bar{u}_n\}$ is bounded and Lemma 16 *(ii)* ensures that $\{u_n\}$ has a convergent subsequence in C. Consequently, I satisfies the (PS) condition and the conclusion follows from [111, Theorem 3.5] and Proposition 1.   □

Remark 11. Condition (106), also of the Ahmad–Lazer–Paul type [1] is, in some sense, "dual" to condition (102).

Example 18. Given $b \in L^1$ such that $\bar{b} < 0$, problem (101) has at least one solution for all $h \in L^1$ such that $|\bar{h}| < |\bar{b}|$.

Indeed, we have in this case $F(t, u) = b(t)[\sqrt{1 + |u|^2} - 1] + c(u)$, and, for any $v \in \mathbb{R}^n \setminus \{0\}$,

$$\bar{b}[\sqrt{1 + |v|^2} - 1] + c(v) + \langle \bar{h}, v \rangle \leq [\bar{b} + |\bar{h}|]|v| - \bar{b} + \sum_{j=1}^n |\mu_j|,$$

with the right-hand member tending to $-\infty$ when $|v| \to \infty$.

Theorem 27. *If*

$$\lim_{|x|\to\infty} [F(t,x) + \langle h(t), x\rangle] = -\infty, \quad uniformly\ in \quad t \in [0, T], \qquad (109)$$

then problem (86) has at least one solution.

Proof. We keep the notations introduced in the proof of Theorem 26. Clearly, (109) implies (106) and from the proof of Theorem 26 it follows that I has the geometry required by the Saddle Point Theorem. To show that I satisfies the (PS) condition, let $\{u_n\} \subset K$ be a (PS)-sequence. If $\{|\bar{u}_n|\}$ is not bounded, we may assume going if necessary to a subsequence, that $|\bar{u}_n| \to \infty$. Using (5) and (109) we deduce that

$$F(t, u_n(t)) + \langle h(t), u_n(t)\rangle \to -\infty, \quad \text{uniformly in } t \in [0, T].$$

This implies

$$\int_0^T [F(t, u_n(t)) - \langle h(t), u_n(t)\rangle] \, dt \to -\infty,$$

contradicting Lemma 16 *(i)*. Hence, $\{\bar{u}_n\}$ is bounded and by Lemma 16 *(ii)*, the sequence $\{u_n\}$ has a convergent subsequence in C. Therefore, I satisfies the (PS) condition. The proof is complete. $\qquad\qquad\qquad\qquad\qquad\qquad\qquad\qquad\qquad\square$

Remark 12. No result corresponding to Theorem 27 holds for the classical case where $(\phi(u'))'$ is replaced by u''. Indeed, if λ_k is a positive eigenvalue of $-u''$ on $[0, T]$ with periodic boundary conditions, and φ_k a corresponding eigenfunction, the problem

$$u'' = -\lambda_k u + \varphi_k(t), \quad u(0) = u(T), \quad u'(0) = u'(T)$$

has no solution, but $-\lambda_k \dfrac{u^2}{2} + \varphi_k(t)u \to -\infty$ uniformly in $[0, T]$ when $|u| \to \infty$.

Example 19. The problem

$$\left(\frac{u'}{\sqrt{1 - u'^2}}\right)' + \frac{u + h(t)}{1 + [u + h(t)]^2} = \cos u, \quad u(0) = u(T), \quad u'(0) = u'(T)$$

has at least one solution for all $h \in C$.

3.5 Multiple Solutions Near Resonance

3.5.1 Introduction and Hypotheses

This subsection, which presents some of the results of [25], is devoted to the existence of multiple solutions for the scalar periodic problems of the form

$$(\phi(u'))' = \lambda|u|^{m-2}u - g(t,u) + h(t), \quad u(0) = u(T), \quad u'(0) = u'(T), \quad (110)$$

where $\phi : (-a, a) \to \mathbb{R}$ satisfies Assumption (H_ϕ) for $n = 1$, $m \geq 2$, $\lambda > 0$, $\overline{h} = 0$ and $g : [0, T] \times \mathbb{R} \to \mathbb{R}$ satisfies suitable conditions. We show in particular the existence of at least three solutions when $G(t, x) = \int_0^x g(t, s)\,ds$ has a polynomial growth of order strictly smaller than m, satisfies a Ahmad–Lazer–Paul condition, and $\lambda > 0$ is sufficiently small. Results of this type, called *multiplicity results near resonance*, have been initiated in the classical case where $(\phi(u'))'$ is replaced by u'' by Schmitt and the author in [82], using bifurcation from infinity and Leray–Schauder degree theory. A variational approach was introduced by Sanchez in [104] to deal with such multiplicity problems, and conditions of type (i)' and (ii)' were introduced by Ma, Ramos and Sanchez in [67, 102] for semilinear and quasilinear Dirichlet problems involving the p-Laplacian. See also [41, 65, 66, 86, 94] for a similar variational treatment of various semilinear or quasilinear equations, systems or inequalities with Dirichlet conditions, [87] for perturbations of p-Laplacian with Neumann boundary conditions, and [63] for periodic solutions of perturbations of the one-dimensional p-Laplacian. The existence of at least two solutions near resonance at a *non-principal* eigenvalue was first obtained in [81] using a topological approach and then for semilinear or quasilinear problems using critical point theory in [42, 57, 110]. This question seems to be meaningless for the singular ϕ considered here because resonance only occurs at $\lambda = 0$.

The main used tools are local minimization results combined with mountain pass techniques in the frame of the Szulkin's critical point theory [111].

Throughout this subsection we assume that the following hypothesis upon g and h hold true.

(H_f) *The functions* $g : [0, T] \times \mathbb{R} \to \mathbb{R}$, $b, h : [0, T] \to \mathbb{R}$ *are continuous; the constant* $m \geq 2$ *is fixed and* λ *is a real positive parameter.*

We denote by G the indefinite integral of g with respect to the second variable defined by

$$G(t, x) := \int_0^x g(t, \xi)\,d\xi, \quad (t, x) \in [0, T] \times \mathbb{R},$$

and assume that G satisfies the following hypotheses :

(i)' *There exists* $k_1, k_2 > 0$ *and* $0 < \sigma < m$ *such that*

$$- l(t) \leq G(t, x) \leq k_1|x|^\sigma + k_2, \quad \text{for all} \quad (t, x) \in [0, T] \times \mathbb{R}, \quad (111)$$

where $l \geq 0$ *and* $l \in L^1$;

(ii)' *One has that either*

$$\lim_{|x| \to \infty} \int_0^T G(t, x)\,dt = +\infty, \quad (112)$$

or the limits $G_{\pm}(t) = \lim\limits_{x \to \pm\infty} G(t,x)$ *exist for all* $t \in [0, T]$ *and*

$$G(t, x) < G_+(t), \quad \text{for all} \quad t \in [0, T], \ x \geq 0,$$
$$G(t, x) < G_-(t), \quad \text{for all} \quad t \in [0, T], \ x \leq 0; \tag{113}$$

(iii)' It holds

$$\overline{h} = 0, \tag{114}$$

We define $\widehat{\mathscr{F}}_\lambda : C \to \mathbb{R}$ by

$$\widehat{\mathscr{F}}_\lambda(u) = \int_0^T \left[\frac{\lambda}{m} |u(t)|^m - G(t, u(t)) + h(t)u(t) \right] dt, \quad u \in C.$$

A standard reasoning shows that $\widehat{\mathscr{F}}_\lambda$ is of class C^1 on C and

$$\widehat{\mathscr{F}}'_\lambda(u)(v) = \int_0^T \left[\lambda |u(t)|^{m-2} u(t) - g(t, u(t)) + h(t) \right] v(t)\, dt, \quad u, v \in C.$$

Then it is clear that $\widehat{I}_\lambda : C \to (-\infty, +\infty]$ defined by

$$\widehat{I}_\lambda = \widehat{\mathscr{F}}_\lambda + \Psi,$$

where Ψ is defined in (88) with $n = 1$, has the structure required by Szulkin's critical point theory. By the results of the beginning of Sect. 2, the search of solutions of problem (110) reduces to finding critical points of the energy functional $\widehat{I}_\lambda$.

We also need in the proof the following inequalities for $u \in K$.

Lemma 17. *Let $p \geq 1$ be a real number. Then*

$$|u(t)|^p \geq |\overline{u}|^p - \frac{pTa}{4} |\overline{u}|^{p-1} \quad \text{for all} \quad u \in K \quad \text{and all} \quad t \in [0, T] \tag{115}$$

and there are constants $\alpha_1, \ \alpha_2 \geq 0$ such that

$$|u(t)|^p \leq |\overline{u}|^p + \alpha_1 |\overline{u}|^{p-1} + \alpha_2 \tag{116}$$

for all $u \in K$ with $|\overline{u}| \geq 1$ and all $t \in [0, T]$.

Proof. The result is trivial for $p = 1$. If $p > 1$, $u \in K$ and $t \in [0, T]$, then, using the convexity of the differentiable function $s \mapsto |s|^p$, we get

$$|u(t)|^p = |\overline{u} + \widetilde{u}(t)|^p \geq |\overline{u}|^p + p|\overline{u}|^{p-2}\overline{u}\,\widetilde{u}(t) \geq |\overline{u}|^p - p|\overline{u}|^{p-1} \frac{Ta}{4}.$$

On the other hand, denoting by $\widetilde{p}$ the smallest integer larger or equal to p and letting $M := Ta/4$, we have, for all $t \in [0, T]$,

$$|u(t)|^p = |\bar{u} + \widetilde{u}(t)|^p \leq (|\bar{u}| + M)^p = |\bar{u}|^p \left(1 + \frac{M}{|\bar{u}|}\right)^p$$

$$\leq |\bar{u}|^p \left(1 + \frac{M}{|\bar{u}|}\right)^{\widetilde{p}} = |\bar{u}|^p \left(1 + \sum_{k=1}^{\widetilde{p}} \frac{\widetilde{p}!}{k!(\widetilde{p} - k)!} \frac{M^k}{|\bar{u}|^k}\right)$$

$$= |\bar{u}|^p + \sum_{k=1}^{\widetilde{p}} \frac{\widetilde{p}!}{k!(\widetilde{p} - k)!} M^k |\bar{u}|^{p-k},$$

and (116) follows easily.   □

3.5.2 Existence of Three Periodic Solutions

The following existence result, inspired from [67, 102] provides a useful tool in obtaining multiple solutions.

Lemma 18. *Assume that condition (114) holds, and that there exists $k_1, k_2 > 0$ and $0 < \sigma < m$ such that*

$$- l(t) \leq G(t, x) \leq k_1 |x|^\sigma + k_2 \quad \text{for all} \quad (t, x) \in [0, T] \times \mathbb{R}_+, \quad (117)$$

with some $l \in L^1, l \geq 0$. If either

$$\lim_{x \to +\infty} \int_0^T G(t, x)\, dt = +\infty, \quad (118)$$

or $G_+(t) := \lim_{x \to +\infty} G(t, x)$ exists for all $t \in [0, T]$ and

$$G(t, x) < G_+(r) \quad \text{for all} \quad t \in [0, T], \ x \geq 0, \quad (119)$$

then there exists $\lambda_+ > 0$ such that problem (110) has at least one solution $u_\lambda > 0$ for any $0 < \lambda < \lambda_+$ which minimize $\widehat{I}_\lambda$ on $C^+ = \{v \in C : v \geq 0\}$. Moreover, u_λ is a local minimum for $\widehat{I}_\lambda$.

Proof. First, notice that, using (5), one has

$$\bar{u} - \frac{Ta}{4} \leq u(t) \leq \bar{u} + \frac{Ta}{4} \quad \text{for all} \quad u \in K, \quad (120)$$

hence

$$\bar{u} \to +\infty \quad \text{as} \quad |u|_\infty \to \infty \quad \text{if} \quad u \in C^+ \cap K. \quad (121)$$

Also, it is clear that

$$|u(t)| \leq |\bar{u}| + \frac{Ta}{4} \quad \text{for all} \quad u \in K \quad \text{and} \quad t \in [0, T]. \tag{122}$$

From (117) it follows that

$$\widehat{I}_\lambda(u) \geq \int_0^T \left[\frac{\lambda}{m} |u(t)|^m - k_1 |u(t)|^\sigma - k_2 - |h|_\infty |u(t)| \right] dt,$$

for all $u \in C^+$. Hence, using (115), (122), (121) and $\sigma < m$, we deduce immediately that

$$\widehat{I}_\lambda(u) \to +\infty \quad \text{whenever} \quad |u|_\infty \to \infty \quad \text{in} \quad C^+, \tag{123}$$

that is $\widehat{I}_\lambda$ is coercive on C^+, and hence bounded from below on C^+. Now, let $\{u_n\} \subset C^+ \cap K$ be a minimizing sequence, i.e. $\widehat{I}_\lambda(u_n) \to \inf_{C^+} \widehat{I}_\lambda$ as $n \to \infty$. From (123), $\{u_n\}$ is bounded in C, and using the fact that $\{u_n\} \subset K$, we infer that $\{u_n\}$ is bounded in $W^{1,\infty}$, compactly embedded in C. Hence $\{u_n\}$ has a subsequence converging in C to some $u_\lambda \in C^+ \cap K$. By the lower semicontinuity of $\widehat{I}_\lambda$ it follows

$$\widehat{I}_\lambda(u_\lambda) = \inf_{C^+} \widehat{I}_\lambda.$$

We *claim* that

$$\bar{u}_\lambda \to +\infty \quad \text{as} \quad \lambda \to 0. \tag{124}$$

Assuming this for the moment, it follows from (120) and (124) that there exists $\lambda_+ > 0$ such that $u_\lambda > 0$ for any $0 < \lambda < \lambda_+$, implying that u_λ is a local minimum for $\widehat{I}_\lambda$. Consequently, u_λ is a critical point of $\widehat{I}_\lambda$, and hence a solution of (110) for any $0 < \lambda < \lambda_+$.

In order to prove the claim, assume first that (118) holds true. Then, consider $M > 0$ and $x_M > 0$ such that

$$\int_0^T G(t, x_M) \, dt > 2M. \tag{125}$$

On the other hand, as $\bar{h} = 0$, one has that for all $\lambda > 0$,

$$\widehat{I}_\lambda(x) = \frac{\lambda T}{m} |x|^m - \int_0^T G(t, x) \, dt \quad (x \in \mathbb{R}). \tag{126}$$

Choosing $\lambda_M > 0$ such that

$$\frac{\lambda_M T}{m} x_M^m < M,$$

and using (125), (126), it follows that

$$\widehat{I}_\lambda(x_M) < -M \quad \text{for all} \quad 0 < \lambda < \lambda_M.$$

Consequently,

$$\inf_{C^+} \widehat{I}_\lambda \to -\infty \quad \text{as} \quad \lambda \to 0,$$

which, together with (120) implies (124), as claimed.

Now, let (119) holds true, and assume also by contradiction that there exists $\lambda_n \to 0$ such that $\{\overline{u}_{\lambda_n}\}$ is bounded. On account of (120) and of the compactness of the embedding of $W^{1,\infty}$ in C, one can assume, going if necessary to a subsequence, that $\{u_{\lambda_n}\}$ converges in C to some $u \in C^+$. Using (119) and Fatou's lemma it follows that

$$\int_0^T G(t, u(t)) \, dt < \int_0^T G_+(t) \, dt \leq \liminf_{s \to \infty} \int_0^T G(t, s + \widetilde{u}(t)) \, dt,$$

which implies the existence of $s_0 > 0$ sufficiently large, with $s_0 + \widetilde{v} \in C^+$ for all $v \in K$, and of $\rho > 0$ such that

$$\int_0^T [G(t, u(t)) - G(t, s_0 + \widetilde{u}(t))] \, dt < -\rho.$$

So, for n sufficiently large, we have

$$\int_0^t [G(t, u_{\lambda_n}(t)) - G(t, s_0 + \widetilde{u}_{\lambda_n}(t))] \, dt < -\rho. \tag{127}$$

On the other hand, using (120), we get

$$\int_0^T \frac{\lambda_n}{m} [|s_0 + \widetilde{u}_{\lambda_n}(t)|^m - |u_{\lambda_n}(t)|^m] \, dt \to 0 \quad \text{as} \quad n \to \infty. \tag{128}$$

Notice that, as $\overline{h} = 0$, for all $\lambda > 0$ and $s \in \mathbb{R}$, one has

$$\widehat{I}_\lambda(s + \widetilde{u}_\lambda) = \int_0^T \Phi(u_\lambda')(t) \, dt + \int_0^T \frac{\lambda}{m} |s + \widetilde{u}_\lambda(t)|^m \, dt$$

$$- \int_0^T G(t, s + \widetilde{u}_\lambda(t)) \, dt - \int_0^T h(t) \widetilde{u}_\lambda(t) \, dt.$$

Then, by (127) and (128) we obtain

$$\widehat{I}_{\lambda_n}(s_0 + \widetilde{u}_{\lambda_n}) < \widehat{I}_{\lambda_n}(u_{\lambda_n}),$$

for n sufficiently large, contradicting the definition of u_{λ_n}. This proves the claim and the proof is complete.   □

The multiplicity result will follow from Lemma 18 and the following version of the mountain pass lemma given in [111].

Lemma 19. *If* $I = \Psi + G$ *satisfies (PS)-condition, 0 is a local minimum of I and if* $I(e) \leq I(0)$ *for some* $e \neq 0$, *then* I *has a critical point different from 0 and e. In particular, if* I *has two local minima, then it has at least a third critical point.*

Theorem 28. *Assume that conditions (114), (111) and either (112) or (113) hold true. Then there exists* $\lambda_0 > 0$ *such that problem (110) has at least three solutions for any* $\lambda \in (0, \lambda_0)$.

Proof. From Lemma 18, it follows that there exists $\lambda_+ > 0$ such that $\widehat{I}_\lambda$ has a local minimum at some $u_{\lambda,1} > 0$ for any $0 < \lambda < \lambda_+$. Using exactly the same strategy, we can find $\lambda_- > 0$ such that $\widehat{I}_\lambda$ has a local minimum at some $u_{\lambda,2} < 0$ for any $0 < \lambda < \lambda_-$. Taking $\lambda_0 = \min\{\lambda_-, \lambda_+\}$ it follows that $\widehat{I}_\lambda$ has two local minima for any $\lambda \in (0, \lambda_0)$. On the other hand, from the proof of Lemma 18, it is easy to see that $\widehat{I}_\lambda$ is coercive on C, implying that $\widehat{I}_\lambda$ satisfies the (PS) condition for any $\lambda > 0$. Hence, from Lemma 19, we infer that $\widehat{I}_\lambda$ has at least three critical points for all $\lambda \in (0, \lambda_0)$ which are solutions of (110).   □

Remark 13.   (i) When g is bounded, it is well known [1] that the Ahmad–Lazer–Paul condition (112) generalizes the Landesman-Lazer condition

$$\int_0^T g^-(t)\, dt < 0 < \int_0^T g_+(t)\, dt,$$

where $g^-(t) = \limsup_{x\to-\infty} g(t,x)$ and $g_+(t) = \liminf_{x\to+\infty} g(t,x)$.
(ii) Condition (113) holds true whenever one has the sign condition

$$xg(t,x) > 0 \quad \text{for all} \quad t \in [0,T] \quad \text{and} \quad x \neq 0.$$

(iii) The condition :
there exists $0 < \theta < m$ *such that*

$$xg(t,x) - \theta G(t,x) \to -\infty \quad \text{as} \quad |x| \to \infty, \text{ uniformly in} \quad t \in [0,T],$$

introduced in [66, 104], together with the sign condition

$$xg(t,x) > 0 \text{ for all } t \in [0,T] \text{ and } |x| \geq x_0$$

for some $x_0 > 0$, imply (111) and (112).

Example 20. Let $m \in \mathbb{N}$ be even and $h \in C$ be with $\overline{h} = 0$. Then there exists $\lambda_0 > 0$ such that the problem

$$\left(\frac{u'}{\sqrt{1-u'^2}}\right)' = \lambda |u|^{m-2}u - \frac{u^{m-1}}{1+u^m} + h(t), \quad u(0) = u(T), \quad u'(0) = u'(T)$$

has at least three solutions for all $\lambda \in (0, \lambda_0)$.

3.6 BV Periodic Solutions of the Forced Pendulum with Curvature Operator

In this section we sketch Obersnel-Omari's recent proof [93] of the existence of at least two solutions for problems of the form

$$\left(\frac{u'}{\sqrt{1+u'^2}}\right)' = f(t, x) + h(t), \quad u(0) = u(T), \quad u'(0) = u'(T), \quad (129)$$

where $f : [0, T] \times \mathbb{R} \to \mathbb{R}$ is supposed, for simplicity, continuous, and satisfies the following periodicity condition, where $F : [0, T] \times \mathbb{R} \to \mathbb{R}$ is the indefinite integral of f defined by

$$F(t, u) = \int_0^u f(t, x) \, dx \quad (t \in [0, T], \ u \in \mathbb{R}).$$

(H_ω) The function F satisfies the ω-periodicity condition

$$F(t, x + \omega) = F(t, x) \quad \text{for all} \quad t \in [0, T], \ x \in \mathbb{R}. \quad (130)$$

This condition is in particular satisfied, with $\omega = 2\pi$, for the *forced pendulum equation with curvature operator*, i.e. for the problem

$$\left(\frac{u'}{\sqrt{1+u'^2}}\right)' + \mu \sin u = h(t), \quad u(0) = u(T), \quad u'(0) = u'(T). \quad (131)$$

We assume in addition that $h \in C$ and that

$$\bar{h} = 0. \quad (132)$$

3.6.1 The Action Functional

The action functional $\mathscr{E}$ associated to problem (129), given by

$$\mathscr{E}(u) = \int_0^T \left[\sqrt{1 + u'^2(t)} + F(t, u(t)) + h(t)u(t)\right] dt,$$

is well defined on the space

$$W_T^{1,1} = \{u \in W^{1,1}([0,T]) : u(0) = u(T)\}$$

and, using the Assumptions (H_ω), (132), and Sobolev inequality, it is not difficult to see that $\mathscr{E}$ is bounded from below. Furthermore, standard reasonings imply that $\mathscr{E}$ is Gateaux-differentiable and lower semi-continuous. Consequently, $\mathscr{E}$ admits a bounded minimizing sequence but, because $W_T^{1,1}$ is not reflexive, the usual extraction of a weakly converging subsequence fails. Indeed, one can construct an example of such sequences having no subsequence converging to some element of $W_T^{1,1}$ even when $f \equiv 0$ (see [93]).

Various considerations lead to the choice of the larger space $BV = BV(0,T)$ of functions having finite total variation, namely such that

$$\int_0^T |Dv| = \sup\{\int_0^T v(t)w'(t)\,dt : w \in C_0^1((0,T)) \quad \text{and} \quad |w|_\infty \le 1\} < +\infty.$$

Here

$$C_0^1(0,T) = \{u \in C^1(0,T) \quad \text{with compact support in} \quad (0,T)\}.$$

BV is a Banach space with respect to the norm

$$|v|_{BV} = \int_0^T |Dv| + |v|_q$$

for any $q \in [1,+\infty)$ fixed. To take in account the periodic boundary condition, the relaxed functional $\mathscr{I} : BV \to \mathbb{R}$ is defined by

$$\mathscr{I}(u) = \int_0^T \sqrt{1 + |Du|^2} + \int_0^T [F(t,u(t)) + h(t)u(t)]\,dt + |u(T^-) - u(0^+)|,$$

where

$$\int_0^T \sqrt{1 + |Du|^2}\,dt = \sup\{\int_0^T [v(t)w_1'(t) + w_2(t)]\,dt : w_1, w_2 \in C_0^1 \text{ and } |w_1^2 + w_2^2|_\infty \le 1\}.$$

Define $\mathscr{J} : BV \to \mathbb{R}$ by

$$\mathscr{J}(u) = \int_0^T \sqrt{1 + |Du|^2} + |u(T^-) - u(0^+)|.$$

It is a nontrivial fact to prove [93] that $\mathscr{J}$ is convex, Lipschitz continuous, and lower semicontinuous with respect to the L^1-convergence. As $\mathscr{F} : BV \to \mathbb{R}$ defined by

$$\mathscr{F}(u) = \int_0^T [F(t, u(t)) + h(t)u(t)]\, dt$$

is of class C^1, we are again in the setting of Szulkin's critical point theory [112] and a critical point of $\mathscr{I}$ is some $u \in BV$ satisfying the differential inequality

$$\mathscr{I}(v) - \mathscr{I}(u) + \int_0^T f(t, u(t))(v(t) - u(t))\, dt \geq 0$$

for all $v \in BV$. Any solution u of this variational inequality belonging to $W_T^{1,1}$ will be a weak solution of (129). The proof of the lower semicontinuity property of $\mathscr{I}$ depends upon the following approximation lemma essentially due to Anzellotti [5].

Lemma 20. *For any given $v \in BV$, there exists a sequence (v_n) in $W_T^{1,1}$ such that*

$$\lim_{n\to\infty} v_n = v \quad in \quad L^1,$$

$$\lim_{n\to\infty} \int_0^T |v_n'(t)|\, dt = \int_0^T [|Dv| + |v(T^-) - v(0^+)|,$$

$$\lim_{n\to\infty} \int_0^T \sqrt{1 + |v_n'(t)|^2}\, dt = \int_0^T \sqrt{1 + |Dv|^2} + |v(T^-) - v(0^+)|.$$

3.6.2 Existence of Two BV Solutions

The proof of existence of a solution depends on the following BV version of Wirtinger inequality [93].

Lemma 21. *For every $v \in BV$ such that $\bar{v} = 0$, one has*

$$|v|_1 \leq \frac{T}{4}\left(\int_0^T |Dv| + |v(T^-) - v(0^+)|\right)$$

and the constant $T/4$ is sharp.

Hence one can prove the following multiplicity result [93].

Theorem 29. *Assume that $h \in L^p$ for some $p > 1$, and assumption (H_ω) holds. Then problem (129) has at least two geometrically distinct solutions if conditions (132) and*

$$\sup\left\{\int_0^T h(t)w(t)\, dt : w \in BV, \int_0^T |Dw| + |w(T^-) - w(0^+)| \leq 1\right\} < 1 \quad (133)$$

hold.

Proof. (sketched). The first solution u_0 is obtained as a global minimum of $\mathscr{I}$ in a rather standard way. Letting $u_1 = u_0 + \omega$ (also a global minimum of $\mathscr{I}$), the second solution is obtained through a modified problem constructed in such a way that any critical point of the association functional $\mathscr{H}$ in BV lies between u_0 and u_1 The functional $\mathscr{H}$ is then extended to a functional $\mathscr{M}$ on L^q where $\frac{1}{p} + \frac{1}{q} = 1$ which is shown to be bounded from below and coercive. Furthermore, $\mathscr{M}$ satisfies Palais–Smale condition in Szulkin's sense. Hence $\mathscr{M}$ has a global minimum u, shown to be a critical point of $\mathscr{H}$ too, so that $u_0(t) \le u(t) \le u_1(t)$. From this information, the existence of a second geometrically distinct critical point follows.   □

Corollary 11. *Assumption (133) holds if either*

$$|h|_\infty < \frac{4}{T}$$

or h has a primitive H such that

$$|H|_\infty < 1.$$

Remark 14. On can compare the result of Corollary 11 with that of Example 7 where the existence of at least two *classical* solutions for

$$\left(\frac{u'}{\sqrt{1+u'^2}}\right)' + \mu \sin u = h(t), \quad u(0) = u(T), \quad u'(0) = u'(T)$$

is proved under the stronger conditions $\overline{h} = 0$, $|h|_\infty < \mu < \frac{1}{2T}$.

4   Hamiltonian Variational Approach for Periodic Solutions

4.1   *Introduction*

Using Lusternik-Schnirelman theory in Hilbert manifolds [95] or variants of it, Chang [34], Rabinowitz [99] and the author [74] have independently obtained results which imply that the problem

$$q'' = \nabla_q F(t,q) + h(t), \quad q(0) = q(T), \quad q'(0) = q'(T) \qquad (134)$$

with $F : [0, T] \times \mathbb{R}^n \to \mathbb{R}$ satisfying assumption

(HF)   *F is continuous, ω_i-periodic with respect to each q_i ($i = 1, 2, \ldots, n$) and such that $\nabla_q F$ exists and is continuous on $[0, T] \times \mathbb{R}^n$, has at least $n + 1$ geometrically distinct solutions for every $h \in L^2$ verifying the Assumption*

(Hh) $\int_0^T h(t)\, dt = 0.$

Because of the periodicity property of F, if $q(t)$ is a solution of (134), the same is true for $(q_1(t)+j_1\omega_1, q_2(t)+j_2\omega_2, \ldots, q_n(t)+j_n\omega_n)$ for any $(j_1, j_2, \ldots, j_n) \in \mathbb{Z}^n$, and hence two solutions q and $\widehat{q}$ of (134) are called *geometrically distinct* if

$$q \not\equiv \widehat{q} \pmod{\omega_j e_j}, j = 1, 2, \ldots, n).$$

This result is an extension of an earlier one of the author and Willem [84] who proved, under the same conditions, the existence of at least two geometrically distinct solutions, using the variant of the mountain pass lemma introduced in [83] to treat the special case where $n = 1$, and in particular the *forced pendulum problem*

$$q'' + \mu \sin q = h(t), \quad q(0) = q(T), \quad q'(0) = q'(T).$$

See [77] for a survey of this problem.

The corresponding existence result for the *relativistic forced pendulum equation*

$$\left(\frac{q'}{\sqrt{1-q'^2}}\right)' + \mu \sin q = h(t), \quad q(0) = q(T), \quad q'(0) = q'(T) \quad (135)$$

has been recently considered by Brezis and the author [32], who proved the existence of at least one solution of (135) when Assumption (Hh) holds, by minimizing the corresponding action functional

$$u \mapsto \int_0^T [1 - \sqrt{1 - q'(t)^2} + A \cos q(t) - h(t)q(t)]\, dt$$

over the closed convex subset made of functions in $W^{1,\infty}$ such that $q(0) = q(T)$ and $|q'|_\infty \le 1$. The main difficulty consisted in showing that such a minimum indeed satisfies (135). The result is obtained in [32] for the more general problem

$$(\phi(q'))' = \partial_q F(t, q) + h(t), \quad q(0) = q(T), \quad q'(0) = q'(T), \quad (136)$$

where $\phi : (-a, a) \to \mathbb{R}$ is an increasing homeomorphism such that $\phi(0) = 0$, F is ω_1-periodic in q, continuous, $\partial_q F$ is continuous and h verifies Assumption (Hh). The same authors in [33] have extended their existence result to the corresponding n-dimensional problem

$$(\phi(q'))' = \nabla_q F(t, q) + h(t), \quad q(0) = q(T), \quad q'(0) = q'(T), \quad (137)$$

when Assumptions (HF) and (Hh) hold, and ϕ satisfies condition (H_ϕ). Very recently, Bereanu and Torres [16] have extended the mountain pass approach of [83] to obtain the existence of at least two geometrically distinct solutions for problem (136). It is not clear if their approach is applicable to system (136) and, would it be the case, the existence of two solutions only would be insured.

The aim of this section which describes some results of [80], is to prove that, under Assumptions (H_ϕ), (HF) and (Hh), problem (136) has at least $n + 1$ geometrically distinct solutions. To do this, we reduce problem (136) to an equivalent *Hamiltonian* system, and apply an abstract result of Szulkin [112] to this system. The advantage of the Hamiltonian formulation with respect to the Lagrangian one used in [32, 33] and presented in Sect. 3 is that the Hamiltonian action functional is now defined on the whole space, so that the Hamiltonian system is trivially its Euler–Lagrange equation, and many standard techniques of critical point theory can be directly applied. The price to pay in the Hamiltonian formalism is that the Hamiltonian action functional is now indefinite, excluding the obtention of existence results by minimization and of multiplicity results through classical Lusternik-Schnirelman category. Although its final result is stated in terms of the classical cuplength of a finite-dimensional manifold, the underlying technique in Szulkin's paper [112] (see also variants in [49, 62]) is a more sophisticated concept of relative category inspired by [47, 48, 103].

4.2   An Equivalent Hamiltonian System and Its Action

4.2.1   Equivalent Hamiltonian System

We introduce the change of variables

$$\nabla\Phi(q') = p$$

which is equivalent to

$$q' = \nabla\Phi^*(p),$$

to transform the problem (136) is the equivalent one

$$q' = \nabla\Phi^*(p), \quad p' = \nabla_q F(t,q) + h(t), \quad q(0) = q(T), \quad p(0) = p(T).$$

$$(138)$$

With the Hamiltonian function $H : [0, T] \times \mathbb{R}^n \times \mathbb{R}^n \to \mathbb{R}$ defined by

$$H(t, p, q) = \Phi^*(p) - F(t,q) - \langle h(t), q \rangle,$$

problem (138) takes the Hamiltonian form

$$p' = -\nabla_q H(t, p, q), \quad q' = \nabla_p H(t, p, q), \quad q(0) = q(T), \quad p(0) = p(T),$$

or, in a more concise way, letting $z = (p, q)$ and introducing the $2n \times 2n$ symplectic matrix

$$J = \begin{pmatrix} 0 & I \\ -I & 0 \end{pmatrix},$$

$$Jz' = \nabla_z H(t,z), \quad z(0) = z(T). \tag{139}$$

We use the same notations $\langle \cdot, \cdot \rangle$ and $|\cdot|$ for the inner product and the corresponding norm in $\mathbb{R}^n$ and in $\mathbb{R}^{2n}$. It is well known that, formally, system (139) is the Euler–Lagrange equation associated to the (action) functional $\mathscr{A}$ defined on a suitable space of T-periodic functions by

$$\mathscr{A}(z) = \int_0^T \left[-\frac{1}{2}\langle Jz'(t), z(t) \rangle + H(t, z(t)) \right] dt,$$

or, in terms of the (p,q) variables and original data, after integrating by parts and using the periodicity, by

$$\mathscr{A}(p,q) = \int_0^T \left[-\langle p(t), q'(t) \rangle + \Phi^*(p(t)) - F(t, q(t)) - \langle h(t), q(t) \rangle \right] dt.$$

4.2.2 The Hamiltonian Action Functional

Define (see e.g. [98]) the space $H_\#^{1/2} := H_\#^{1/2}(0, T; \mathbb{R}^{2n})$ as the space of functions $z \in L^2(0, T; \mathbb{R}^{2n})$ with Fourier series $z(t) = \sum_{k \in \mathbb{Z}} e^{k\omega t J} z_k$ ($\omega = \frac{2\pi}{T}$), such that $z_k \in \mathbb{R}^{2n}$ ($k \in \mathbb{Z}$) and

$$|z|_{1/2}^2 := \sum_{k \in \mathbb{Z}} (1 + |k|)|z_k|^2 < +\infty.$$

With the corresponding inner product

$$(z|w) := \sum_{k \in \mathbb{Z}} (1 + |k|)\langle z_k, w_k \rangle,$$

$H_\#^{1/2}$ is a Hilbert space such that $H_\#^1(0, T; \mathbb{R}^{2n}) \subset H_\#^{1/2} \subset L^s(0, T; \mathbb{R}^{2n})$ for any $s \in [1, +\infty)$. We have also, by easy computations based on Fourier series and use of Cauchy-Schwarz inequality, for z smooth,

$$\left| \int_0^T [-\langle Jz'(t), w(t) \rangle] dt \right| \leq C|z|_{1/2}|w|_{1/2},$$

so that the bilinear form defined in the left-hand member can be extended to $H_\#^{1/2}$ as a continuous quadratic form $B(z, w)$, and the linear self-adjoint operator $A : H_\#^{1/2} \to H_\#^{1/2}$ defined through Riesz's representation theorem by the relation

$$(Az|w) = B(z, w) \quad (z, w \in H_\#^{1/2}) \tag{140}$$

is continuous. In terms of Fourier series,

$$(Az|w) = 2\pi \sum_{k \in \mathbb{Z}} k \langle z_k, w_k \rangle$$

and hence

$$(Az|z) = 2\pi \sum_{k \in \mathbb{Z}} k |z_k|^2. \qquad (141)$$

It is easily seen that the spectrum of A is made of the eigenvalues $\lambda_k = 2\pi \frac{k}{1+|k|}$ ($k \in \mathbb{Z}$), each of multiplicity $2n$, and of the elements -2π, 2π in the essential spectrum. Therefore, if we let, with $E(\lambda_k)$ the eigenspace associated to λ_k,

$$H^0 = \ker A = E(\lambda_0) \simeq \mathbb{R}^{2n},$$

$$H^- = \overline{\{\cup_{k \leq -1} E(\lambda_k)\}}, \qquad H^+ = \overline{\{\cup_{k \geq 1} E(\lambda_k)\}},$$

then $H_{\#}^{1/2} = H^- \oplus H^0 \oplus H^+$ (orthogonal sum with respect to $(\cdot|\cdot)$ and to L^2), and, using (141), we have, for $z^- \in H^-$, $z^+ \in H^+$,

$$(Az^-|z^-) = 2\pi \sum_{k \leq -1} \frac{k}{1 + |k|}(1 + |k|)|z_k|^2$$

$$\leq -\pi \sum_{k \leq -1} (1 + |k|)|z_k|^2 = -\pi |z^-|_{1/2}^2,$$

$$(Az^+|z^+) = 2\pi \sum_{k \geq 1} \frac{k}{1 + |k|}(1 + |k|)|z_k|^2$$

$$\geq \pi \sum_{k \geq 1}(1 + |k|)|z_k|^2 = \pi |z^+|_{1/2}^2.$$

Furthermore the subspaces H^- and H^+ are invariant for A.

Finally, using estimate (6), it is well known [98] that the assumptions (HΦ) and (HF) imply that $\mathscr{A}$ is of class C^1 on $H_{\#}^{1/2}$ and that any critical point $(\widehat{p}, \widehat{q})$ of the functional

$$\mathscr{A}(p,q) = -\frac{1}{2}(A(p,q)|(p,q)) + \int_0^T [\Phi^*(p(t)) - F(t,q(t)) - \langle h(t), q(t) \rangle] \, dt$$

satisfies the Euler equation

$$(A(\widehat{p},\widehat{q})|(p,q)) + \int_0^T [\langle \nabla \Phi^*(\widehat{p}(t)), p(t) \rangle - \langle \nabla_q F(t,\widehat{q}(t)) - h(t), q(t) \rangle] \, dt = 0,$$

for all $(p,q) \in H_{\#}^{1/2}$. A classical reasoning shows then that $(\widehat{p}, \widehat{q})$ is a (Carathéodory) solution of (139) (see e.g. [98]).

4.3 Multiplicity of Periodic Solutions

4.3.1 Szulkin's Theorem

If X is a closed smooth manifold of dimension n, let $\Lambda^k(X)$ denote the vector space of all smooth differential k-forms on X. Taking for coboundary operator the exterior differential $d : \Lambda^k(X) \to \Lambda^{k+1}(X)$, one can define De Rham cohomology $H^*(X)$ through the vector spaces

$$Z^k(X) = \{\omega \in \Lambda^k(X) : d\omega = 0\}, \quad B^k(X) = dC^{k-1}(X),$$

by $H^k(X) = Z^k(X)/B^k(X)$. If $\omega_1 \in \Lambda^{k_1}(X)$, $\omega_2 \in \Lambda^{k_2}(X)$, then $\omega_1 \wedge \omega_2 \in \Lambda^{k_1+k_2}(X)$, and the easily checked fact that the exterior differential of the wedge product of two cocycles is a cocycle and the wedge product of a cocycle and a coboundary is a coboundary implies that the exterior product induces on $H^*(X)$ a product $\cup$, the cup product, operating as follows

$$\cup : H^{k_1}(X) \times H^{k_2}(X) \mapsto H^{k_1+k_2}(X).$$

Then the *cuplength* of X is the greatest number of elements of non-zero degree in $H^*(X)$ with non vanishing cup product, namely the largest integer m for which there exists $\alpha_j \in H^{k_j}(X)$, $1 \leq j \leq m$, such that $k_1, \ldots, k_m \geq 1$ and $\alpha_1 \cup \ldots \cup \alpha_m \neq 0$ in $H^{k_1+\ldots+k_m}(X)$. For the n-dimensional torus $\mathbb{T}^n$, cuplength($\mathbb{T}^n$) $= n$.

Let E be a real Hilbert space with inner product $(\cdot|\cdot)$ and norm $\|\cdot\|$, and V^d a compact d-dimensional C^2-manifold without boundary. Let $L : E \to E$ be a bounded linear self-adjoint operator to which there corresponds an orthogonal decomposition $E = E^- \oplus E^0 \oplus E^+$ into invariant subspaces, with $E^0 = \ker L$, and a number $\varepsilon > 0$ such that

$$\langle Lx^+, x^+ \rangle \geq \varepsilon \|x^+\|^2 \quad (x^+ \in E^+), \quad \langle Lx^-, x^- \rangle \leq -\varepsilon \|x^-\|^2 \quad (x^- \in E^-).$$

The following result is due to Szulkin [112]

Lemma 22. *Let $\Psi \in C^1(E \times V^d, \mathbb{R})$ be given by $\Psi(x,v) = \frac{1}{2}(Lx|x) - \psi(x,v)$, where ψ' is compact. Suppose that $\psi'(E \times V^d)$ is a bounded set, E^0 is finite dimensional and, if $\dim E^0 > 0$, $\psi(x^0,v) \to -\infty$ (or $\psi(x^0,v) \to +\infty$) as $\|x^0\| \to \infty$, $x^0 \in E^0$. Then Φ has at least cuplength $(V^d) + 1$ critical points.*

4.3.2 Existence of Multiple Periodic Solutions

Lemma 22 applied to a suitable reformulation of $\mathscr{A}$ will give our multiplicity theorem.

Theorem 30. *If $\Phi : \overline{B}(a) \to \mathbb{R}$ satisfies Assumption (HΦ) and $F : [0, T] \times \mathbb{R}^n \to \mathbb{R}$ satisfies Assumption (HF), then, for every $h \in L^s$ ($s > 1$) verifying assumption (Hh), problem (136) has at least $n + 1$ geometrically distinct solutions.*

Proof. Assumptions (HF) and (Hh) imply that, for any $(j_1, \ldots, j_n) \in \mathbb{Z}^n$,

$$\mathscr{A}(p, q_1 + \omega_1, \ldots, q_n + \omega_n)$$

$$= \frac{1}{2}(A(p,q)|(p,q)) + \int_0^T [\Phi^*(p(t)) - F(t, q_1(t) + \omega_1, \ldots, q_n(t) + \omega_n)$$

$$- \langle h(t), q(t) \rangle - \sum_{k=1}^n h_k(t)\omega_k] \, dt$$

$$= \frac{1}{2}(A(p,q)|(p,q)) + \int_0^T [\Phi^*(p(t)) - F(t, q(t)) - \langle h(t), q(t) \rangle] \, dt = \mathscr{A}(p, q).$$

To each critical point $(\widehat{p}, \widehat{q})$ of $\mathscr{A}$ on $H_\#^{1/2}$, corresponds the orbit

$$(\widehat{p}, \widehat{q}_1 + j_1\omega_1, \ldots, \widehat{q}_n + j_n\omega_n) \quad ((j_1, \ldots, j_n) \in \mathbb{Z}^n)$$

of critical points, which can be considered as a single critical point lying on the manifold $E \times V^n$, with V^n the n-torus $\mathbb{T}^n = \mathbb{R}^n/(\omega_1\mathbb{Z}, \ldots, \omega_n\mathbb{Z})$, and

$$E = \{(p, q) \in H_\#^{1/2} : \overline{q} = 0\}.$$

Denoting by $L : E \to E$ the restriction to E of A given in (140), we have $E = H^- \oplus E^0 \oplus H^+$, where $E^0 \simeq \mathbb{R}^n = \{(p, 0) \in \mathbb{R}^{2n} : p \in \mathbb{R}^n\} = \ker L$. Hence, $\mathscr{A}$ has the equivalent expression

$$\frac{1}{2}(L(p, \widetilde{q})|(p, \widetilde{q}) + \int_0^T [\Phi^*(p(t)) - F(t, \overline{q} + \widetilde{q}(t)) - \langle h(t), \widetilde{q}(t) \rangle] \, dt,$$

namely

$$\Psi(x, v) = \frac{1}{2}\langle L(p, \widetilde{q}), (p, \widetilde{q}) \rangle - \psi(p, \widetilde{q}; \overline{q})$$

requested by Szulkin's lemma with $x = (p, \widetilde{q})$, $v = \overline{q}$, considered as an element of V^n, and

$$\psi(p, \widetilde{q}; \overline{q}) := \int_0^T [F(t, \overline{q} + \widetilde{q}(t)) + \langle h(t), \widetilde{q}(t) \rangle - \Phi^*(p(t))] \, dt.$$

Therefore, for any $v, \widetilde{w}, \overline{w}$, we have

$$(\psi'(p, \widetilde{q}; \overline{q})|(v, \widetilde{w}; \overline{w}))$$

$$= \int_0^T [\langle \nabla_q F(t, \overline{q} + \widetilde{q}(t)), \overline{w} + \widetilde{w} \rangle + \langle h(t), \widetilde{w}(t) \rangle - \langle \phi^{-1}(p(t)), v(t) \rangle] \, dt.$$

Because $\nabla_q F(t, \cdot)$ and ϕ^{-1} have a bounded range, ψ' has a bounded range, and ψ' is compact using the compact embedding of $H_\#^{1/2}$ in L^s for any $s \geq 1$. On the other hand, because of (6) and the fact that any $(p, \widetilde{q}) \in E^0$ has the form $(p^0, 0)$ with $p^0 \in \mathbb{R}^n$, we have, for $|p^0| \to \infty$,

$$\psi(p^0, \overline{q}) = \int_0^T [-F(t, \overline{q}) - \Phi^*(p^0)]\, dt = -T[\overline{F(\cdot, \overline{q})} + T\Phi^*(p^0)] \to -\infty.$$

All the assumptions of Lemma 22 are satisfied, and Ψ has at least cuplength $(\mathbb{T}^n) + 1 = n + 1$ critical points, i.e. $\mathscr{A}$ has at least $n + 1$ geometrically distinct critical points.   □

Example 21. For any $h \in L^s$ ($s > 1$) such that $\overline{h} = 0$ and S defined in (99), the problem

$$\left(\frac{q'}{\sqrt{1 - |q'|^2}}\right)' + S(q) = h(t), \quad q(0) = q(T), \quad q'(0) = q'(T)$$

has at least $n + 1$ geometrically distinct solutions.

In particular, for any $\mu \in \mathbb{R}$ and $h \in L^s$ ($s > 1$) such that $\overline{h} = 0$, the forced relativistic pendulum problem

$$\left(\frac{q'}{\sqrt{1 - q'^2}}\right)' + \mu \sin q = h(t), \quad q(0) = q(T), \quad q'(0) = q'(T)$$

has at least 2 geometrically distinct solutions.

Remark 15. A similar Hamiltonian approach has been recently used by Manásevich and Ward [71] to give an alternative proof to the result of Brezis and the author on the relativistic forced pendulum [32]. The existence of the corresponding critical point for the associated Hamiltonian action is obtained using Rabinowitz' saddle point theorem [98].

References

1. S. Ahmad, A.C. Lazer, J.L. Paul, Elementary critical point theory and perturbations of elliptic boundary value problems. Indiana Univ. Math. J. **25**, 933–944 (1976)
2. H. Amann, A. Ambrosetti, G. Mancini, Elliptic equations with noninvertible Fredholm linear part and bounded nonlinearities. Math. Z. **158**, 179–194 (1978)
3. A. Ambrosetti, G. Prodi, On the inversion of some differentiable mappings with singularities between Banach spaces. Ann. Mat. Pura Appl. **93**, 231–247 (1973)
4. A. Ambrosetti, P. Rabinowitz, Dual variational methods in critical point theory and applications. J. Funct. Anal. **14**, 349–381 (1973)
5. G. Anzellotti, The Euler equation for functionals with linear growth. Trans. Am. Math. Soc. **290**, 483–501 (1985)
6. J. Benedikt, P. Girg, P. Takáč, On the Fredholm alternative for the p-Laplacian at higher eigenvalues (in one dimension). Nonlinear Anal. **72**, 3091–3107 (2010)

7. P. Benevieri, J.M. do Ó, E. Souto de Medeiros, Periodic solutions for nonlinear systems with mean curvature-like operators. Nonlinear Anal. **65**, 1462–1475 (2006)
8. P. Benevieri, J.M. do Ó, E. Souto de Medeiros, Periodic solutions for nonlinear equations with mean curvature-like operators. Appl. Math. Lett. **20**, 484–492 (2007)
9. C. Bereanu, J. Mawhin, Nonlinear Neumann boundary value problems with ϕ-Laplacian operators. An. Stiint. Univ. Ovidius Constanta **12**, 73–92 (2004)
10. C. Bereanu, J. Mawhin, Boundary-value problems with non-surjective ϕ-laplacian and one-sided bounded nonlinearity. Adv. Differ. Equat. **11**, 35-60 (2006)
11. C. Bereanu, J. Mawhin, Existence and multiplicity results for some nonlinear problems with singular ϕ-laplacian. J. Differ. Equat. **243**, 536–557 (2007)
12. C. Bereanu, J. Mawhin, Boundary value problems for some nonlinear systems with singular ϕ-Laplacian. J. Fixed Point Theor. Appl. **4**, 57–75 (2008)
13. C. Bereanu, J. Mawhin, Multiple periodic solutions of ordinary differential equations with bounded nonlinearities and ϕ-Laplacian. NoDEA Nonlinear Differ. Equat. Appl. **15**, 159–168 (2008)
14. C. Bereanu, J. Mawhin, Periodic solutions of nonlinear perturbations of ϕ-Laplacian with possibly bounded ϕ. Nonlinear Anal. **68**, 1668–1681 (2008)
15. C. Bereanu, J. Mawhin, Nonhomogeneous boundary value problems for some nonlinear equations with singular ϕ-Laplacian. J. Math. Anal. Appl. **352**, 218–233 (2009)
16. C. Bereanu, P. Torres, Existence of at least two periodic solutions of the forced relativistic pendulum. Proc. Am. Math. Soc. **140**, 2713–2719 (2012)
17. C. Bereanu, P. Jebelean, J. Mawhin, Non-homogeneous boundary value problems for ordinary and partial differential equations involving singular ϕ-Laplacians. Matemática Contemporânea **36**, 51–65 (2009)
18. C. Bereanu, P. Jebelean, J. Mawhin, Radial solutions for some nonlinear problems involving mean curvature operators in Euclidian and Minkowski spaces. Proc. Am. Math. Soc. **137**, 161–169 (2009)
19. C. Bereanu, P. Jebelean, J. Mawhin, Radial solutions for systems involving mean curvature operators in Euclidian and Minkowski spaces, in *Mathematical Models in Engineering*, ed. by A. Cabada, E. Liz, J.J. Nieto. Biology and Medicine (American Institute of Physics, New York, 2009), pp. 50–59
20. C. Bereanu, P. Jebelean, J. Mawhin, Periodic solutions of pendulum-like perturbations of singular and bounded ϕ-Laplacians. J. Dynam. Differ. Equat. **22**, 463–471 (2010)
21. C. Bereanu, P. Jebelean, J. Mawhin, Radial solutions for Neumann problems involving mean curvature operators in Euclidean and Minkowski spaces. Math. Nachr. **283**, 379–391 (2010)
22. C. Bereanu, P. Jebelean, J. Mawhin, Radial solutions for Neumann problems with ϕ-Laplacians and pendulum-like nonlinearities. Discrete Contin. Dynam. Syst. A **28**, 637–648 (2010)
23. C. Bereanu, P. Jebelean, J. Mawhin, Variational methods for nonlinear perturbations of singular ϕ-Laplacians. Rend. Lincei Mat. Appl. **22**, 89–111 (2011)
24. C. Bereanu, P. Jebelean, J. Mawhin, Radial solutions of Neumann problems involving mean extrinsic curvature and periodic nonlinearities. Calculus of Variations and Partial Differential Equations (to appear)
25. C. Bereanu, P. Jebelean, J. Mawhin, Multiple solutions for Neumann and periodic problems with singular ϕ-Laplacian. J. Funct. Anal. **261**, 3226–3246 (2011)
26. M.S. Berger, *Nonlinearity and Functional Analysis* (Academic, New York, 1977)
27. P.A. Binding, P. Drábek, Y.X. Huang, The range of the p-Laplacian. Appl. Math. Lett. **10**, 77–82 (1997)
28. M.S. Berger, M. Schechter, On the solvability of semi-linear gradient operator equations. Adv. Math. **25**, 97–132 (1977)
29. P.A. Binding, P. Drábek, Y.X. Huang, On the Fredholm alternative for the p-Laplacian. Proc. Am. Math. Soc. **125**, 3555–3559 (1997)

30. D. Bonheure, P. Habets, F. Obersnel, P. Omari, Classical and non-classical solutions of a prescribed curvature equation. J. Differ. Equat. **243**, 208–237 (2007)
31. D. Bonheure, P. Habets, F. Obersnel,P. Omari, Classical and non-classical positive solutions of a prescribed curvature equation with singularities. Rend. Istit. Mat. Univ. Trieste **39**, 63–85 (2007)
32. H. Brezis, J. Mawhin, Periodic solutions of the forced relativistic pendulum. Differ. Integr. Equat. **23**, 801–810 (2010)
33. H. Brezis, J. Mawhin, Periodic solutions of Lagrangian systems of relativistic oscillators. Comm. Appl. Anal. **15**, 235–250 (2011)
34. K.C. Chang, On the periodic nonlinearity and the multiplicity of solutions. Nonlinear Anal. **13**, 527–537 (1989)
35. J. Chu, J. Lei, M. Zhang, The stability of the equilibrium of a nonlinear planar system and application to the relativistic oscillator. J. Differ. Equat. **247**, 530–542 (2009)
36. J.A. Cid, P.J. Torres, Solvability of some boundary value problems with ϕ-Laplacian operators. Discrete Contin. Dynam. Syst. A **23**, 727–732 (2009)
37. C.V. Coffman, W.K. Ziemer, A prescribed mean curvature problem on domains without radial symmetry. SIAM J. Math. Anal. **22**, 982–990 (1991)
38. C. De Coster, P. Habets, *Two-Point Boundary Value Problems. Lower and Upper Solutions* (Elsevier, Amsterdam, 2006)
39. K. Deimling, *Nonlinear Functional Analysis* (Springer, Berlin, 1985)
40. M. del Pino, P. Drábek, R. Manásevich, The Fredholm alternative at the first eigenvalue for the one-dimensional p-Laplacian. J. Differ. Equat. **151**, 386–419 (1999)
41. P. De Nápoli, M.C. Mariani, Three solutions for quasilinear equations in $\mathbb{R}^N$, in *Proceedings USA-Chile Workshop on Nonlinear Analysis* (Southwest Texas State University, Texas, 2001), pp. 131–140
42. F.O. de Paiva, E. Massa, Semilinear elliptic problems near resonance with a nonprincipal eigenvalue. J. Math. Anal. Appl. **342**, 638–650 (2008)
43. P. Drábek, P. Takáč, A counterexample to the Fredholm alternative for the p-Laplacian. Proc. Am. Math. Soc. **127**, 1079–1087 (1999)
44. P. Drábek, P. Girg, R. Manásevich, Generic Fredholm alternative-type results for the one dimensional p-Laplacian. NoDEA Nonlinear Differ. Equat. Appl. **8**, 285–298 (2001)
45. C. Fabry, J. Mawhin, M. Nkashama, A multiplicity result for periodic solutions of forced nonlinear second order ordinary differential equations. Bull. Lond. Math. Soc. **18**, 173–180 (1986)
46. L. Ferracuti, F. Papalini, Boundary value problems for strongly nonlinear multivalued equations involving differenti ϕ-Laplacians. Adv. Differ. Equat. **14**, 541–566 (2009)
47. G. Fournier, M. Willem, Multiple solutions of the forced double pendulum equation. Ann. Inst. Henri-Poincaré. Anal. Non Linéaire **5** (suppl.), 259–281 (1989)
48. G. Fournier, M. Willem, Relative category and the calculus of variations, in *Variational Problems*, ed. by H. Berestycki, J.M. Coron, I. Ekeland (Birkhäuser, Basel, 1990), pp. 95–104
49. G. Fournier, D. Lupo, M. Ramos, M. Willem, Limit relative category and critical point theory, in *Dynamics Reported*, vol. 3 (Springer, Berlin, 1994), pp. 1–24
50. M. Furi, M.P. Pera, On the existence of an unbounded connected set of solutions for nonlinear equations in Banach spaces. Atti Accad. Naz. Lincei, Rend. Cl. Sci. Mat. Fis. Natur. **47**, 31–38 (1979)
51. M. García-Huidobro, R. Manásevich, F. Zanolin, Strongly nonlinear second-order ODE's with unilateral conditions. Differ. Integr. Equat. **6**, 1057–1078 (1993)
52. M. García-Huidobro, R. Manásevich, F. Zanolin, A Fredholm-like result for strongly nonlinear second order ODE's. J. Differ. Equat. **114**, 132–167 (1994)
53. P. Girg, Neumann and periodic boundary-value problems for quasilinear ordinary differential equations with a nonlinearity in the derivatives. Electronic J. Differ. Equat. **2000-63**, 1–28 (2000)

182 J. Mawhin

54. P. Habets, P. Omari, Positive solutions of an indefinite prescribed mean curvature problem on a general domain. Adv. Nonlinear Stud. **4**, 1–14 (2004)
55. P. Habets, P. Omari, Multiple positive solutions of a one-dimensional prescribed mean curvature problem. Comm. Contemp. Math. **9** 701–730 (2007)
56. P. Jebelean, J. Mawhin, Periodic solutions of singular nonlinear perturbations of the ordinary p-Laplacian. Adv. Nonlinear Stud. **2**, 299–312 (2002)
57. X.F. Ke, C.L. Tang, Multiple solutions for semilinear elliptic equations near resonance at higher eigenvalues. Nonlinear Anal. **74**, 805–813 (2011)
58. T. Kusahara, H. Usami, A barrier method for quasilinear ordinary differential equations of the curvature type. Czech. Math. J. **50**, 185–196 (2000)
59. A. C. Lazer, S. Solimini, On periodic solutions of nonlinear differential equations with singularities. Proc. Am. Math. Soc. **99**, 109–114 (1987)
60. J. Leray, J. Schauder, Topologie et équations fonctionnelles. Ann. Ec. Norm. Sup. **51**, 45–78 (1934)
61. W.S. Li, Z.L. Liu, Exact number of solutions of a prescribed mean curvature equation. J. Math. Anal. Appl. **367**, 486–498 (2010)
62. J.Q. Liu, A generalized saddle point theorem. J. Differ. Equat. **82**, 372–385 (1989)
63. H.S. Lü, On the existence of multiple periodic solutions for the p-Laplacian. Indian J. Pure Appl. Math. **35**, 1185–1199 (2004)
64. J. Ma, C.L. Tang, Periodic solutions for some nonautonomous second-order systems. J. Math. Anal. Appl. **275**, 482–494 (2002)
65. T.F. Ma, M.L. Pelicer, Perturbations near resonance for the p-Laplacian in $\mathbb{R}^N$. Abstract Appl. Anal. **7:6**, 323–334 (2002)
66. T.F. Ma, L. Sanchez, Three solutions of a quasilinear elliptic problem near resonance. Math. Slovaca **47**, 451–457 (1997)
67. T.F. Ma, M. Ramos, L. Sanchez, Multiple solutions for a class of nonlinear boundary value problems near resonance, a variational approach. Nonlinear Anal. **30**, 3301–3311 (1997)
68. R. Manásevich, J. Mawhin, Periodic solutions for nonlinear systems with p-Laplacian-like operators. J. Differ. Equat. **145**, 367–393 (1998)
69. R. Manásevich, J. Mawhin, Boundary value problems for nonlinear perturbations of vector p-Laplacian-like operators. J. Korean Math. Soc. **37**, 665–685 (2000)
70. R. Manásevich, P. Takáč, On the Fredholm alternative for the p-Laplacian in one dimension. Proc. Lond. Math. Soc. (3) **84**, 324–342 (2002)
71. R. Manásevich, J.R. Ward Jr., On a result of Brezis and Mawhin. Proc. Am. Math. Soc. **140**, 531–539 (2012)
72. I. Massabó, J. Pejsachowicz, On the connectivity properties of the solution set of parametrized families of compact vector fields. J. Funct. Anal. **59**, 151–166 (1984)
73. J. Mawhin, in *Topological Degree Methods in Nonlinear Boundary Value Problems.* CBMS Series No. 40 (American Mathematical Society, Providence, 1979)
74. J. Mawhin, Forced second order conservative systems with periodic nonlinearity. Ann. Inst. Henri-Poincaré Anal. Non Linéaire **5** (suppl.), 415–434 (1989)
75. J. Mawhin, Topological degree and boundary value problems for nonlinear differential equations, in *Topological Methods in Ordinary Differential Equations*, CIME, Montecatini Terme, 1991, ed. by M. Furi, P. Zecca. LNM 1537 (Springer, Berlin, 1993), pp. 74–142
76. J. Mawhin, Leray-Schauder degree: A half century of extensions and applications. Topological Meth. Nonlinear Anal. **14**, 195–228 (1999)
77. J. Mawhin, Global results for the forced pendulum equations, in *Handbook on Differential Equations*, ed. by A. Cañada, P. Drábek, A. Fonda. Ordinary Differential Equations, vol. 1 (Elsevier, Amsterdam, 2004), pp. 533–589
78. J. Mawhin, The periodic Ambrosetti-Prodi problem for nonlinear perturbations of the p-Laplacian. J. Eur. Math. Soc. **8**, 375–388 (2006)
79. J. Mawhin, Periodic solutions of the forced pendulum: classical vs relativistic. Le Matematiche **65**, 97–107 (2010)
80. J. Mawhin, Multiplicity of solutions of variational systems involving ϕ-Laplacians with singular ϕ and periodic nonlinearities. Discrete Contin. Dynam. Syst. **32**, 4015–4026 (2012)

81. J. Mawhin, K. Schmitt, Landesman-Lazer type problems at an eigenvalue of odd multiplicity. Results Math. **14**, 138–146 (1988)
82. J. Mawhin, K. Schmitt, Nonlinear eigenvalue problems with the parameter near resonance. Ann. Polon. Math. **60**, 241–248 (1990)
83. J. Mawhin, M. Willem, Multiple solutions of the periodic boundary value problem for some forced pendulum-type equations. J. Differ. Equat. **52**, 264–287 (1984)
84. J. Mawhin, M. Willem, Variational methods and boundary value problems for vector second order differential equations and applications to the pendulum equation, in *Nonlinear Analysis and Optimisation*, ed. by C. Vinti (Bologna, 1982). LNM 1107 (Springer, Berlin, 1984), pp. 181–192
85. J. Mawhin, M. Willem, *Critical Point Theory and Hamiltonian Systems* (Springer, New York, 1989)
86. D. Motreanu, V.V. Motreanu, N.S. Papageorgiou, Positive solutions and multiple solutions at non-resonance, resonance and near resonance for hemivariational inequalities with p-Laplacian. Trans. Am. Math. Soc. **360**, 2527–2545 (2008)
87. D. Motreanu, V.V. Motreanu, N.S. Papageorgiou, Nonlinear Neumann problems near resonance. Indiana Univ. Math. J. **58**, 1257–1279 (2009)
88. D.G. Northcott, Some inequalities between periodic functions and their derivatives. J. Lond. Math. Soc. **14**, 198–202 (1939)
89. E.S. Noussair, Ch.A. Swanson, J.F. Yang, A barrier method for mean curvature problems. Nonlinear Anal. **21**, 631–641 (1993)
90. F. Obersnel, Classical and non-classical sign changing solutions of a one-dimensional autonomous prescribed curvature equation. Adv. Nonlinear Stud. **7**, 1–13 (2007)
91. F. Obersnel, P. Omari, Existence and multiplicity results for the prescribed mean curvature equation via lower and upper solutions. Differ. Integr. Equat. **22**, 853–880 (2009)
92. F. Obersnel, P. Omari, On a result of C.V. Coffman and W.K. Ziemer about the prescribed mean curvature equation. Quad. Mat. Univ. Trieste **593**, 1–10 (2009)
93. F. Obersnel, P.P. Omari, Multiple bounded variation solutions of a periodically perturbed sine-curvature equation. Comm. Contemp. Math. **13**, 863–883 (2011)
94. Z.Q. Ou, C.M. Tang, Existence and multiplicity results for some elliptic systems at resonance. Nonlinear Anal. **71**, 2660–2666 (2009)
95. R.S. Palais, Ljusternik Schnirelmann theory on Banach manifolds. Topology **5**, 115–132 (1966)
96. H.J. Pan, One-dimensional prescribed mean curvature equation with exponential nonlinearity. Nonlinear Anal. **70**, 999–1010 (2009)
97. H.J. Pan, R.X. Xing, Time maps and exact multiplicity results for one-dimensional prescribed mean curvature equations. Nonlinear Anal. **74**, 1234–1260 (2011)
98. P. Rabinowitz, in *Minimax Methods in Critical Point Theory with Applications to Differential Equations*. CBMS Regional Conf. No. 65 (American Mathematical Society, Providence, 1986)
99. P. Rabinowitz, On a class of functionals invariant under a Z_n action. Trans. Am. Math. Soc. **310**, 303–311 (1988)
100. I. Rachunková, M. Tvrdy, Periodic problems with ϕ-Laplacian involving non-ordered lower and upper solutions. Fixed Point Theor. **6**, 99–112 (2005)
101. I. Rachunková, M. Tvrdý, Periodic singular problems with quasilinear differential operator. Math. Bohemica **131**, 321–336 (2006)
102. M. Ramos, L. Sanchez, A variational approach to multiplicity in elliptic problems near resonance. Proc. Roy. Soc. Edinb. Sect. A **127**, 385–394 (1997)
103. M. Reeken, Stability of critical points under small perturbations. Part I : Topological theory. Manuscripta Math. **7**, 387–411 (1972)
104. L. Sanchez, Boundary value problems for some fourth order ordinary differential equations. Applicable Anal. **38**, 161–177 (1990)
105. J. Schauder, Der Fixpunktsatz in Funktionalräumen. Studia Math. **2**, 171–180 (1930)

106. M. Schechter, Periodic non-autonomous second-order dynamical systems. J. Differ. Equat. **223**, 290–302 (2006)
107. J.T. Schwartz, *Nonlinear Functional Analysis* (Gordon and Breach, New York, 1969)
108. J. Serrin, Positive solutions of a prescribed mean curvature problem, in *Calculus of Variations and Partial Differential Equations*, ed. by S. Hildebrandt, D. Kinderlehrer, M. Miranda. LNM 1340 (Springer, Berlin, 1988), pp. 248–255
109. M. Struwe, in *Variational Methods*. Applications to Nonlinear Partial Differential Equations and Hamiltonian Systems, 2nd edn. (Springer, Berlin, 1996)
110. H.M. Suo, C.L. Tang, Multiplicity results for some elliptic systems near resonance with a nonprincipal eigenvalue. Nonlinear Anal. **73**, 1909–1920 (2010)
111. A. Szulkin, Minimax principles for lower semicontinuous functions and applications to nonlinear boundary value problems. Ann. Inst. H. Poincaré Anal. Non Linéaire **3**, 77–109 (1986)
112. A. Szulkin, A relative category and applications to critical point theory for strongly indefinite functionals. Nonlinear Anal. **15**, 725–739 (1990)
113. C.L. Tang, Periodic solutions of non-autonomous second order systems with γ-quasisubadditive potential. J. Math. Anal. Appl. **189**, 671–675 (1995)
114. C.L. Tang, Periodic solutions of non-autonomous second order systems. J. Math. Anal. Appl. **202**, 465–469 (1996)
115. C.L. Tang, Some existence results for periodic solutions of non-autonomous second order systems. Acad. Roy. Belg. Bull. Cl. Sci. (6) **8**, 13–19 (1997)
116. C.L. Tang, Periodic solutions for nonautonomous second order systems with sublinear nonlinearity. Proc. Am. Math. Soc. **126**, 3263–3270 (1998)
117. C.L. Tang, X.P. Wu, Periodic solutions for second order systems with not uniformly coercive potential. J. Math. Anal. Appl. **259**, 386–397 (2001)
118. C.L. Tang, X.P. Wu, Notes on periodic solutions of subquadratic second order systems. J. Math. Anal. Appl. **285**, 8–16 (2003)
119. C.L. Tang, X.P. Wu, A note on periodic solutions of nonautonomous second-order systems. Proc. Am. Math. Soc. **132**, 1295–1303 (2004)
120. Y. Tian, G.S. Zhang, W.G. Ge, Periodic solutions for a quasilinear non-autonomous second-order system. J. Appl. Math. Comput. **22**, 263–271 (2006)
121. P.J. Torres, Periodic oscillations of the relativistic pendulum with friction. Phys. Lett. A **372**, 6386–6387 (2008)
122. P.J. Torres, Nondegeneracy of the periodically forced Liénard differential equation with ϕ-Laplacian. Comm. Contemp. Math. **13**, 283–292 (2011)
123. G. Villari, Soluzioni periodiche di una classe di equazione differenziali del terz'ordine. Ann. Mat. Pura Appl. **73**, 103–110 (2966)
124. Z.Y. Wang, J.H. Zhang, Periodic solutions of non-autonomous second order systems with p-Laplacian. Electron. J. Differ. Equat. **2009-17**, 1–12 (2009)
125. J.R. Ward Jr., Periodic solutions of ordinary differential equations with bounded nonlinearities. Topological Meth. Nonlinear Anal. **19**, 275–282 (2002)
126. X.P. Wu, Periodic solutions for nonautonomous second-order systems with bounded nonlinearity. J. Math. Anal. Appl. **230**, 135–141 (1999)
127. X.P. Wu, C.L. Tang, Periodic solutions of a class of nonautonomous second order systems. J. Math. Anal. Appl. **236**, 227–235 (1999)
128. X.J. Yang, On the Fredholm alternative for the p-Laplacian. Appl. Math. Comput. **153**, 537–556 (2004)
129. X.J. Yang, The Fredholm alternative for the one-dimensional p-Laplacian. J. Math. Anal. Appl. **299**, 494–507 (2004)
130. F.K. Zhao, X. Wu, Existence and multiplicity of periodic solution for non-autonomous second-order systems with linear nonlinearity. Nonlinear Anal. **60**, 325–335 (2005)

Non-autonomous Functional Differential Equations and Applications

Sylvia Novo and Rafael Obaya

Abstract This chapter deals with the applications of dynamical systems techniques to the study of non-autonomous, monotone and recurrent functional differential equations. After introducing the basic concepts in the theory of skew-product semiflows and the appropriate topological dynamics techniques, we study the long-term behavior of relatively compact trajectories by describing the structure of minimal and omega-limit sets, as well as the attractors. Both the cases of finite and infinite delay are considered. In particular, we show the relevance of uniform stability in this study. Special attention is also paid to the almost periodic case, in which the presence of almost periodic and almost automorphic dynamics is analyzed. Some applications of these techniques to the study of neural networks, compartmental systems and certain biochemical control circuit models are shown.

1 Introduction

In this work we study the long-term behaviour of the solutions of non-autonomous ordinary differential equations (ODEs for short) and non-autonomous functional differential equations (FDEs for short). In order to unify our theory, we frequently assume a general expression for the FDEs which contains the ODEs models as a particular case. We investigate the structure and the qualitative behaviour of the omega-limit sets of relatively compact trajectories, as well as some properties for the minimal sets they contain. This information becomes essential to understand the local or global behaviour of the solutions.

Some recurrence properties on the temporal variation of the FDEs are assumed, and therefore, their solutions induce a skew-product semiflow with a minimal flow

S. Novo (✉) · R. Obaya
Departamento de Matemática Aplicada, E.I. Industriales, Paseo del Cauce 59,
Universidad de Valladolid, 47011 Valladolid, Spain
e-mail: sylnov@wmatem.eis.uva.es; rafoba@wmatem.eis.uva.es

A. Capietto et al., *Stability and Bifurcation Theory for Non-Autonomous Differential
Equations*, Lecture Notes in Mathematics 2065, DOI 10.1007/978-3-642-32906-7_4,
© Springer-Verlag Berlin Heidelberg 2013

on the base. As it is common in the analysis of non-autonomous dynamical systems, our phase space is a positively invariant closed subset of $\Omega \times X$, where Ω is a compact metric space with a continuous flow describing the dynamics associated to the time-variation of the equation, and X is the Banach or Fréchet space representing the state space. In the periodic case Ω is just a circle, but our formulation includes more general cases as almost-periodic and almost automorphic temporal variations of the vector field, with significant interest in practical applications, in which the structure of Ω becomes more complicated. Thus we study the trajectories of a complete family of FDEs and we transfer information from one equation to the others using topological and ergodic methods. In general, a collective property is easier to recognize than those verified by a single differential equation; this explains some of the advantages of this collective formulation.

In addition, special attention is paid to the study of cooperative FDEs, whose trajectories preverse in the future the order of the initial data. This property, which implies the monotonicity of the semiflow, is essential in most of the results and has important dynamical implications.

Basically, this work is divided in three parts of increasing complexity. The first part analyzes the presence of almost periodic and almost automorphic dynamics in the skew-product semiflows induced by non-autonomous families of ODEs. In particular, we will recall that almost automorphy is a fundamental notion in the study of almost periodic differential equations.

The theory of almost periodicity, including almost periodic differential equations, has given a strong impetus to the development of the topological and smooth dynamical systems for the last century. The notion of almost automorphy, as a generalization of almost periodicity, was introduced by Bochner in 1955 in the context of differential geometry. Roughly speaking the presence of almost periodic dynamics represents a regular behaviour, while almost automorphic minimal sets can exhibit a big complexity, as sensitive dependence with respect to initial data, existence of several ergodic measures or chaos. For example, almost automorphic symbolic minimal flows may admit positive topological entropy as shown by Markley and Paul [61].

Probably motivated by its abstract origin, almost automorphic dynamics was not originally considered in the study of almost periodic differential equations. Later, its importance was motivated by the examples, given by Levitan and Zhikov [57] and Johnson [46], of scalar almost periodic linear equations with almost automorphic but not almost periodic solutions. This means that the solutions of almost periodic differential equations could exhibit more complexity than the original terms of the equation. From here it is not hard to construct an example of a scalar almost periodic differential equation with bounded solutions whose omega-limit sets contain two minimal sets. This represents again a significant difference between the autonomous and the non-autonomous dynamics.

Analogously, it is proved in Johnson [48] that the Riccati equation provided by the non-uniformly hyperbolic almost periodic linear equations constructed by Millionščikov [65] and Vinograd [115] admits almost automorphic but not almost periodic solutions. These linear systems are disconjugate and define the first point of the spectrum of the corresponding self-adjoint Schrödinger operators. This fact establishes important connections between the dynamical, spectral and ergodic theories and admits generalizations in several directions. In the same line of results the relevance of the almost automorphic dynamics in the Floquet theory of two-dimensional almost periodic linear ODEs was shown in Johnson [45].

The references by Alonso and Obaya [1] and Novo et al. [71] study the existence of almost periodic and almost automorphic solutions for scalar convex or concave almost periodic differential equations. This theory explains a saddle-node bifurcation problem, where the almost automorphic dynamics could appear. The same kind of problems was considered by Ortega and Tarallo [86] for the second order case. The presence of almost periodic and almost automorphic dynamics in the Lagrangian flow induced by a disconjugate linear Hamiltonian system is analyzed in Johnson et al. [51,52], extending the previous two-dimensional results.

Shen and Yi [97–100] have developed similar dynamical techniques for the study of parabolic partial differential equations. In particular, in the 1-space dimension case, they study the ergodic and topological structure of minimal sets, characterize the number of minimal sets of an omega-limit set, and prove that every minimal is an almost automorphic extension of the base.

The second part of the work, included in section 3, is devoted to the study of almost periodic and almost automorphic dynamics in non-autonomous FDEs with finite delay. When the delay is small the theory of Driver [21] proves the existence of special solutions which satisfy a supplementary ODE. Although the delay can stabilize or destabilize previous solutions of ODEs (see Kuang [56], Smith [106] among others), Alonso et al. [2] prove, in the scalar case, that a small delay is harmless in the sense that the minimal sets are those defined by families of ODEs.

In the case of FDEs with fixed delay we introduce general conditions which ensure the extensibility of the trajectories in the minimal sets and state different properties which imply the monotonicity or eventual strong monotonicity of the corresponding semiflow. We give a version of a theorem of Shen and Yi [101] which proves, in the almost periodic and cooperative case, that the linearly stable minimal sets are almost automorphic assuming the irreducibility of the linearized FDE. More precisely, these minimal sets define an N-copy of an almost automorphic extension of the base $(\Omega, \sigma, \mathbb{R})$.

A large number of mathematical models in applied sciences present some monotonicity properties with respect to the state argument. The rest of this chapter studies monotone semiflows induced by cooperative FDEs. The paper by Novo et al. [72] introduces a topological version of the concept of semi-equilibrium given by Arnold and Chueshov [5, 6] in the measurable case. Under natural compactness conditions a semicontinuous semi-equilibrium provides the existence of a minimal set which is an almost automorphic extension of the base; if this minimal set is uniformly stable then it is a copy of the base. The concept of strong semi-equilibrium and its connection with the upper and lower solutions of FDEs is also analyzed.

When the increasing rate of the vector field defining the cooperative FDEs decreases (or increases) as the state argument increases, the model exhibits concave (or convex) nonlinearities. There are also well-known phenomena for which only positive state arguments make sense and the dynamics can be essentially described by a sublinear vector field. Convex, concave and sublinear monotone semiflows have been extensively studied in the literature (see Hirsch [34, 35], Smith [103], Takáč [108], Krause and Ranft [55], Arnold and Chuesov [5, 6], Chuesov [18], Zhao [121], Novo and Obaya [70], Novo et al. [72–74] and the references therein). More recently Núñez et al. [80,82] describe all possible dynamical scenarios generated by monotone sublinear or concave skew-product semiflows, and they provide a more accurate description of the dynamics for two-dimensional systems [81,83].

In subsection 3.3 we analyze some of the scenarios more relevant in applications. We begin with the dynamics in the concave case, describing two different situations which allow us to prove the existence, on the positively invariant region determined by a semicontinuous sub-equilibrium, of a minimal set given by an exponentially stable copy of the base. The first one requires the vector field to be concave, the semi-equilibrium to be strong and the existence of a bounded trajectory above the graph of the semi-equilibrium. In the second situation the vector field is strongly concave and the bounded trajectory must be strongly above the graph of the sub-equilibrium. In the sublinear case, the existence of a minimal set given by a global asymptotically stable copy of the base is guaranteed by the assumption of the strong sublinearity of the vector field and the existence of a strongly positive bounded semiorbit. We finish the section with the description of relevant applications of these results to the study of non-autonomous cyclic feedback systems, as well as cellular neural networks.

Finally, section 4 deals with the study of the dynamical properties of the semiflow induced by recurrent families of FDEs with infinite delay. Before introducing this skew-product semiflow we review the standard lifting theorem, proved by Sacker and Sell [91] for almost periodic differential equations, in a more general setting: a skew-product semiflow $(\Omega \times X, \tau, \mathbb{R}^+)$ where $(\Omega, \sigma, \mathbb{R})$ is a minimal flow on a compact metric space and X is a complete metric space. We state that a uniformly stable compact and positively invariant set admitting a backward orbit for every point has a flow extension, which is fiber distal and uniformly stable when t goes to $-\infty$. In addition, if the set is uniformly asymptotically stable we show that it is an N-copy of the base flow. As a consequence, the omega-limit set of a uniformly stable trajectory is a uniformly stable minimal set admitting a fiber distal extension, and it is a uniformly asymptotically stable N-cover of the base if the trajectory is uniformly asymptotically stable.

In the case of FDEs with infinite delay, we consider the phase space $BU \subset C((-\infty, 0], \mathbb{R}^m)$ of bounded and uniformly continuous functions with the supremum norm, and we assume the continuous differentiability of the vector field with respect to the state variable; the standard theory provides existence, uniqueness and continuous dependence of the solutions (see [33]). In this setting every bounded trajectory is relatively compact for the compact-open topology. We also assume the vector field to be continuous for the metric topology on bounded sets, which implies the continuity of the semiflow for this topology. In particular, the restriction of the semiflow to the omega-limit set of a bounded trajectory is continuous, when the compact-open topology is taken in BU, and admits a flow extension. Then, we apply the above general results on the structure of compact invariant sets to contracting semiflows, i.e., semiflows generated by FDEs with infinite delay admitting a Lyapunov function. When the Lyapunov function is strict we prove the existence of a unique minimal set globally asymptotically stable. We also remark some important points of difficulty appearing in the application of this theory in our non-autonomous formulation.

Next we study the presence of almost periodic and almost automorphic dynamics for monotone skew-product semiflows induced by a cooperative family of FDEs with infinite delay. Many important results in the theory of monotone dynamical systems deduced in the last decades require strong monotonicity. However, this condition never holds when we consider FDEs with infinite delay in BU with the standard ordering, because every trajectory always remembers its complete past or history. For this reason, we describe recent results with significative dynamical meaning requiring only the monotonicity of the semiflow. We extend the theory proved in section 3 for finite delay, deducing the presence of almost automorphic dynamics from the existence of a semicontinuos semi-equilibrium which satisfies additional compactness conditions. If the base $(\Omega, \sigma, \mathbb{R})$ is almost periodic these methods ensure the existence of almost automorphic minimal sets which in some cases become copies of the base. Moreover, in the above dynamical scenario we assume that the trajectories are bounded and uniformly stable on bounded sets and prove that their omega-limit sets are copies of the base. This provides a recurrent and infinite delay version of previous results proved by Jiang and Zhao [44] for distal FDEs with finite delay. We finish the chapter with applications of this theory to the study of the long-term behaviour of the solutions of some compartmental systems with infinite delay.

2   Basic Notions and Results

2.1   Flows Over Compact Metric Spaces

Let Ω be a compact metric space. A real *continuous flow* $(\Omega, \sigma, \mathbb{R})$ is defined by a continuous mapping $\sigma : \mathbb{R} \times \Omega \to \Omega$, $(t, \omega) \mapsto \sigma(t, \omega)$ satisfying

(i) $\sigma_0 = \mathrm{Id}$;
(ii) $\sigma_{t+s} = \sigma_t \circ \sigma_s$ for each $s, t \in \mathbb{R}$,

where $\sigma_t(\omega) = \sigma(t, \omega)$ for all $\omega \in \Omega$ and $t \in \mathbb{R}$. The set $\{\sigma_t(\omega) \mid t \in \mathbb{R}\}$ is called the *orbit* or the *trajectory* of the point ω.

We say that a subset $\Omega_1 \subset \Omega$ is σ-*invariant* if $\sigma_t(\Omega_1) = \Omega_1$ for every $t \in \mathbb{R}$. A mapping $f : \Omega \to \mathbb{R}$ is σ-*invariant* if it is constant along the trajectories, i.e., $f(\sigma_t(\omega)) = f(\omega)$ for all $\omega \in \Omega$ and $t \in \mathbb{R}$. A subset $\Omega_1 \subset \Omega$ is called *minimal* if it is compact, σ-invariant and it has no other nonempty compact σ-invariant subset but itself. Every compact and σ-invariant set contains a minimal subset; in particular it is easy to prove that a compact σ-invariant subset is minimal if and only if every trajectory is dense. We say that the continuous flow $(\Omega, \sigma, \mathbb{R})$ is *recurrent* or *minimal* if Ω is minimal.

If $\omega_0 \in \Omega$, the subset $\{\sigma_t(\omega_0) \mid t \geq t_0\} \subset \Omega$ is relatively compact for each $t_0 > 0$, then its *omega-limit set* can be defined by

$$\bigcap_{s \geq t_0} \text{closure}\{\sigma(t + s, \omega_0) \mid t \geq 0\},$$

which is a compact and invariant subset. Analogously, given $\omega_0 \in \Omega$ the set $\{\sigma_t(\omega_0) : t \leq -t_0\} \subset \Omega$ is relatively compact for each $t_0 > 0$, we can consider its *alpha-limit set*, defined by

$$\bigcap_{s \leq -t_0} \text{closure}\{\sigma(t + s, \omega_0) \mid t \leq 0\}.$$

Both omega-limit set and alpha-limit set contain minimal subsets.

Let d be a metric on Ω. We say that the flow $(\Omega, \sigma, \mathbb{R})$ is *distal* when, for each pair ω_1, ω_2 of different elements of Ω, the orbits keep at a positive distance, that is, there is a $\delta > 0$ such that $\mathsf{d}(\sigma_t(\omega_1), \sigma_t(\omega_2)) > \delta$ for every $t \in \mathbb{R}$. Equivalently, $\inf\{\mathsf{d}(\sigma_t(\omega_1), \sigma_t(\omega_2)) \mid t \in \mathbb{R}\} = 0$ if and only if $\omega_1 = \omega_2$.

The flow $(\Omega, \sigma, \mathbb{R})$ is said to be *almost periodic* when for every $\varepsilon > 0$ there is a $\delta > 0$ such that, if $\omega_1, \omega_2 \in \Omega$ with $\mathsf{d}(\omega_1, \omega_2) < \delta$ then $\mathsf{d}(\sigma_t(\omega_1), \sigma_t(\omega_2)) < \varepsilon$ for every $t \in \mathbb{R}$; equivalently, the flow $(\Omega, \sigma, \mathbb{R})$ is almost periodic if the family $\{\sigma_t\}_{t \in \mathbb{R}}$ is equicontinuous. If $(\Omega, \sigma, \mathbb{R})$ is almost periodic, it is distal. The converse is not true; even if $(\Omega, \sigma, \mathbb{R})$ is minimal and distal, it does not need to be almost periodic.

We say that $\omega_1, \omega_2 \in \Omega$ form a *proximal pair* if $\inf\{\mathsf{d}(\sigma_t(\omega_1), \sigma_t(\omega_2)) \mid t \in \mathbb{R}\} = 0$, otherwise the pair is said to be *distal*. It is said that the points ω_1 and ω_2 are a *positively* (resp. *negatively*) *proximal pair* if $\inf\{\mathsf{d}(\sigma_t(\omega_1), \sigma_t(\omega_2)) \mid t \geq 0\} = 0$ (resp. $\inf\{\mathsf{d}(\sigma_t(\omega_1), \sigma_t(\omega_2)) \mid t \leq 0\} = 0$). For the basic properties on almost periodic and distal flows we refer the reader to Ellis [22] and Sacker and Sell [91].

Given another continuous flow $(Y, \Psi, \mathbb{R})$, a *flow homomorphism* from $(Y, \Psi, \mathbb{R})$ into $(\Omega, \sigma, \mathbb{R})$ is a continuous mapping $\pi : Y \to \Omega$ such that, for every $y \in Y$ and $t \in \mathbb{R}$, $\pi(\Psi(t, y)) = \sigma(t, \pi(y))$. If π is also surjective, then it is called a *flow epimorphism*; in this case, Ω is a *factor* of Y, and Y is an *extension* of Ω. If π is a flow epimorphism and there exists $k \geq 1$ such that $\text{card}(\pi^{-1}(\omega)) = k$ for all $\omega \in \Omega$, then it is said that the flow $(Y, \Psi, \mathbb{R})$ is a *k-cover* or a *k-copy* of $(\Omega, \sigma, \mathbb{R})$. If $k = 1$, then the flows are isomorphic; in particular, they have the same topological properties. In such a case, we will simply say that they are covers or copies. As for homomorphisms between distal flows, now we present a relevant result (see [91, 101]).

Theorem 2.1. *Let* $(\Omega, \sigma, \mathbb{R})$ *be a minimal and distal flow, and consider a homomorphism between distal flows* $\pi : (Y, \Psi, \mathbb{R}) \to (\Omega, \sigma, \mathbb{R})$. *If there is an* $\omega \in \Omega$ *such that* $\mathrm{card}(\pi^{-1}(\omega)) = N$ *for some* $N \in \mathbb{N}$, *then*

(i) Y *is an* N-*copy of* Ω;
(ii) $(Y, \Psi, \mathbb{R})$ *is almost periodic if and only if* $(\Omega, \sigma, \mathbb{R})$ *is almost periodic.*

Let $\pi : (Y, \Psi, \mathbb{R}) \to (\Omega, \sigma, \mathbb{R})$ be a flow epimorphism, and suppose that $(Y, \Psi, \mathbb{R})$ is a minimal flow (then, so is $(\Omega, \sigma, \mathbb{R})$, because, given $\omega = \pi(y)$ and $\omega_0 = \pi(y_0)$, there exists $\{t_n\}_n \subset \mathbb{R}$ such that $\Psi_{t_n}(y_0) \to y$ as $n \to \infty$, and, due to the continuity of π and its being a homomorphism, we have that $\pi(\Psi_{t_n}(y_0)) = \sigma_{t_n}(\omega_0) \to \omega$ as $n \to \infty$). $(Y, \Psi, \mathbb{R})$ is said to be an *almost automorphic extension* of $(\Omega, \sigma, \mathbb{R})$ if there exists $\omega \in \Omega$ such that $\mathrm{card}(\pi^{-1}(\omega)) = 1$. Furthermore, $(Y, \Psi, \mathbb{R})$ is said to be a *proximal extension* of $(\Omega, \sigma, \mathbb{R})$ if, whenever $\pi(y_1) = \pi(y_2)$ for some y_1, $y_2 \in Y$, then they are a proximal pair. An almost automorphic extension is always a proximal extension (see [112]). From this last remark together with statement (i) of Theorem 2.1, it is deduced that, if $(Y, \Psi, \mathbb{R})$ is a minimal and almost periodic flow which is an almost automorphic extension of an almost periodic flow $(\Omega, \sigma, \mathbb{R})$, then it must be a copy of $(\Omega, \sigma, \mathbb{R})$.

A point $\omega_0 \in \Omega$ is said to be an *almost automorphic point* if, given any sequence $\{s_n\}_n \subset \mathbb{R}$, we can find a subsequence $\{t_n\}_n$ of it such that the limits $\lim_{n\to\infty} \sigma_{t_n}(\omega_0) = \omega_1$ and $\lim_{n\to\infty} \sigma_{-t_n}(\omega_1) = \omega_0$ exist. The flow $(\Omega, \sigma, \mathbb{R})$ is *almost automorphic* when there is an almost automorphic point which has a dense orbit. An almost automorphic flow is always minimal, that is, actually all the orbits are dense. Almost automorphic minimal flows were first introduced and studied by Veech [112–114]. The theorem known as *Veech almost automorphic structure theorem* says that a flow is almost automorphic if and only if it is an almost automorphic extension of an almost periodic (minimal) flow (see [112]). A relation between almost automorphic and Levitan almost periodic points (see [57]) in a class of compact minimal flows can be found in Miller [63].

If $(Y, \Psi, \mathbb{R})$ is an almost automorphic flow and $(\Omega, \sigma, \mathbb{R})$ is an almost periodic (and minimal) flow satisfying that there exists a flow epimorphism $p : (Y, \Psi, \mathbb{R}) \to (\Omega, \sigma, \mathbb{R})$ such that $\mathrm{card}(p^{-1}(\omega)) = 1$ for some $\omega \in \Omega$, then the subset of Y formed by all the almost automorphic points in Y is given by $\{y \in Y \mid p^{-1}(p(y)) = \{y\}\}$, and it is a residual set (see Remark 2.6 in [101], part I).

We recall that a subset of a topological space E is said to be *residual* if its complementary is of first category in the sense of Baire, that is, its complementary is given by the union of countably many nowhere dense subsets of E.

A Borel measure on Ω will be a finite regular measure defined on the Borel sets. Let μ be a normalized Borel measure on Ω, μ is σ-*invariant* (or *invariant under* σ) if $\mu(\sigma_t(\Omega_1)) = \mu(\Omega_1)$ for every Borel subset $\Omega_1 \subset \Omega$ and every $t \in \mathbb{R}$. It is σ-*ergodic* (or *ergodic under* σ) if, in addition, $\mu(\Omega_1) = 0$ or $\mu(\Omega_1) = 1$ for every σ-invariant subset $\Omega_1 \subset \Omega$.

We denote by $\mathcal{M}_{\mathrm{inv}}(\Omega, \sigma)$ the set of positive and normalized σ-invariant measures on Ω. The Krylov-Bogoliubov theorem (see [68]) asserts that $\mathcal{M}_{\mathrm{inv}}(\Omega, \sigma)$ is nonempty when Ω is a compact metric space. The extremal points of the convex and weakly compact set $\mathscr{M}_{\text{inv}}(\Omega, \sigma)$ are the σ-ergodic measures, from which it is deduced that also the set of σ-ergodic measures is nonempty. The decomposition of the flow $(\Omega, \sigma, \mathbb{R})$ into ergodic components and the construction and representation theorems of σ-invariant measures from σ-ergodic measures are well known (see [59, 89]).

We say that $(\Omega, \sigma, \mathbb{R})$ is *uniquely ergodic* (u.e.) if it has a unique normalized invariant measure which is then necessarily ergodic. If $(\Omega, \sigma, \mathbb{R})$ is u.e. it is not necessarily minimal; however, if $(\Omega, \sigma, \mathbb{R})$ is u.e. and $\mu(U) > 0$ for every non-empty open set U, then $(\Omega, \sigma, \mathbb{R})$ is minimal. An almost periodic and minimal flow $(\Omega, \sigma, \mathbb{R})$ is always u.e. but an almost automorphic minimal one can be non-uniquely ergodic and can admit positive topological entropy (see [61]).

2.2 Almost Periodic and Almost Automorphic Dynamics

In order to find a link between non-autonomous differential equations with some recurrence in time and the theory of dynamical systems, we recall the basic definitions and results for the class of almost periodic and almost automorphic functions. We will give a brief explanation about the way this kind of equations give rise to skew-product flows or semiflows using the so-called hull as a base flow, which in turn will have some recurrence properties as well.

The concept of almost periodic function came up in the 1920s as an extension of the notion of periodicity. Authors like Bohr [10, 11], Favard [25], Besicovitch [7] and Bochner [9] studied exhaustively the properties of these functions. The book by Fink [26] is a detailed and well written reference on this topic.

Several equivalent definitions of almost periodic function may be found in the literature. Thus, in order to study harmonic functions, it is better to choose the characterization (as adopted by Corduneanu [20]) saying that a function is almost periodic whenever it can be approximated uniformly by a sequence of trigonometric polynomials on the whole real line, whereas, if our aim is to study differential equations, the preferred definition is the one introduced by Bohr, which is in the end the most frequently chosen one, (as seen in [4, 7]). A subset S of $\mathbb{R}$ is said to be *relatively dense* if there exists $l > 0$ such that every interval of length l intersects S. A complex function f, defined and continuous on $\mathbb{R}$, is *almost periodic* if, for all $\varepsilon > 0$, the set

$$T(f, \varepsilon) = \{s \in \mathbb{R} \mid |f(t+s) - f(t)| < \varepsilon \text{ for all } t \in \mathbb{R}\}$$

is relatively dense. The set $T(f, \varepsilon)$ is called *ε-translation set of f*. Almost periodic functions are bounded and uniformly continuous on $\mathbb{R}$. The set formed by all these functions is an algebra over $\mathbb{C}$, which is invariant by translations and closed under conjugation and uniform limits. Moreover, if f is almost periodic and $|f(t)| \geq m > 0$ for all $t \in \mathbb{R}$, then the function $1/f$ is almost periodic as well. Besides, if f is almost periodic and differentiable, then f' is almost periodic if and only if it is uniformly continuous on $\mathbb{R}$. As for integration, if a primitive of an almost periodic function is bounded, then it is also almost periodic.

The concept of almost periodicity can be extended to continuous functions taking values in a complete metric space (E, d) in a straightforward way: for each $\varepsilon > 0$, the set

$$T(f, \varepsilon) = \{s \in \mathbb{R} \mid \mathsf{d}(f(t + s), f(t)) < \varepsilon \text{ for all } t \in \mathbb{R}\}$$

must be relatively dense in $\mathbb{R}$. The reference [4] contains a study about almost periodic functions taking values in a Banach space and their relation with the theory of functional equations.

Bochner introduced another equivalent definition in terms of sequences (adopted for instance in [26]): a continuous function f is almost periodic if, given any sequence $\{\alpha_n\}_n \subset \mathbb{R}$, we can find a subsequence $\{\alpha_{n_j}\}_j$ of the previous one such that $\lim_{j \to \infty} f(t + \alpha_{n_j})$ exists uniformly on $\mathbb{R}$.

Besides, Bochner pointed out that, in order to simplify the proofs involving almost periodic functions, a property satisfied by such functions with respect to a group G could be used (see [9]); when $G = \mathbb{R}$, this property can be stated as follows: given a complex function f, defined and continuous on $\mathbb{R}$, and given any sequence $\{\alpha_n\}_n$ of real numbers, we can find a subsequence $\{\alpha_{n_j}\}_j$ in such a manner that the following limits exist pointwise on $\mathbb{R}$:

$$\lim_{j \to \infty} f(t + \alpha_{n_j}) = g(t), \qquad \lim_{j \to \infty} g(t - \alpha_{n_j}) = f(t)$$

for some function g. All the functions satisfying that property, whether they are almost periodic or not, are said to be *almost automorphic*. As we explained before, the fundamental properties of these functions with respect to groups, together with almost automorphic abstract minimal flows, were studied by Veech. In principle, the function g does not need to be continuous. If the function g is continuous for all sequences, then we say that f is *almost automorphic in the sense of Bohr*. From now on, we will assume that almost automorphic functions are almost automorphic in the sense of Bohr, so that almost automorphic functions are bounded and uniformly continuous on $\mathbb{R}$ (see [112]). Almost periodic functions are always almost automorphic, but the converse is not true; several examples can be found in the foregoing references.

One can define Fourier series both for almost periodic and almost automorphic functions valued in a Banach space but the one for an almost periodic function is unique and converges uniformly in terms of Bochner–Fejer summation, while the one for an almost automorphic function is in general non-unique and its Bochner–Fejer sum only converges pointwise. However, one can define the *frequency module* $\mathcal{M}(f)$ of an almost automorphic function f in the usual way as the smallest Abelian group containing a Fourier spectrum (the set of Fourier exponents associated with a Fourier series), and it has been shown that such a frequency module is uniquely defined [101]. In the above sense both almost periodic and almost automorphic functions can be viewed as natural generalizations to the periodic ones in t he strongest and the weakest sense respectively.

In the early 1940s, Fréchet defined and studied the concept of asymptotic almost periodicity. A function f continuous on $\mathbb{R}^+ = [0, \infty)$ is said to be *asymptotically almost periodic* if it can be represented as $f = f_1 + f_2$, where f_1 is an almost periodic function, and f_2 vanishes pointwise as $t \to \infty$. In fact, that representation is unique.

The relation between almost periodic functions and almost periodic flows is quite simple (see [22,26,68]). First, if $(\Omega, \sigma, \mathbb{R})$ is an almost periodic continuous real flow, then all the trajectories $t \in \mathbb{R} \mapsto \sigma(t, \omega) \in \Omega$ define almost periodic functions taking values in the compact metric space Ω. It is said that an element ω of a continuous real flow $(\Omega, \sigma, \mathbb{R})$ is an *almost periodic point* if, for any $\varepsilon > 0$, the set

$$T(\omega, \varepsilon) = \{s \in \mathbb{R} \mid \mathsf{d}(\sigma(s, \omega), \omega) < \varepsilon\}$$

is relatively dense in $\mathbb{R}$; such points are sometimes referred to as points with a *recurrent* orbit (see [68]). This condition is equivalent to the fact that the closure of the trajectory of such point, closure$\{\sigma(t, \omega) \mid t \in \mathbb{R}\}$, is a minimal subset for the flow. Note that, if the flow is minimal, then all its points are almost periodic. As a consequence, the flow $(\Omega, \sigma, \mathbb{R})$ can be decomposed as the disjoint union of a family of minimal subsets if and only all its points are almost periodic. Clearly, if the trajectory of ω, $t \in \mathbb{R} \mapsto \sigma(t, \omega) \in \Omega$, is an almost periodic function, then ω is an almost periodic point; moreover, in this case, the closure of its orbit is an almost periodic minimal set which coincides with both the omega-limit and alpha-limit sets of ω. Specifically, almost periodic flows are decomposed as a disjoint union of almost periodic and minimal flows.

As for almost automorphic flows, we know that there is an almost automorphic point with a dense orbit. If a point $\omega \in \Omega$ is almost automorphic, then its trajectory, $t \in \mathbb{R} \mapsto \sigma(t, \omega) \in \Omega$, is an almost automorphic function taking values in Ω (as before, the definition can be extended to this case in a natural manner). However, now there is no need for all the points to be almost automorphic, though all the points in a residual subset of Ω are (as we remarked in subsection 2.1). In fact, an almost automorphic minimal flow become almost periodic only if every point is almost automorphic (see [112]).

Conversely, let us check how to obtain almost periodic and almost automorphic flows from functions with analogous properties.

Definition 2.2. A function $f : \mathbb{R} \times \mathbb{R}^n \to \mathbb{R}^m$ is said to be *admissible* if, for every compact subset $K \subset \mathbb{R}^n$, f is bounded and uniformly continuous on $\mathbb{R} \times K$. Besides, if f is of class C^r $(r \geq 1)$ in $x \in \mathbb{R}^n$ and f and all its partial derivatives with respect to x up to order r are admissible, then we will say that f is either C^r-*admissible* or *admissible of class* C^r. A function $f \in C(\mathbb{R} \times \mathbb{R}^n, \mathbb{R}^m)$ is *uniformly almost automorphic* (resp. *almost periodic*) if it is admissible and almost automorphic (resp. almost periodic) in $t \in \mathbb{R}$.

Given an admissible function $f \in C(\mathbb{R} \times \mathbb{R}^n, \mathbb{R}^m)$, we consider the family of time translated functions $\{f_s \mid s \in \mathbb{R}\}$, where $f_s(t, x) = f(t + s, x)$ for all $s, t \in \mathbb{R}$, and all $x \in \mathbb{R}^n$. Hence, we can define the *hull* of f, which will be denoted by Ω or $H(f)$, as the closure within the space $C(\mathbb{R} \times \mathbb{R}^n, \mathbb{R}^m)$ of the set of time translated functions for the compact-open topology, that is, the topology of uniform convergence over compact subsets. Thanks to Arzelà-Ascoli's theorem, we can assure that the space $H(f)$ is compact and, furthermore, metrizable. Moreover, a continuous real flow is induced over the hull in a natural way, just by considering the mapping $\sigma : \mathbb{R} \times \Omega \to \Omega$, $(s, h) \mapsto h_s$, h translated a time s, that is, there is a flow over the hull defined by translation.

The next result assures that the initial function f admits a unique continuous extension to the hull and shows how the properties of recurrence of f are translated to the hull (see e.g. [101]).

Theorem 2.3. *Let* $f \in C(\mathbb{R} \times \mathbb{R}^n, \mathbb{R}^m)$ *be an admissible function. The following statements hold:*

  (i) *All the functions* $h \in H(f)$ *are admissible; in fact, if* f *is admissible of class* C^r, *so are all the functions* $h \in H(f)$.
  (ii) *There exists a unique function* $F \in C(H(f) \times \mathbb{R}^n, \mathbb{R}^m)$ *which extends* f, *in the sense that* $F(f_t, x) = f(t, x)$ *for all* $t \in \mathbb{R}$ *and* $x \in \mathbb{R}^n$; *besides, if* f *is* C^r-*admissible, then* F *is of class* C^r *in* x.
  (iii) *The flow* $(H(f), \sigma)$ *is almost automorphic (resp. almost periodic) if* f *is uniformly almost automorphic (resp. almost periodic).*

It is convenient to point out that the function F is defined specifically by $F(h, x) = h(0, x)$, $(h, x) \in H(f) \times \mathbb{R}^n$. The construction of the flow on the hull is often used when dealing with differential equations, as we will see in the next subsection. In particular, systematic studies of almost automorphic dynamics in differential equations were made in the 90s in a series of works by Shen and Yi [97–101].

2.3 Some Important ODEs Examples

Let $f : \mathbb{R} \times \mathbb{R}^m \to \mathbb{R}^m$ be a C^r-admissible function such that the flow $(H(f), \sigma)$ is minimal, and consider its unique continuous extension to the hull $\Omega = H(f)$, $F : \Omega \times \mathbb{R}^m \to \mathbb{R}^m$, which, according to Theorem 2.3, is a function of class C^r in $x \in \mathbb{R}^m$. In particular, if the initial equation is given by a uniformly almost periodic or almost automorphic function, then we are in the foregoing context. This way, from a system of non-autonomous ordinary differential equations

$$x' = f(t, x),$$

we can obtain a family of differential equations with indexes in the hull

$$x'(t) = F(\omega{\cdot}t, x(t)), \quad \omega \in \Omega, \tag{2.1}$$

where the flow on Ω is denoted by $\omega{\cdot}t = \sigma(t,\omega)$. Note that, fixing $\omega = f$, we get the original system, i.e., $x'(t) = f(t,x(t))$.

According to the standard theory of existence, uniqueness, and continuation of solutions for this kind of equations (see e.g. Hale [31]), these families of systems give rise to a local flow of skew-product type

$$\tau\colon \mathcal{U} \subset \mathbb{R} \times \Omega \times \mathbb{R}^m \longrightarrow \Omega \times \mathbb{R}^m, \quad (t,\omega,x) \mapsto (\omega{\cdot}t, u(t,\omega,x)), \qquad (2.2)$$

where $u(t,\omega,x)$ is the value of the solution of the system corresponding to ω with initial value $x(0) = x$ at time t, for t in the interval where the solution is defined. Thanks to the classical theorems of continuous dependence with respect to the initial values, u inherits the same regularity, C^r, with respect to x.

The use of this technique, that is, of including a non-autonomous system within a family of systems linked to one another by means of the flow on the hull, is focused to the application of the methods and results of the theory of skew-product flows to the new problem, where the solutions of the systems have been considered as a part of the trajectories of a dynamical system. It is noteworthy that, in the new family of systems generated from a given system, there are just their translated systems as well as their limits, so that the flow associated to this family is a good representation of the dynamics of the initial system and, in particular, the asymptotic behavior of its bounded solutions. Such a formulation was originated in Miller [64] and Sell [95].

The importance of the presence of almost automorphic dynamics in the study of almost periodic differential equations was motivated by the examples given by Levitan and Zhikov [57] and Johnson [46] of scalar almost periodic equations with almost automorphic but not almost periodic solutions. Ortega and Tarallo [88] describe a qualitative property, which is satisfied by the above examples, and it provides almost automorphic solutions of almost periodic linear systems. The linear case with Levitan almost periodic coefficient is studied by Caraballo and Cheban [12, 13].

Similar dynamical phenomena occur in the Riccati equations obtained from the non-uniformly hyperbolic two-dimensional linear systems constructed by Millionščikov [65] and Vinograd [115]. We now review the importance of some of these examples in the study of the almost automorphic dynamics, as shown in [101].

Before, we recall that in the scalar and almost periodic case, $m = 1$, every minimal set M is almost automorphic (see [101] for a more general version of this result, valid for scalar parabolic partial differential equations with the Neumann boundary condition).

Lemma 2.4. *Let $m = 1$ and let M be a minimal set of (2.2). Then (M,τ) is an almost automorphic extension of the base flow (Ω,σ), and hence an almost automorphic minimal set if f is uniformly almost periodic.*

Proof. We define $x_i\colon \Omega \to \mathbb{R}$, $i = 1,2$ by $x_1(\omega) = \inf\{x \in \mathbb{R} \mid (\omega,x) \in M\}$ and $x_2(\omega) = \sup\{x \in \mathbb{R} \mid (\omega,x) \in M\}$. It is easy to check that x_1 is lower semi-continuous and x_2 is upper semi-continuous. As a consequence (see [17]), there is a residual set $\Omega_0 \subset \Omega$ of continuity points for x_1 and x_2. In fact, since the flow is scalar, and then monotone, it can be shown that

$$x_i(\omega{\cdot}t) = u(t, \omega, x_i(\omega)), \quad t \in \mathbb{R}, \; \omega \in \Omega, \; i = 1, 2,$$

and that Ω_0 is an invariant set.

Let $\pi: \Omega \times \mathbb{R} \to \Omega$ be the natural projection and $\omega \in \Omega_0$. Next we check that $\mathrm{card}(M \cap \pi^{-1}(\omega)) = 1$. Since both $(\omega, x_1(\omega)), (\omega, x_2(\omega)) \in M$, which is minimal, let $t_n \uparrow \infty$ be such that $\lim_{n \to \infty} \tau(t_n, \omega, x_1(\omega)) = (\omega, x_2(\omega))$, that is, $\lim_{n \to \infty} \omega{\cdot}t_n = \omega$ and $\lim_{n \to \infty} u(t_n, \omega, x_1(\omega)) = x_2(\omega)$. Moreover, since $u(t_n, \omega, x_1(\omega)) = x_1(\omega{\cdot}t_n)$ and ω is a continuity point of x_1, we conclude that $x_1(\omega) = x_2(\omega)$. As a consequence, $\mathrm{card}(M \cap \pi^{-1}(\omega)) = 1$, and hence, M is an almost automorphic extension of the base flow (Ω, σ), as stated. $\qquad\Box$

For escalar linear homogeneous equations, each minimal set is a copy of the base.

Proposition 2.5. *Let* $m = 1$, $F(\omega, x) = d(\omega)x$ *and let* M *be a minimal set of* (2.2).

(i) *(M, τ) is a copy of the base flow (Ω, σ), that is, $M = \{(\omega, c(\omega)) \mid \omega \in \Omega\}$ for a continuous map $c: \Omega \to \mathbb{R}$.*

(ii) *There are infinitely many minimal subsets if and only if there is a continuous map $D: \Omega \to \mathbb{R}$ such that $\int_0^t d(\omega{\cdot}s)\,ds = D(\omega{\cdot}t) - D(\omega)$.*

Proof. (i) $\Omega \times \{0\}$ is a minimal subset, which is a copy of the base. Let us assume that there is another one $M \subset \Omega \times \mathbb{R}^+$, and assume by contradiction that there are $(\omega, x_1), (\omega, x_2) \in M$ with $0 < x_1 < x_2$. Take $\lambda = x_2/x_1 > 1$. It is easy to check that the map $\Omega \times \mathbb{R} \to \Omega \times \mathbb{R}$, $(\omega, x) \mapsto (\omega, \lambda x)$ takes invariant subsets into invariant subsets. Moreover $(\omega, \lambda x_1) = (\omega, x_2)$ and hence it takes M to itself. As a consequence, $(\omega, \lambda^n x_1) \in M$ for each $n \in \mathbb{N}$, which contradicts that M is bounded, and we deduce that $M = \{(\omega, c(\omega)) \mid \omega \in \Omega\}$.

(ii) If there are infinitely many, let $M \neq \Omega \times 0$. Then $c(\omega) > 0$ for each $\omega \in \Omega$ and from $\frac{d}{dt}c(\omega{\cdot}t) = d(\omega{\cdot}t)c(\omega{\cdot}t)$ we deduce that $D(\omega) = \log c(\omega)$ satisfies $\int_0^t d(\omega{\cdot}s)\,ds = D(\omega{\cdot}t) - D(\omega)$, as stated. Conversely, if there is a continuous map D satisfying this relation, it is immediate to check that $M_\alpha = \{(\omega, \alpha\, e^{D(\omega)}) \mid \omega \in \Omega\}$ is a different minimal subset for each $\alpha \in \mathbb{R}$. $\qquad\Box$

The next subsection shows that this result is not true for the nonhomogeneous case.

2.3.1 Existence of an Almost Automorphic but Non-almost Periodic Minimal Set

Based on a previous example by Conley and Miller [19], Jonhson constructed in [46] a linear almost periodic scalar ordinary differential equation

$$x' + a(t)\, x = b(t) \tag{2.3}$$

satisfying the following properties:

   (i) $a(t)$ and $b(t)$ are uniform limits of 2^n-periodic continuous functions $a_n(t)$ and $b_n(t)$ respectively.
  (ii) $\lim_{t \to \infty} \int_0^t a(s) \, ds = \infty$.
 (iii) If $x_0(t)$ is the solution of (2.3) with $x_0(0) = 0$, then $|x_0(t)| \leq 1$, and for $n \geq 4$, $x_0(2^n) = 1/5$ if n is odd, and $x_0(2^n) = 0$ if n is even.

As before, we take Ω the hull of the uniformly almost periodic function f given by $f(t, x) = -a(t)x + b(t)$, $(t, x) \in \mathbb{R}^2$, and the corresponding family of differential equations with indexes in the hull (2.1) and its induced skew-product flow (2.2).

Since $x_0(t)$ is a bounded solution with $x_0(0) = 0$, the omega-limit set of the point $(f, 0) \in \Omega \times \mathbb{R}$ contains a minimal set $M \subset \Omega \times \mathbb{R}$ for the flow, which is almost automorphic by Lemma 2.4. The uniqueness of the minimal set follows from (ii) because the existence of two different minimal sets would contradict the fact that the solutions of $x' + a(t) x = 0$ tend to 0 at $+\infty$.

Johnson showed that (2.3) admits no almost periodic solutions and there is one of the equations in the hull with an almost automorphic but non-almost periodic solution $x(t, \widetilde{\omega}, \widetilde{x})$. As a consequence, the unique minimal set M is almost automorphic but not almost periodic because the trajectory $\{\tau(t, \widetilde{\omega}, \widetilde{x}) = (\widetilde{\omega} \cdot t, x(t, \widetilde{\omega}, \widetilde{x})) \mid t \in \mathbb{R}\}$ is not almost periodic. In addition, Johnson showed in [47] that M is uniquely ergodic.

2.3.2 An Omega-Limit Set Which Contains Two Minimal Sets

A modification of the previous example provides an example of a skew-product scalar flow with an omega-limit set which contains two minimal sets. First note that the family of equations in the hull corresponding to (2.3) could have be written in the form

$$x' + A(\omega \cdot t) x = B(\omega \cdot t), \quad \omega \in \Omega, \tag{2.4}$$

for continuous functions $A, B \in C(\Omega, \mathbb{R})$. Let M be, as before, the unique minimal set for the induced skew-product flow. We take $y_0 \in \mathbb{R}$ such that $(\omega, y_0) \notin M$ for all $\omega \in \Omega$. The change of variables $z = 1/(x - y_0)$ takes (2.4) to

$$z' = A(\omega \cdot t) z + (B(\omega \cdot t) - A(\omega \cdot t) y_0) z^2, \quad \omega \in \Omega, \tag{2.5}$$

and we will denote by $\hat{\tau}$ the induced local skew-product flow for this family. Clearly, $M_1 = \{(\omega, 0) \mid \omega \in \Omega\} = \Omega \times \{0\}$ and $M_2 = \{(\omega, 1/(x - y_0) \mid (\omega, x) \in M\}$ are two different minimal sets for this flow.

We claim that there is an omega-limit set which contains M_1 and M_2. Since $\int_0^t a(s) \, ds$ is unbounded in $t \in \mathbb{R}$, also $\int_0^t A(\omega \cdot s) \, ds$ is unbounded in $t \in \mathbb{R}$ for each $\omega \in \Omega$, and it can be shown (see [47, 53]) that there is a point $\hat{\omega} \in \Omega$ (in fact a residual set) such that

$$\sup_{t\geq 0} \int_0^t A(\hat{\omega}\cdot s)\, ds = +\infty, \quad \inf_{t\geq 0} \int_0^t A(\hat{\omega}\cdot s)\, ds = -\infty, \tag{2.6}$$

and (2.4) for $\hat{\omega}$ has a unique bounded solution $x(t,\hat{\omega},\hat{x})$. If we choose α such that $\left| x(t,\hat{\omega},\hat{x}) + \alpha \exp(-\int_0^t A(\hat{\omega}\cdot s)\, ds) - y_0 \right| \geq \varepsilon > 0$ for each t, the function

$$z(t,\hat{\omega},\hat{z}) = \frac{1}{x(t,\hat{\omega},\hat{x}) + \alpha \exp(-\int_0^t A(\hat{\omega}\cdot s)\, ds) - y_0}$$

is a bounded solution of (2.5) for $\hat{\omega}$. Moreover from (2.6) there are sequences $\{t_n\}_n$ and $\{s_n\}_n \uparrow \infty$ such that $\lim_{n\to\infty} z(t_n,\hat{\omega},\hat{z}) = 0$ and $\lim_{n\to\infty} z(s_n,\hat{\omega},\hat{z}) = z_0$ with $\lim_{n\to\infty} \hat{\omega}\cdot s_n = \omega_0$ and $(\omega_0, z_0) \in M_2$. Hence the omega-limit set of $(\hat{\omega}, \hat{z})$, i.e., the closure of the set $\{\hat{\tau}(t,\hat{\omega},\hat{z}) = (\hat{\omega}\cdot t, z(t,\hat{\omega},\hat{z})) \mid t \in \mathbb{R}\}$ contains M_1 and M_2. Note that M_2 is also almost automorphic but non-almost periodic.

2.3.3 Existence of Non-uniquely Ergodic Minimal Sets

The almost automorphic but non-almost periodic minimal set of subsection 2.3.1 is uniquely ergodic. An idea of constructing examples of non-uniquely ergodic almost automorphic minimal sets is suggested by Johnson in [45,48] by studying the skew-product flow induced in the real projective bundle by a family of two-dimensional linear systems whose Sacker–Sell spectrum (see [92]) is a nondegenerate closed interval. We review the application of this technique to the non-uniformly hyperbolic family of systems obtained from the quasi-periodic Vinograd example [115]:

$$x' = \begin{pmatrix} 0 & 1 + a(\omega\cdot t) \\ 1 - a(\omega\cdot t) & 0 \end{pmatrix} x, \quad \omega = (\omega_1, \omega_2) \in \mathbb{T}^2, \tag{2.7}$$

where the flow on the base, which is a two torus, is given by a frequency vector $(1,\alpha)$ for an irrational number α, i.e., $\omega\cdot t = (\omega_1 + t, \omega_2 + \alpha t)$, and hence is minimal and almost periodic.

In polar coordinates (r, θ) $(\theta = \arg x)$, (2.7) take the form $r' = r\sin(2\theta)$ and

$$\theta' = -a(\omega\cdot t) + \cos(2\theta), \quad \omega \in \mathbb{T}^2, \tag{2.8}$$

and this family of scalar equations for the angular coordinate induces a skew-product flow on the projective bundle $\Sigma_p = \mathbb{T}^2 \times \mathbb{P}^1$.

Moreover, the function $a(\omega)$ is constructed as the limit of a nondecreasing sequence of positive functions $a_n(\omega)$, satisfying that for each $n \geq 1$ the system

$$x' = \begin{pmatrix} 0 & 1 + a_n(\omega\cdot t) \\ 1 - a_n(\omega\cdot t) & 0 \end{pmatrix} x, \quad \omega \in \mathbb{T}^2 \tag{9$_{\omega,n}$}$$

- Has two Lyapunov exponents $\beta_n, -\beta_n$ with $\beta_n > 1/2$, and its Sacker–Sell spectrum is $\{-\beta_n, \beta_n\}$.
- The angular equation

$$\theta' = -a_n(\omega_0 \cdot t) + \cos(2\theta), \tag{$10)_n$}$$

for $\omega_0 = (0,0)$ has two solutions $\theta_n^1(t)$ and $\theta_n^2(t)$ with

$$-\frac{\pi}{4} < \theta_n^1(t) < \theta_{n+1}^1(t) < \theta_{n+1}^2(t) < \theta_n^2(t) < \frac{\pi}{4}, \quad t \in \mathbb{R},$$

and $0 < \inf_{t \in \mathbb{R}} |\theta_n^1(t) - \theta_n^2(t)| = \delta_n \to 0$ as $n \to \infty$.
- The sets $M_n^1 = \text{closure}\{(\omega_0 \cdot t, \theta_n^1(t)) \mid t \in \mathbb{R}\}$ and $M_n^2 = \text{closure}\{(\omega_0 \cdot t, \theta_n^2(t)) \mid t \in \mathbb{R}\}$ are disjoint almost period minimal sets for the skew-product flow induced on the projective bundle Σ_p, i.e., copies of the base: $M_n^1 = \{(\omega, h_n^1(\omega)) \mid \omega \in \mathbb{T}^2\}$ and $M_n^2 = \{(\omega, h_n^2(\omega)) \mid \omega \in \mathbb{T}^2\}$ with

$$-\frac{\pi}{4} < h_n^1(\omega) \leq h_{n+1}^1(\omega) < h_{n+1}^2(\omega) \leq h_n^2(\omega) < \frac{\pi}{4}.$$

As a consequence, in the limit, as studied in Johnson [48]:

- The family (2.7) is non-uniformly hyperbolic: the positive Lyapunov exponent is $\beta \geq 1/2$ and the family of systems does not have an exponential dichotomy.
- The Sacker–Sell spectrum of (2.7) is a nondegenerate interval containing $[-\frac{1}{2}, \frac{1}{2}]$.
- The set $\Omega_1 = \{\omega \in \mathbb{T}^2 \mid h^1(\omega) = h^2(\omega)\}$ is a residual set of null measure, where $h^i(\omega) = \lim_{n \to \infty} h_n^i(\omega), i = 1, 2$, and $J = \{(\omega, \varphi) \in \mathbb{T}^2 \times \mathbb{P}^1 \mid h^1(\omega) \leq \varphi \leq h^2(\omega)\}$ is a compact invariant set which contains a unique minimal set M satisfying our assertions: it is a non-uniquely ergodic minimal set which is almost automorphic but non-almost periodic. There are two different invariant measures concentrated on the sets $\{(\omega, x_i(\omega)) \mid \omega \in \mathbb{T}^2\} \subset M, i = 1, 2$, where, as in the proof of Lemma 2.4, x_1 and x_2 are defined by $x_1(\omega) = \inf\{x \in \mathbb{P}^1 \mid (\omega, x) \in M\}$, $x_2(\omega) = \sup\{x \in \mathbb{P}^1 \mid (\omega, x) \in M\}$, and they coincide a.e. with $h^1(\omega)$ and $h^2(\omega)$ respectively.

Similar assertions are obtained for the complex projective flow induced by the Riccati equations associated to the family (2.7) with $z = x_2/x_1$

$$z' = 1 - a(\omega \cdot t) - (1 + a(\omega \cdot t)) z^2, \quad \omega \in \mathbb{T}^2. \tag{2.11}$$

We refer the reader to Jorba et al. [53] for the relation of these minimal sets with the occurrence of strange non-chaotic attractors (SNA).

These systems correspond with the first point of the spectrum for the spectral problem defined by the associated self-adjoint Schrödinger operators. In fact, the construction of real almost automorphic minimal sets is possible for the first or the last point of each interval gap in which the Schrödinger linear systems have positive

Lyapunov exponent (see [49,53]). Bjerklöv [8] shows the existence of quasi periodic non-uniformly hyperbolic second order Schrödinger equations whose projective flow is minimal. The previous construction is not possible in this situation. Other situations showing the presence of almost automorphic dynamics in almost periodic ODEs can be found in Yi [119] and Huang and Yi [39].

2.3.4 Bifurcation in the Scalar, Coercive and Concave Case

In some cases, the appearance of almost automorphic dynamics as collision of minimal sets, can also be understood as a bifurcation process, as shown in Novo et al. [71]. The result generalizes to the almost periodic case, the Ambrosetti–Prodi type result showed by Mawhin [62] in the periodic case, concerning the existence and number of periodic solutions of a one-parameter scalar, coercive and convex (or concave) equation with periodic coefficients. The discrete-time analogue has been recently published by Nguyen et al. [69].

For each $\alpha \in \mathbb{R}$, we consider the family of differential equations

$$x' = g(\omega{\cdot}t, x) + \alpha , \quad \omega \in \Omega , \tag{12$_\alpha$}$$

with (Ω, σ) a minimal flow defined on a compact metric space and, as usual, represent $\omega{\cdot}t = \sigma(t, \omega)$ for each $t \in \mathbb{R}$ and $\omega \in \Omega$, and the continuous function $g \colon \Omega \times \mathbb{R} \to \mathbb{R}$ satisfies:

- g is concave in x, that is,

$$g(\omega, \lambda\, x_1 + (1 - \lambda)\, x_2) \geq \lambda\, g(\omega, x_1) + (1 - \lambda)\, g(\omega, x_2)$$

  for each $\omega \in \Omega$, $0 \leq \lambda \leq 1$, $x_1, x_2 \in \mathbb{R}$.
- g is strictly concave in x for some $\omega_0 \in \Omega$, that is,

$$g(\omega_0, \lambda\, x_1 + (1 - \lambda)\, x_2) > \lambda\, g(\omega_0, x_1) + (1 - \lambda)\, g(\omega_0, x_2)$$

  for each $0 < \lambda < 1$ and $x_1 \neq x_2$.
- g is C^1 with respect to x and $\displaystyle\lim_{x \to \pm\infty} g(\omega, x) = -\infty$.

This family induces a local skew-product flow on $\Omega \times \mathbb{R}$

$$\tau_\alpha \colon \mathscr{U} \subset \mathbb{R} \times \Omega \times \mathbb{R} \longrightarrow \Omega \times \mathbb{R}, \quad (t, \omega, x_0) \mapsto \tau_\alpha(t, \omega, x_0) = (\omega{\cdot}t, x(t, \omega, x_0, \alpha)),$$

where $x(t, \omega, x_0, \alpha)$ is the solution of $(12)_\alpha$ evaluated along the trajectory of ω with initial value x_0, and t belongs to its maximal interval of definition (t_-, t_+) (note that, although dropped from the notation, t_+ and t_- depend on ω, x_0 and α). It is well known that if $x(t, \omega, x_0, \alpha)$ remains bounded, then it is defined for every $t \in \mathbb{R}$. We consider the set of bounded solutions

$$B_\alpha = \left\{ (\omega, x_0) \in \Omega \times \mathbb{R} \mid \sup_{t \in \mathbb{R}} |x(t, \omega, x_0, \alpha)| < \infty \right\},$$

and we denote $\pi : B_\alpha \mapsto \Omega$ as the natural projection. Let us assume that $B_\alpha \neq \emptyset$ and let $(\omega_0, x_0) \in B_\alpha$. Thus, closure$\{(\omega_0 \cdot t, x(t, \omega_0, x_0, \alpha)) \mid t \in \mathbb{R}\} \subset B_\alpha$ and the minimal character of (Ω, σ) provides $\pi^{-1}(\omega) \cap B_\alpha \neq \emptyset$ for every $\omega \in \Omega$. Therefore, the map τ_α defines a global flow on B_α. Moreover, as shown in [71],

$$B_\alpha = \{(\omega, x) \in \Omega \times \mathbb{R} \mid x_1(\omega, \alpha) \leq x \leq x_2(\omega, \alpha)\},$$

where $x_1(\omega, \alpha) = \inf\{x \mid (\omega, x) \in B_\alpha\}$ and $x_2(\omega, \alpha) = \sup\{x \mid (\omega, x) \in B_\alpha\}$. The following result was proved in [71] for the convex case.

Theorem 2.6. *There exists a value of the parameter $\alpha^* \in \mathbb{R}$ such that*

(i) *If $\alpha > \alpha^*$, then $B_\alpha \neq \emptyset$ and it contains two minimal subsets which are copies of the base*

$$M_{i,\alpha} = \{(\omega, x_i(\omega, \alpha)) \mid \omega \in \Omega\}, \quad i = 1, 2.$$

Every intermediate trajectory in B_α moves from the lower top to the upper top as t goes from $-\infty$ to $+\infty$ and if $\alpha_1 > \alpha_2 > \alpha^*$ then $B_{\alpha_2} \subset B_{\alpha_1}$.

(ii) *If $\alpha = \alpha^*$, then $B_\alpha \neq \emptyset$ and it contains a unique minimal subset M_α^* which is an almost automorphic extension of (Ω, σ) and then almost automorphic if Ω is almost periodic.*

(iii) *If $\alpha^* > \alpha$, then $B_\alpha = \emptyset$, i.e., there are no bounded solutions.*

In the almost periodic case, the previous theorem implies that each of the equations of the family $(12)_\alpha$ has exactly two a.p. solutions if $\alpha > \alpha^*$, at most one a.p. solution if $\alpha = \alpha^*$ and no bounded solution if $\alpha < \alpha^*$. This, as stated above, generalizes Mawhin's result for the periodic case.

The next result, also proved in [71], explains the possible situations in the case $\alpha = \alpha^*$. Let us fix an ergodic measure m on Ω. Denote by $\gamma_i(\alpha)$ the Lyapunov exponents with respect to m given by the formula

$$\gamma_i(\alpha) = \int_\Omega \frac{\partial g}{\partial x}(\omega, x_i(\omega, \alpha)) \, dm, \quad i = 1, 2.$$

We know that $\gamma_2(\alpha) < 0 < \gamma_1(\alpha)$ for each $\alpha > \alpha^*$. We denote

$$x_1(\omega, \alpha^*) = \lim_{\alpha \to (\alpha^*)^+} x_1(\omega, \alpha), \quad x_2(\omega, \alpha^*) = \lim_{\alpha \to (\alpha^*)^+} x_2(\omega, \alpha).$$

Proposition 2.7. *Let α^* be the value of the parameter obtained in Theorem 2.6. Then, there exists a unique minimal subset $M_{\alpha^*} \subset B_{\alpha^*}$, which is an almost* *automorphic extension of the base (Ω, σ) and it is not hyperbolic. Besides, one of the following cases holds:*

(c.1) $x_1(\omega, \alpha^*) = x_2(\omega, \alpha^*)$ *for every* $\omega \in \Omega$. *In this situation* $B_{\alpha^*} = M_{\alpha^*}$ *is a copy of the base, there exists a unique ergodic measure concentrated on* B_{α^*} *and projecting onto* m, *and the Lyapunov exponent is null. If the base* (Ω, σ) *is a.p., then so is* M_{α^*};

(c.2) $x_1(\omega, \alpha^*) = x_2(\omega, \alpha^*)$ *for almost every* $\omega \in \Omega$ *with respect to* m. *As in the previous case, there exists a unique ergodic measure concentrated on* B_{α^*} *and projecting onto* m, *and the Lyapunov exponent is null. If* (Ω, σ) *is a.p., then* M_{α^*} *is an a.a. minimal set, which is a.p. if and only if it is a copy of the base;*

(c.3) $x_1(\omega, \alpha^*) = x_2(\omega, \alpha^*)$ *only on a set of null measure. Then, there exist two unique ergodic measures concentrated on* B_{α^*} *and projecting onto* m, *and the Lyapunov exponents are not zero:* $\gamma_2(\alpha^*) < 0 < \gamma_1(\alpha^*)$. *In this case if* (Ω, σ) *is a.p., then* M_{α^*} *is an a.a. minimal set which is not a.p.*

It can be checked that all the above situations hold. An example for situation (c.1) is the periodic case. An example of (c.2) for the convex case is constructed in Sanz [93] starting from Johnson's example (2.3). Finally, the Riccati equation (2.11) associated to Vinograd's example provides an illustration of situation (c.3) with $\alpha^* = 0$. The last two examples illustrate how an almost automorphic, and not a.p. minimal set is obtained as collision of almost periodic minimal sets.

The references Johnson and Mantellini [50], Fabbri et al. [23] and Núñez and Obaya [78] study other bifurcation problems for scalar almost periodic differential equations which confirm the relevance of the almost automorphic dynamics in this theory.

2.4 Ordered Banach Spaces: Monotone Skew-Product Semiflows

In the one-dimensional ODE case, the induced flow is obviously monotone, i.e., ordered initial states lead to ordered subsequent states. In general, this is no longer true, and in addition, we are interested in the study of functional differential equations in which the trajectories are not defined backwards and we obtain a semiflow.

Let E be a complete metric space and $\mathbb{R}^+ = \{t \in \mathbb{R} \mid t \geq 0\}$. A *semiflow* $(E, \Phi, \mathbb{R}^+)$ is determined by a continuous map $\Phi : \mathbb{R}^+ \times E \to E$, $(t, x) \mapsto \Phi(t, x)$ which satisfies

(i) $\Phi_0 = \text{Id}$,
(ii) $\Phi_{t+s} = \Phi_t \circ \Phi_s$ for all $t, s \in \mathbb{R}^+$,

where $\Phi_t(x) = \Phi(t, x)$ for each $x \in E$ and $t \in \mathbb{R}^+$. The set $\{\Phi_t(x) \mid t \geq 0\}$ is the *semiorbit* of the point x. A subset E_1 of E is *positively invariant* (or just Φ-invariant) if $\Phi_t(E_1) \subset E_1$ for all $t \geq 0$. A semiflow $(E, \Phi, \mathbb{R}^+)$ admits a *flow extension* if there exists a continuous flow $(E, \widetilde{\Phi}, \mathbb{R})$ such that $\widetilde{\Phi}(t, x) = \Phi(t, x)$

for all $x \in E$ and $t \in \mathbb{R}^+$. A compact and positively invariant subset admits a flow extension if the semiflow restricted to it admits one.

Write $\mathbb{R}^- = \{t \in \mathbb{R} \mid t \leq 0\}$. A *backward orbit* of a point $x \in E$ in the semiflow $(E, \Phi, \mathbb{R}^+)$ is a continuous map $\psi : \mathbb{R}^- \to E$ such that $\psi(0) = x$ and for each $s \leq 0$ it holds that $\Phi(t, \psi(s)) = \psi(s + t)$ whenever $0 \leq t \leq -s$. If for $x \in E$ the semiorbit $\{\Phi(t, x) \mid t \geq 0\}$ is relatively compact, we can consider the *omega-limit set* of x,

$$\mathcal{O}(x) = \bigcap_{s \geq 0} \text{closure}\{\Phi(t + s, x) \mid t \geq 0\},$$

which is a nonempty compact connected and Φ-invariant set. Namely, it consists of the points $y \in E$ such that $y = \lim_{n \to \infty} \Phi(t_n, x)$ for some sequence $t_n \uparrow \infty$. It is well-known that every $y \in \mathcal{O}(x)$ admits a backward orbit inside this set. Actually, a compact positively invariant set M admits a flow extension if every point in M admits a unique backward orbit which remains inside the set M (see [101], part II).

A compact positively invariant set M for the semiflow $(E, \Phi, \mathbb{R}^+)$ is *minimal* if it does not contain any other nonempty compact positively invariant set than itself. If E is minimal, we say that the semiflow is minimal.

A semiflow is *of skew-product type* when it is defined on a vector bundle and has a triangular structure; more precisely, a semiflow $(\Omega \times X, \tau, \mathbb{R}^+)$ is a *skew-product* semiflow over the product space $\Omega \times X$, for a compact metric space (Ω, d) and a complete metric space (X, d), if the continuous map τ is as follows:

$$\tau: \mathbb{R}^+ \times \Omega \times X \longrightarrow \Omega \times X, \quad (t, \omega, x) \mapsto (\omega{\cdot}t, u(t, \omega, x)), \qquad (2.13)$$

where $(\Omega, \sigma, \mathbb{R})$ is a real continuous flow $\sigma : \mathbb{R} \times \Omega \to \Omega$, $(t, \omega) \mapsto \omega{\cdot}t$, called the *base flow*. The skew-product semiflow (2.13) is *linear* if $u(t, \omega, x)$ is linear in x for each $(t, \omega) \in \mathbb{R}^+ \times \Omega$.

Let $K \subset \Omega \times X$ be a compact positively invariant set such that every point of K admits a backward orbit. We introduce the lifting flow associated to the semiflow $(K, \tau, \mathbb{R}^+)$ (see [101] and the references therein). Since every point of K admits a backward orbit, hence an entire orbit (although not necessarily unique), we consider $\widehat{K}$ the set of entire orbits of $(K, \tau, \mathbb{R}^+)$, that is,

$$\widehat{K} = \{\phi \in C(\mathbb{R}, K) \mid \tau(t, \phi(s)) = \phi(t + s), \quad t \geq 0, s \in \mathbb{R}\}.$$

Note that, if $\phi \in \widehat{K}$, we have that $\phi(t) = (\omega{\cdot}t, u(t, \omega, x))$ for each $t \geq 0$, where $\phi(0) = (\omega, x) \in K$. The set $\widehat{K}$ is compact with respect to the compact-open topology on $C(\mathbb{R}, K)$, which is metrizable. For each $\phi \in \widehat{K}$ and $t \in \mathbb{R}$, the translated orbit $\phi_t(s) = \phi(t + s)$, $s \in \mathbb{R}$, also belongs to $\widehat{K}$. Therefore, the map

$$\widehat{\tau}: \mathbb{R} \times \widehat{K} \longrightarrow \widehat{K}, \quad (t, \phi) \mapsto \phi_t$$

defines a flow $(\widehat{K}, \widehat{\tau}, \mathbb{R})$, called the lifting flow associated to $(K, \tau, \mathbb{R}^+)$, which is isomorphic to a skew-product flow as $\widehat{K} \simeq \{(\omega, \phi) \mid \phi \in \widehat{K}, \; \phi(0) = (\omega, x)\} \subset \Omega \times \widehat{K}$. For simplicity of notation we do not repeat the first component, and sometimes we will refer to $(\widehat{K}, \widehat{\tau}, \mathbb{R})$ as the corresponding skew-product flow.

Now, we introduce some definitions concerning the stability of the trajectories. A forward orbit $\{\tau(t, \omega_0, x_0) \mid t \geq 0\}$ of the skew-product semiflow (2.13) is said to be *uniformly stable* if for every $\varepsilon > 0$ there is a $\delta(\varepsilon) > 0$, called the *modulus of uniform stability*, such that, if $s \geq 0$ and $\mathsf{d}(u(s, \omega_0, x_0), x) \leq \delta(\varepsilon)$ for certain $x \in X$, then for each $t \geq 0$,

$$\mathsf{d}(u(t + s, \omega_0, x_0), u(t, \omega_0 \cdot s, x)) = \mathsf{d}(u(t, \omega_0 \cdot s, u(s, \omega_0, x_0)), u(t, \omega_0 \cdot s, x)) \leq \varepsilon\,.$$

A forward orbit $\{\tau(t, \omega_0, x_0) \mid t \geq 0\}$ of the skew-product semiflow (2.13) is said to be *uniformly asymptotically stable* if it is uniformly stable and there is a $\delta_0 > 0$ with the following property: for each $\varepsilon > 0$ there is a $t_0(\varepsilon) > 0$ such that, if $s \geq 0$ and $\mathsf{d}(u(s, \omega_0, x_0), x) \leq \delta_0$, then

$$\mathsf{d}(u(t + s, \omega_0, x_0), u(t, \omega_0 \cdot s, x)) \leq \varepsilon \quad \text{for each } t \geq t_0(\varepsilon)\,.$$

Next we introduce the basic definitions and preliminary results of the theory of *monotone dynamical systems*, that is, dynamical systems on an ordered metric space X which have the property that ordered initial states lead to ordered subsequent states. We refer the reader to Smith [105], Amann [3] and Krasnoselskii et al. [54] for more details.

We say that X is a *strongly ordered* Banach space if there is a closed convex cone, that is, a nonempty closed subset $X_+ \subset X$ satisfying

(i) $X_+ + X_+ \subset X_+$,   (ii) $\mathbb{R}^+ X_+ \subset X_+$,   (iii) $X_+ \cap (-X_+) = \{0\}$

with nonempty interior $\mathrm{Int} X_+ \neq \emptyset$. The *strong ordering* on X is defined as follows:

$$x < y \quad \Longleftrightarrow \quad y - x \in X_+\,;$$
$$x < y \quad \Longleftrightarrow \quad y - x \in X_+ \text{ and } x_1 \neq x_2\,;$$
$$x \ll y \quad \Longleftrightarrow \quad y - x \in \mathrm{Int} X_+\,.$$

The positive cone X_+ is said to be *normal* if the norm of the Banach space X is *semimonotone*, i.e., there is a positive constant $k > 0$ such that $0 \leq x \leq y$ implies $\|x\| \leq k \|y\|$. A norm of X is called *monotone* if $0 \leq x \leq y$ implies $\|x\| \leq \|y\|$.

The skew-product semiflow $(\Omega \times X, \tau, \mathbb{R}^+)$ is *monotone* if

$$u(t, \omega, x) \leq u(t, \omega, y) \quad \text{for each } t \geq 0,\ \omega \in \Omega \text{ and } x, y \in X \text{ with } x \leq y\,;$$

it is *strongly monotone* if

$$u(t, \omega, x) \ll u(t, \omega, y) \quad \text{for each } t > 0,\ \omega \in \Omega \text{ and } x, y \in X \text{ with } x < y\,,$$

and it is *eventually strongly monotone* if there is a $t_0 > 0$ such that

$$u(t, \omega, x) \ll u(t, \omega, y) \quad \text{for each } t > t_0,\ \omega \in \Omega \text{ and } x, y \in X \text{ with } x < y\,,$$

whenever they are defined.

3 Non-autonomous FDEs with Finite Delay

Throughout this section we will study the monotone skew-product semiflow induced by a finite-delay functional differential equation, as we explain in what follows. In this setting, the strongly ordered Banach space is the set $X = C([-r, 0], \mathbb{R}^m)$, whose ordering relies on the usual one of $\mathbb{R}^m$,

$$v \leq w \iff v_j \leq w_j \quad \text{for } j = 1, \ldots, m,$$

$$v < w \iff v \leq w \quad \text{and} \quad v_j < w_j \quad \text{for some } j \in \{1, \ldots, m\},$$

$$v \ll w \iff v_j < w_j \quad \text{for } j = 1, \ldots, m,$$

where v_j represents the j-th component of any point $v \in \mathbb{R}^m$.

The subset $X_+ = \{x \in X \mid x(s) \geq 0 \text{ for each } s \in [-r, 0]\}$ is a normal positive cone in X. Since its interior is nonempty, this cone induces a strong order relation on X,

$$x \leq y \iff x(s) \leq y(s) \quad \text{for each } s \in [-r, 0],$$

$$x < y \iff x \leq y \quad \text{and} \quad x \neq y,$$

$$x \ll y \iff x(s) \ll y(s) \quad \text{for each } s \in [-r, 0].$$

The spaces $\mathbb{R}^m$ and X will be respectively endowed with the maximum norm $\|v\| = \max_{j=1,\ldots,m} |v_j|$ and with the supremum norm $\|x\|_\infty = \sup_{t \in [-r,0]} \|x(t)\|$, both of them monotone. As usual, given an interval $I \subset \mathbb{R}$, a point $t \in \mathbb{R}$ with $[t - r, t] \subset I$, and a continuous function $z : I \to \mathbb{R}^m$, z_t will denote the element of X given by $z_t(s) = z(t + s)$ for $s \in [-r, 0]$.

Our starting point is the non-autonomous finite-delay FDE

$$z' = f(t, z_t), \tag{3.14}$$

defined by a function $f : \mathbb{R} \times X \to \mathbb{R}^m$ satisfying the following conditions:

(C1) f is C^1-admissible; i.e., f is C^1 in the variable x, and the functions

$$\mathbb{R} \times X \to \mathbb{R}^m, (t, x) \mapsto f(t, x), \qquad \mathbb{R} \times X \to L(X, \mathbb{R}^m), (t, x) \mapsto f_x(t, x)$$

are admissible. As usual, $L(X, \mathbb{R}^m)$ represents the set of linear maps from X to $\mathbb{R}^m$. We recall that given a Banach space Y, a map $g : \mathbb{R} \times X \to Y$ is *admissible* if the family $\{g(t, \cdot) \mid t \in \mathbb{R}\}$ is equicontinuous at every $x_0 \in X$, and for each $x_0 \in X$, $\{g(t, x_0) \mid t \in \mathbb{R}\}$ is a relatively compact subset of Y;

(C2) f takes $\mathbb{R} \times B$ into a bounded set of $\mathbb{R}^m$ for any bounded subset B of X.

As in the ODEs case, let Ω be the *hull* of f, defined as the closure in the topology of uniform convergence on compact sets of the set of time-translated maps $\{f_t \mid t \in \mathbb{R}\}$, with $f_t(s, x) = f(t + s, x)$. Property (C1) and the separability of X guarantee that Ω is a compact metric space (see [33]). It is possible to define a real continuous flow on Ω, as $\sigma : \mathbb{R} \times \Omega \rightarrow \Omega$, $(t, \omega) \mapsto \omega \cdot t$ with $\omega \cdot t(s, x) = \omega(t + s, x)$. It is also known that each $\omega \in \Omega$ is also a C^1-admissible function and f has a unique extension to a continuous function $F : \Omega \times X \rightarrow \mathbb{R}^m$, $(\omega, x) \mapsto \omega(0, x)$. Thus we obtain the family of finite-delay equations

$$z' = F(\omega \cdot t, z_t), \quad \omega \in \Omega. \tag{3.15}$$

Note that the element of this family corresponding to $\omega = f$ is our initial equation (3.14). A recurrence property is also assumed on f, namely,

(C3) $(\Omega, \sigma, \mathbb{R})$ is a minimal flow.

This is satisfied, for instance, when f is a *uniformly almost periodic* or, more generally, a *uniformly almost automorphic function*, i.e., when it is admissible and almost periodic or almost automorphic in $t \in \mathbb{R}$ (see [101]).

Fix now an element $(\omega, x) \in \Omega \times X$. Condition (C1) ensures the existence of a unique maximal solution $z : [-r, \beta) \rightarrow \mathbb{R}^m$ of the initial value problem given by the Eq. (3.15) corresponding to ω and by the initial condition $z|_{[-r,0]} = x$, which in addition varies continuously with respect to the initial data (see e.g. Hale [31] or Hale and Verduyn Lunel [32]). We represent this maximal solution by $z(t, \omega, x)$. In this context, maximality means that the solution cannot be continued to the right of β. Therefore, the family (3.15) induces a *local* skew-product semiflow

$$\tau : \mathbb{R}^+ \times \Omega \times X \longrightarrow \Omega \times X, \quad (t, \omega, x) \mapsto (\omega \cdot t, u(t, \omega, x)), \tag{3.16}$$

where $u(t, \omega, x)$ is the element of X defined by $u(t, \omega, x)(s) = z(t + s, \omega, x)$ for each $s \in [-r, 0]$. Since f is C^1 in x we deduce that u is C^1 in x, and $u_x(t, \omega, x)v$ is continuous in $t > 0$, $(\omega, x) \in \Omega \times X$ and $v \in X$.

If $K \subset \Omega \times X$ is a compact positively invariant set, we can define a linear skew-product semiflow called the *linearized skew-product semiflow* of (3.16)

$$L : \mathbb{R}^+ \times K \times X \longrightarrow K \times X, \quad (t, (\omega, x), v) \mapsto (\tau(t, \omega, x), u_x(t, \omega, x)v).$$

We note that u_x satisfies the following semi-cocycle property

$$u_x(t + s, \omega, x) = u_x(t, \tau(s, \omega, x)) u_x(s, \omega, x), \quad s, t \in \mathbb{R}^+, \quad (\omega, x) \in K.$$

Definition 3.1. For $(\omega, x) \in K$, we define the *Lyapunov exponent* $\lambda(\omega, x)$ as

$$\lambda(\omega, x) = \limsup_{t \to \infty} \frac{\ln \|u_x(t, \omega, x)\|}{t}.$$

The number $\lambda_K = \sup_{(\omega,x) \in K} \lambda(\omega, x)$ is called the *upper Lyapunov exponent on K*. If $\lambda_K \leq 0$, then K is said to be *linearly stable*.

We refer the reader to Chicone and Latushkin [16] for a systematic treatment of recent results in the area of linear differential equations on Banach spaces and infinite dimensional dynamical systems, in terms of spectral properties of the associated evolution semigroup.

On its turn, condition (C2) guarantees that if $z(t, \omega, x)$ is a bounded solution, then it is defined in $[-r, \infty)$; hence $u(t, \omega, x)$ exists for all $t \geq 0$ and the forward orbit $\{\tau(t, \omega, x) \mid t \geq 0\}$ is relatively compact in $\Omega \times X$.

Next, we introduce a topological tool that we call the *section map*, which will prove to be useful in the sequel. Given a compact and positively invariant set $K \subset \Omega \times X$, let us introduce the projection set of K into the fiber space

$$K_X = \{x \in X \mid \text{ there exists } \omega \in \Omega \text{ such that } (\omega, x) \in K\} \subseteq X.$$

From the compactness of K it is immediate to show that also K_X is a compact subset of X. Let $\mathscr{P}_c(K_X)$ denote the set of closed subsets of K_X, endowed with the Hausdorff metric ρ, that is, for any two sets $A, B \in \mathscr{P}_c(K_X)$,

$$\rho(A, B) = \sup\{\alpha(A, B),\ \alpha(B, A)\},$$

where $\alpha(A, B) = \sup\{r(a, B) \mid a \in A\}$ and $r(a, B) = \inf\{\mathsf{d}(a, b) \mid b \in B\}$. Then, define the so-called *section map*

$$\Omega \longrightarrow \mathscr{P}_c(K_X), \quad \omega \mapsto K_\omega = \{x \in X \mid (\omega, x) \in K\}. \tag{3.17}$$

Due to the minimality of Ω and the compactness of K, the set K_ω is nonempty for every $\omega \in \Omega$; besides, the map is trivially well-defined.

Lemma 3.2. *There exists a residual set $\Omega_0 \subseteq \Omega$ of continuity points for the section map* (3.17) *associated to a compact and positively invariant set $K \subset \Omega \times X$. Moreover, if each point in K admits a backward extension then Ω_0 is positively invariant.*

Proof. We refer the reader to Lemma 3.2 in Novo et al. [76] for the first part of the proof. For the second part of the statement, let $\omega \in \Omega_0$ and $t_0 > 0$, we claim that $\omega \cdot t_0 \in \Omega_0$, that is, $\lim_{n \to \infty} \rho(K_{\omega_n}, K_{\omega \cdot t_0}) = 0$ for each sequence $\omega_n \to \omega \cdot t_0$. First of all note that $\rho(A, B) < \varepsilon$ if and only if $A \subset N(\varepsilon, B)$ and $B \subset N(\varepsilon, A)$ where $N(\varepsilon, A) = \{x \mid \exists y \in A \text{ with } \mathsf{d}(x, y) < \varepsilon\}$. Moreover, since each point in K admits a backward extension, for each $x \in K_{\omega \cdot t_0}$ we can find a point $y \in K_\omega$ such that $u(t_0, \omega, y) = x$, and analogously for each point $x_n \in K_{\omega_n}$, there is $y_n \in K_{\omega_n \cdot (-t_0)}$ with $u(t_0, \omega_n \cdot (-t_0), y_n) = x_n$. From these facts together with the uniform continuity of τ_{t_0} in $\Omega \times K_X$ and $\lim_{n \to \infty} \rho(K_{\omega_n \cdot (-t_0)}, K_\omega) = 0$ it is easy to finish the proof. $\qquad \square$

Lemma 2.4 in subsection 2.3 proves that minimal sets of skew-product flows generated by recurrent non-autonomous scalar ODEs are a.a. extensions of the base flow. This result is no longer true for general scalar delay equations. As shown in

Alonso et al. [2], the result is true provided that the delay is small enough. The proof is based on a result on the existence of special solutions by Driver [21].

More specifically, if instead of (C1) we assume

(C1)* f is an admissible function satisfying a global Lipschitz condition:

$$|f(t, x_1) - f(t, x_2)| \le L\,|x_1 - x_2| \quad \text{for all } x_1, x_2 \in X \,, \ t \in \mathbb{R}\,,$$

a certain constant $L > 0$, and $Lre < 1$,

then for each $\omega \in \Omega$, $t_0 \in \mathbb{R}$ and $v_0 \in \mathbb{R}^m$ there exists a unique solution of (3.15) defined on the real line, which we denote by $z(\omega, t_0, v_0)(t)$, such that

$$z(\omega, t_0, v_0)(t_0) = v_0 \quad \text{and} \quad \sup_{t \le t_0} |z(\omega, t_0, v_0)(t)|\,e^{t/r} < \infty\,.$$

The solutions $z(\omega, t_0, v_0)(t)$ are the so-called *special solutions*. As shown in [2] for the scalar case, although the proof remains valid for our case, $z(\omega, t_0, v_0)(t)$ is the unique solution of the Cauchy problem

$$\begin{cases} v' = G(\omega{\cdot}t, v) \\ v(t_0) = v_0 \end{cases} \tag{3.18}$$

where $G(\omega, v) = F(\omega, z(\omega, 0, v)_0)$, which is also a globally Lipschitzian function in v. As a consequence, we prove the following result.

Theorem 3.3. *Under assumptions* (C1)*, (C2) *and* (C3), *let M be a minimal set for the skew-product semiflow* (3.16) *induced by the family* (3.15). *Then*

(i) The semiflow $(M, \tau, \mathbb{R}^+)$ admits a flow extension.
(ii) In the scalar case, $m = 1$, (M, τ) is an almost automorphic extension of the base flow (Ω, σ), and hence an almost automorphic minimal set if f is uniformly almost periodic.

Proof. (i) By Proposition 2.1, part II in [101], every $(\omega, x) \in M$ admits a backward orbit, and by Theorem 2.3, part II in [101], M admits a flow extension if and only if every point admits a unique backward extension. Therefore, let us assume by contradiction that there is a point $(\omega, x) \in M$ with two backward extensions $\{(\omega{\cdot}t, x_t) \mid t \in \mathbb{R}\}$ and $\{(\omega{\cdot}t, y_t)) \mid t \in \mathbb{R}\}$. We assume without loss of generality that $t = 0$ is the first time in which the two semiorbits coincide. Then, $x_t(s) = y_t(s) = z(t + s, \omega, x)$ for each $t \ge 0$ and $s \in [-1, 0]$, and $x_t \neq y_t$ for $t < 0$. Moreover, $x_t(0)$ and $y_t(0)$ are solutions of (3.15) satisfying

$$x_0(0) = x(0) \quad \text{and} \quad \sup_{t \le 0} |x_t(0)|e^{t/r} < \infty\,,$$

$$y_0(0) = x(0) \quad \text{and} \quad \sup_{t \le 0} |y_t(0)|e^{t/r} < \infty\,.$$

Then, since the special solution $z(\omega, 0, x(0))(t)$ is unique it must coincide for each $t \in \mathbb{R}$ both with the functions $x_t(0)$ and $y_t(0)$, contradicting that $x_t \neq y_t$ for $t < 0$.

(ii) Since, as shown in (i), all the solutions with initial data in M are special solutions, the proof, done in Theorem 3.2 in [2], follows from Lemma 2.4 after constructing an isomorphism between the restriction of the flow induced by the family (3.18) to $M_0 = \{(\omega, x(0)) \mid (\omega, x) \in M\}$, and the flow (M, τ).   □

However, even though frequently small delays can be neglected, small delays can also have large effects, as remarked in Kuang [56]. In particular, the small delay critical value that destabilizes the local stability of a steady state for some delayed logistic equations is shown.

In order to get the semiflow to be monotone, we also assume on f a *quasimonotone condition* of Kamke type (see [105]),

(C4) If $x \leq y$ and $x_j(0) = y_j(0)$ holds for some $j \in \{1, \ldots, m\}$, then $f_j(t, x) \leq f_j(t, y)$ for each $t \in \mathbb{R}$.

It is easily seen that this condition is simultaneously satisfied by each function in Ω. Property (C4) has important consequences for the semiflow $(\Omega \times X, \tau, \mathbb{R}^+)$ given by (3.16): as checked e.g. in [105],

$$u(t, \omega, x) \leq u(t, \omega, y) \quad \text{for } t \geq 0, \ \omega \in \Omega \text{ and } x, y \in X \text{ with } x \leq y,$$
$$u(t, \omega, x) \ll u(t, \omega, y) \quad \text{for } t \geq 0, \ \omega \in \Omega \text{ and } x, y \in X \text{ with } x \ll y,$$
(3.19)

whenever they are defined. In particular, $(\Omega \times X, \tau, \mathbb{R}^+)$ is a monotone semiflow.

3.1 Cooperative and Irreducible Systems of Finite Delay Equations

In this section we consider the case in which (3.14) takes the form

$$z'(t) = f(t, z(t), z(t-1)),$$
(3.20)

where $f : \mathbb{R} \times \mathbb{R}^m \times \mathbb{R}^m \to \mathbb{R}^m$, $(t, v, w) \mapsto f(t, v, w)$ is a C^1-admissible and uniformly almost periodic or uniformly almost automorphic function, and the family on the hull is
$$z'(t) = F(\omega \cdot t, z(t), z(t-1)), \quad \omega \in \Omega.$$
(3.21)

Note that this case is included in the general previous formulation because (3.20) can be expressed as $z'(t) = g(t, z_t)$ with $g : \mathbb{R} \times X \to \mathbb{R}^m$, $(t, x) \mapsto f(t, x(0), x(-1))$ and the phase space is $X = C([-1, 0], \mathbb{R}^m)$.

The induced skew-product semiflow is

$$\tau : \mathbb{R}^+ \times \Omega \times X \longrightarrow \Omega \times X, \quad (t, \omega, x) \mapsto (\omega \cdot t, u(t, \omega, x)),$$
(3.22)

where, as before, $u(t, \omega, x)(s) = z(t + s, \omega, x)$ for $s \in [-1, 0]$.

As explained before, from (C2), a point $(\omega_0, x_0) \in \Omega \times X$ with a bounded semi-trajectory $\{(\omega_0 \cdot t, u(t, \omega_0, x_0)) \mid t \geq 0\}$ provides a relatively compact forward orbit; hence the omega-limit set $\mathcal{O}(\omega_0, x_0)$ exists, and contains at least a minimal set M. As we will recall in section 4.1, uniform stability implies that the restriction of the semiflow to M admits a flow extension. The next result provides another condition for this fact based only on F. Since $F \colon \Omega \times \mathbb{R}^m \times \mathbb{R}^m \to \mathbb{R}^m$, $(\omega, v, w) \mapsto F(\omega, v, w)$, as usual, F_w denotes the matrix of partial derivatives with respect to w.

Proposition 3.4. *We consider* $(\omega_0, x_0) \in \Omega \times X$ *with a bounded semitrajectory* $\{(\omega_0 \cdot t, u(t, \omega_0, x_0)) \mid t \geq 0\}$, *and let* M *be a minimal set* $M \subset \mathcal{O}(\omega_0, x_0)$. *Assume that* $\det F_w(\omega, x(0), x(-1)) \neq 0$ *for each* $(\omega, x) \in M$. *Then the semiflow* $(M, \tau, \mathbb{R}^+)$ *admits a flow extension.*

Proof. As in Theorem 3.3, let us assume by contradiction that there is $(\omega^*, x^*) \in M$ with two backward extensions $\{(\omega^* \cdot t, x_t) \mid t \in \mathbb{R}\}$ and $\{(\omega^* \cdot t, y_t)) \mid t \in \mathbb{R}\}$. We assume without loss of generality that $t = 0$ is the first time in which the two semiorbits coincide. Then, $x_t(s) = y_t(s) = z(t + s, \omega^*, x^*)$ for each $t \geq 0$ and $s \in [-1, 0]$ and $x_t \neq y_t$ for $t < 0$. As a consequence, if we denote $h(t) = x_t(0) - y_t(0)$,

$$h'(t) = F(\omega^* \cdot t, x_t(0), x_t(-1)) - F(\omega^* \cdot t, y_t(0), y_t(-1)) = 0$$

for each $t \geq -1$. Since $t = 0$ is the first time in which the two semiorbits coincide, we can find a sequence $t_n \uparrow 0$ with $-1 \leq t_n < 0$, $x_{t_n}(0) = y_{t_n}(0)$ but $x_{t_n}(-1) \neq y_{t_n}(-1)$. Moreover, from $h'(t_n) = 0$ and the mean value theorem, we deduce that

$$\begin{aligned} 0 &= F(\omega^* \cdot t_n, x_{t_n}(0), x_{t_n}(-1)) - F(\omega^* \cdot t_n, y_{t_n}(0), y_{t_n}(-1)) \\ &= D(t_n)\,(x_{t_n}(-1) - y_{t_n}(-1)) \end{aligned}$$

where $D(t_n) = [d_{ij}(t_n)]$ satisfies

$$d_{ij}(t_n) = \frac{\partial F_i}{\partial w_j}(\omega^* \cdot t_n, x_{t_n}(0), s_{i,n} x_{t_n}(-1) + (1 - s_{i,n}) y_{t_n}'(-1))$$

for some $s_{i,n} \in (0, 1)$, $i = 1, \ldots, m$. Consequently, for each n, $\det D(t_n) = 0$, which contradicts, since $x_{t_n}(-1)$ and $y_{t_n}(-1)$ converge to $x^*(-1)$ as n goes to ∞, that $\det F_w(\omega^*, x^*(0), x^*(-1)) \neq 0$. $\square$

Now we provide sufficient conditions for the induced semiflow (3.22) to be eventually strongly monotone, where as before, $u(t, \omega, x)(s) = z(t + s, \omega, x)$ for $s \in [-1, 0]$.

We recall that an $m \times m$ matrix $A = [a_{ij}]$ is irreducible if for any nonempty proper subset I of $N_m = \{1, \ldots, m\}$, there are $i \in I$ and $j \in J = N_m - I$ such that $a_{ij} \neq 0$.

Definition 3.5. The delay system (3.20) is:

- *Cooperative* if for each $v, w \in \mathbb{R}^m$, $t \in \mathbb{R}$,

$$\frac{\partial f_i}{\partial v_j}(t, v, w) \geq 0 \quad \text{for each } i \neq j \quad \text{and} \quad \frac{\partial f_i}{\partial w_j}(t, v, w) \geq 0 \quad \text{for } i, j = 1, 2, \ldots m.$$

- *Irreducible* with respect to $z(t)$ if there is a continuous function $\delta: [0, \infty) \to (0, \infty)$ such that if two nonempty subsets I, J form a partition of $N_m = \{1, \ldots, m\}$, then for any $v, w \in \mathbb{R}^m$, $t \in \mathbb{R}$, there are $i \in I$ and $j \in J = N_m - I$ with

$$\left| \frac{\partial f_i}{\partial v_j}(t, v, w) \right| \geq \delta(\max(\|v\|, \|w\|)) > 0.$$

- *Irreducible* with respect to $z(t-1)$ if there is a continuous function $\delta: [0, \infty) \to (0, \infty)$ such that if two nonempty subsets I, J form a partition of $N_m = \{1, \ldots, m\}$, then for any $v, w \in \mathbb{R}^m$, $t \in \mathbb{R}$, there are $i \in I$ and $j \in J = N_m - I$ with

$$\left| \frac{\partial f_i}{\partial w_j}(t, v, w) \right| \geq \delta(\max(\|v\|, \|w\|)) > 0.$$

- *Irreducible* with respect to $z(t) + z(t-1)$ if there is a continuous function $\delta: [0, \infty) \to (0, \infty)$ such that if two nonempty subsets I, J form a partition of $N_m = \{1, \ldots, m\}$, then for any $v, w \in \mathbb{R}^m$, $t \in \mathbb{R}$, there are $i \in I$ and $j \in J = N_m - I$ with

$$\left| \frac{\partial f_i}{\partial v_j}(t, v, w) \right| + \left| \frac{\partial f_i}{\partial w_j}(t, v, w) \right| \geq \delta(\max(\|v\|, \|w\|)) > 0.$$

- *Strongly increasing* if there is a continuous function $\delta: [0, \infty) \to (0, \infty)$ such that

$$\left| \frac{\partial f_i}{\partial w_j}(t, v, w) \right| \geq \delta(\max(\|v\|, \|w\|)) \quad \text{for } v, w \in \mathbb{R}^m, \, t \in \mathbb{R} \text{ and } i, j = 1, 2, \ldots m.$$

It is easy to see that if (3.20) is a cooperative system, an irreducible system or a strongly increasing system then so are the systems (3.21).

Proposition 3.6. *Consider* (3.20) *and its induced skew-product semiflow* (3.22).

(i) *If* (3.20) *is cooperative then* (3.22) *is monotone.*

(ii) *If* (3.20) *is cooperative and irreducible with respect to* $z(t)$ *and* $x < y$, *then* $u(1, \omega, x) = u(1, \omega, y)$ *or* $u(t, \omega, x) \ll u(t, \omega, y)$ *for each* $t \geq 2$ *whenever they are defined.*

(iii) *If* (3.20) *is cooperative and irreducible with respect to* $z(t-1)$ *and* $x < y$, *then* $u(t, \omega, x) \ll u(t, \omega, y)$ *for each* $t \geq m + 1$ *whenever they are defined, i.e.,* (3.22) *is eventually strongly monotone.*

(iv) *If (3.20) is cooperative and irreducible with respect to $z(t) + z(t-1)$ and $x < y$, then $u(1, \omega, x) = u(1, \omega, y)$ or $u(t, \omega, x) \ll u(t, \omega, y)$ for each $t \geq m + 1$ whenever they are defined.*

(v) *If (3.20) is cooperative and strongly increasing and $x < y$, then $u(t, \omega, x) \ll u(t, \omega, y)$ for each $t \geq 2$, whenever they are defined, i.e., (3.22) is eventually strongly monotone.*

Proof. (i) Follows from Theorem 5.1.1 in [105] because if (3.20) is cooperative then the quasimonotone condition (C4) holds.

(ii) We fix $\omega \in \Omega$, $x < y$ and denote $h(t) = z(t, \omega, y) - z(t, \omega, x)$. From (i), $h(t) \geq 0$ for each $t \geq -1$, whenever defined. Moreover,

$$h'(t) = F(\omega \cdot t, z(t, \omega, y), z(t-1, \omega, y)) - F(\omega \cdot t, z(t, \omega, x), z(t-1, \omega, x))$$
$$= A(t) h(t) + B(t) h(t-1)$$

$$(3.23)$$

with

$$A(t) = \int_0^1 F_v(\omega \cdot t, s\, z(t, \omega, y) + (1-s)\, z(t, \omega, x), z(t-1, \omega, y))\, ds,$$

$$B(t) = \int_0^1 F_w(\omega \cdot t, z(t, \omega, x), s\, z(t-1, \omega, y) + (1-s)\, z(t-1, \omega, x))\, ds.$$

Assume that $u(1, \omega, x) \neq u(1, \omega, y)$. Then, there is a $t_1 \in [0, 1)$ such that $h(t_1) = z(t_1, \omega, y) - z(t_1, \omega, x) > 0$. From the assumptions, system (3.23) satisfies the quasimonotone condition (C4) and all the entries of $B(t)$ are positive. Then again the comparison Theorem 5.1.1 in [105] shows that for each $t \geq t_1$ we have $h(t) \geq z(t, t_1, v_1)$, the solution of $z' = A(t) z$ with initial data $v_1 = h_{t_1}$ at time t_1. Moreover, also from the assumptions, $A(t)$ is an irreducible matrix and hence Theorem 4.1.1 in [105] proves that $z(t, t_1, v_1) \gg 0$ for each $t > t_1$, and hence $z(t, \omega, y) - z(t, \omega, x) \gg 0$ for $t > t_1$ which implies that $u(t, \omega, x) \ll u(t, \omega, y)$ for each $t \geq 2$, as stated.

(iii) We fix $\omega \in \Omega$ and $x < y$. With the same notation as above, since $y - x > 0$ there is a $t_0 \in [-1, 0)$ such that $h(t_0) = y(t_0) - x(t_0) > 0$. We assume without loss of generality that $h_1(t_0) > 0$. Now, $B(t_0 + 1)$ is an irreducible matrix and we can find $j_1 \in \{2, \ldots, m\}$ such that $b_{j_1,1}(t_0 + 1) \geq \delta > 0$.

If $h_{j_1}(t_0 + 1) > 0$, a comparison argument and Lemma 5.1.3 in [105] proves that $h_{j_1}(t) > 0$ for each $t \geq t_0 + 1$, and hence for each $t \geq 1$. If $h_{j_1}(t_0 + 1) = 0$, since all the entries of $B(t)$ are positive, $a_{i,j}(t) \geq 0$ for $i \neq j$, $b_{j_1,1}(t_0 + 1) > 0$ and $h_1(t_0) > 0$, we deduce from (3.23) that $h'_{j_1}(t_0 + 1) > 0$. Consequently, $h_{j_1}(t) > 0$ for t in a right neighborhood of $t_0 + 1$ and, as before $h_{j_1}(t) > 0$ for each $t \geq 1$ whenever defined.

Analogously, in a second step, by choosing $j_2 \in \{1, \ldots, m\} - \{j_1\}$ such that $b_{j_2,j_1}(t_1 + 1) \geq \delta > 0$ for some fixed $t_1 \in [1, 2)$, we deduce that $h_{j_2}(t) > 0$ for $t \geq 2$. In a recursive way, we obtain $\{j_1, \ldots, j_m\} = \{1, \ldots, m\}$ such that $h_{j_k}(t) > 0$ for $t \geq k$, and we conclude that $h(t) \gg 0$ for $t \geq m$, and consequently $u(t, \omega, x) \ll u(t, \omega, y)$ for each $t \geq m + 1$ whenever they are defined, as stated.

(iv) We fix $\omega \in \Omega$ and $x < y$. With the same notation as above, assume that $u(1, \omega, x) \neq u(1, \omega, y)$. Then, as in (ii), there is a $t_1 \in [0, 1)$ such that $h(t_1) = z(t_1, \omega, y) - z(t_1, \omega, x) > 0$, i.e., there is $j_1 \in \{1, \ldots, m\}$ such that $h_{j_1}(t_1) > 0$, and as in (iii) we deduce that $h_{j_1}(t) > 0$ for $t \geq t_1$ and hence for each $t \geq 1$. Since now $A(t_1 + 1) + B(t_1 + 1)$ is an irreducible matrix, we can find $j_2 \in \{1, \ldots, m\} - \{j_1\}$ such that $a_{j_2, j_1}(t_1 + 1) + b_{j_2, j_1}(t_1 + 1) \geq \delta > 0$. Then, as in (iii), we conclude that $h_{j_2}(t) > 0$ for $t \geq 2$ and, in a recursive way, the existence of $\{j_1, \ldots, j_m\} = \{1, \ldots, m\}$ such that $h_{j_k}(t) > 0$ for $t \geq k$, to finish the proof.

(v) We fix $\omega \in \Omega$ and $x < y$. With the same notation as above, since $y - x > 0$ there is a $t_0 \in [-1, 0)$ such that $h(t_0) = y(t_0) - x(t_0) > 0$. As a consequence, a similar argument to the one in (iii), but now taking into account that all the entries of $B(t)$ are strictly positive, yields to $h(t) \gg 0$ for $t \geq 1$ and hence $u(t, \omega, x) \ll u(t, \omega, y)$ for each $t \geq 2$ whenever they are defined, as stated.

$\square$

The next result characterizes, in the almost periodic and cooperative case, the linearly stable minimal subsets (see Definition 3.1) admitting flow extension. We refer to [101] for the definition and properties of the frequency module of an almost automorphic function, introduced in subsection 2.2.

Theorem 3.7. *Consider* (3.20) *and its induced skew-product semiflow* (3.22). *If*

(a) f *is* C^2-*admissible and uniformly almost periodic;*
(b) (3.20) *is cooperative;*
(c) K *is a linearly stable minimal subset, i.e.,* $\lambda_K \leq 0$, *and the semiflow* $(K, \tau, \mathbb{R}^+)$ *admits a flow extension;*
(d) $F_w(\omega, x(0), x(-1))$ *is an irreducible matrix for each* $(\omega, x) \in K$;

then, the flow $(K, \tau, \mathbb{R})$ *is almost automorphic and there is an integer* $N \geq 1$ *such that for each almost automorphic point* $(\omega, x) \in K$, *the frequency module of the almost automorphic function* $z(t, \omega, x)$ *satisfies* $N \mathcal{M}(z(t, \omega, x)) \subset \mathcal{M}(f)$.

Proof. We consider $(\omega, x) \in K$ and the linearized equation

$$y'(t) = A(t) \, y(t) + B(t) \, y(t - 1), \tag{3.24}$$

with $A(t) = F_v(\omega \cdot t, z(t, \omega, x), z(t - 1, \omega, x))$, $B(t) = F_w(\omega \cdot t, z(t, \omega, x), z(t - 1, \omega, x))$. From (b) and (d) the system (3.24) is cooperative and irreducible with respect to $y(t - 1)$ and, from Proposition 3.6 (iii) the induced semiflow, which coincides with the linearized skew-product semiflow

$$L : \mathbb{R}^+ \times K \times X \longrightarrow K \times X, \quad (t, (\omega, x), v) \mapsto (\tau(t, \omega, x), u_x(t, \omega, x) v),$$

is eventually strongly monotone. In particular $u_x(t, \omega, x) v \gg 0$ for each $(\omega, x) \in K$, $v > 0$ and $t \geq m + 1$, that is, $u_x(m + 1, \omega, x)$ is a strongly positive operator. Since, in addition, $u_x(m + 1, \omega, x)$ is compact and $(K, \tau, \mathbb{R}^+)$ admits a flow extension, Theorem 4.4, part II in [101] provides a continuous separation for K (see Definition 4.6, part II in [101] and Poláčik–Tereščák [90]).

Next, from Lemma 3.2 there exists a positively invariant residual set $\Omega_0 \subseteq \Omega$ of continuity points for the section map (3.17) associated to K. We claim that for each $\omega \in \Omega_0$ there are no ordered points in $K_\omega = \{x \in X \mid (\omega, x) \in K\}$. First we check that there are not strongly ordered points. Assume by contradiction that there are two strongly ordered points (ω_0, x_0), (ω_0, y_0) with $x_0 \ll y_0$ and $\omega_0 \in \Omega_0$, and let (ω_0, z_0) be the maximal point satisfying $x_0 \ll z_0$. Since K is minimal and $(K, \tau, \mathbb{R}^+)$ admits a flow extension, there is a sequence $s_n \downarrow -\infty$ such that

$$\lim_{n \to \infty} (\omega \cdot s_n, u(s_n, \omega_0, z_0)) = (\omega_0, x_0). \tag{3.25}$$

Moreover, ω_0 is a continuity point of the section map, and we can find a sequence $(\omega_0 \cdot s_n, x_n) \in K$ such that $\lim_{n \to \infty} (\omega_0 \cdot s_n, x_n) = (\omega_0, z_0)$. Again, the flow extension implies that $x_n = u(s_n, \omega_0, z_n)$ for some $(\omega_0, z_n) \in K$, and hence

$$\lim_{n \to \infty} (\omega_0 \cdot s_n, u(s_n, \omega_0, z_n)) = (\omega_0, z_0). \tag{3.26}$$

Since $x_0 \ll z_0$, from (3.25) and (3.26) we deduce that there is an n_0 such that $u(s_{n_0}, \omega_0, z_0)) \ll u(s_{n_0}, \omega_0, z_{n_0})$, and the monotonicity of the flow yields to $z_0 \ll z_{n_0}$ contradicting that z_0 was maximal.

Next, we assume that there are two ordered points (ω_0, x_0), $(\omega_0, y_0) \in K$ with $x_0 < y_0$. Then, $u_x(m+1, \omega_0, x_0) (y_0 - x_0) \gg 0$ and $u_x(m+1, \omega_0, y_0) (y_0 - x_0) \gg 0$ because $u_x(m + 1, \omega, x)$ is a strongly positive operator. Since, in addition,

$$u(m+1, \omega_0, y_0) - u(m+1, \omega_0, x_0) = \int_0^1 u_x(m+1, \omega_0, s\, y_0 + (1-s)\, x_0)\, (y_0 - x_0)\, ds\,,$$

we conclude that $u(m + 1, \omega_0, y_0) - u(m + 1, \omega_0, x_0) \gg 0$, contradicting that there are no strongly ordered pairs in $K_{\omega_0 \cdot (m+1)}$.

Finally, Theorem 4.5, part II in [101] and similar arguments to those in Theorem 6.5, part III in [101] finish the proof. □

As Takáč showed in [109] the case $N > 1$ can occur. To clarify this comment in the simplest case, $N = 2$, we recall the four-dimensional π-periodic system of ODEs

$$\begin{aligned}
z_1' &= z_4 + 4 z_1 (\cos^2 t - z_1^2) \\
z_2' &= z_1 + 4 z_2 (\sin^2 t - z_2^2) \\
z_3' &= z_2 + 4 z_3 (\cos^2 t - z_3^2) \\
z_4' &= z_3 + 4 z_4 (\sin^2 t - z_4^2)\,.
\end{aligned} \tag{3.27}$$

It is cooperative and strongly irreducible and, hence, the induced flow is strongly monotone. It has a global attractor, and $(\cos t, \sin t, -\sin t, \cos t)$ is a hyperbolic 2π-periodic solution. Its omega-limit provides a 2-copy of the base flow, defined over the circumference $\mathbb{R}/\pi\mathbb{Z}$.

Similarly, the four-dimensional π-periodic system with finite delay

$$z_1' = z_4(t - 2\pi) + 4z_1(\cos^2 t - z_1^2)$$

$$z_2' = z_1(t - 2\pi) + 4z_2(\sin^2 t - z_2^2)$$

$$z_3' = z_2(t - 2\pi) + 4z_3(\cos^2 t - z_3^2)$$

$$z_4' = z_3(t - 2\pi) + 4z_4(\sin^2 t - z_4^2),$$

which is cooperative and irreducible with respect ro $z(t-1)$, induces an eventually strongly monotone skew-product semiflow, and the omega-limit set of the 2π-periodic solution $(\cos t, \sin t, -\sin t, \cos t)$ provides a 2-copy of the base flow, defined over the circumference $\mathbb{R}/\pi\mathbb{Z}$.

Based on these examples and Johnson's example (3), Shen and Yi [101] construct an example of a cooperative and strongly irreducible ODE system which exhibits the subharmonic phenomena indicated in Theorem 3.7 with $N > 1$, and in addition, since the solution is almost automorphic but not almost periodic, the omega-limit set is not a N-copy of the base flow.

In the rest of the section we will impose additional hypotheses under which the omega-limit sets are 1-copies of the base flow.

3.2 Semicontinuous Equilibria and Almost Automorphic Extensions

In this subsection, we consider the monotone skew-product semiflow (3.16) induced by Eq. (3.14) satisfying Hypotheses (C1)–(C4). However, some of the results apply to a general monotone skew-product semiflow over $\Omega \times X$ for a strongly ordered Banach space X and a minimal real continuous flow $(\Omega, \sigma, \mathbb{R})$ over a compact metric space Ω.

Definition 3.8. A measurable map $a : \Omega \to X$ such that $u(t, \omega, a(\omega))$ is defined for any $t \geq 0$ is

(a) An *equilibrium* if $a(\omega \cdot t) = u(t, \omega, a(\omega))$ for any $\omega \in \Omega$ and $t \geq 0$,
(b) A *super-equilibrium* if $a(\omega \cdot t) \geq u(t, \omega, a(\omega))$ for any $\omega \in \Omega$ and $t \geq 0$, and
(c) A *sub-equilibrium* if $a(\omega \cdot t) \leq u(t, \omega, a(\omega))$ for any $\omega \in \Omega$ and $t \geq 0$.

We will call *semi-equilibrium* to either a super or a sub-equilibrium.

The following dynamical interpretation of the concept of a super and a sub-equilibrium appeared in Novo et al. [72] in a topological framework.

Definition 3.9. A super-equilibrium (resp. sub-equilibrium) $a : \Omega \to X$ is *semicontinuous* if the following properties hold:

(1) $\Gamma_a = \mathrm{closure}_X\{a(\omega) \mid \omega \in \Omega\}$ is a compact subset of X, and
(2) $C_a = \{(\omega, x) \mid x \leq a(\omega)\}$ (resp. $C_a = \{(\omega, x) \mid x \geq a(\omega)\}$) is a closed subset of $\Omega \times X$.

An equilibrium is *semicontinuous* in any of these cases.

A semicontinuous semi-equilibrium does always have a residual subset of continuity points, as it is derived from the next result, proved in [72].

Proposition 3.10. *Let $a\colon \Omega \to X$ be a map satisfying (1) and (2) in Definition 3.9. Then, it is continuous over a residual subset $\Omega_0 \subset \Omega$.*

A semicontinuous semi-equilibrium provides a minimal set which is an almost automorphic extension of the base if a relatively compact trajectory exists.

Proposition 3.11. *Let $a : \Omega \to X$ be a semicontinuous semi-equilibrium and assume that there is an $\omega_0 \in \Omega$ such that $\mathrm{closure}_X\{u(t, \omega_0, a(\omega_0)) \mid t \geq 0\}$ is a compact subset of X. Then:*

(i) *The omega-limit set $\mathcal{O}(\omega_0, a(\omega_0))$ contains a unique minimal set M which is an almost automorphic extension of the base flow.*
(ii) *The minimal lifting skew-product flow $(\widehat{M}, \widehat{\tau}, \mathbb{R})$, introduced in subsection* Ordered Banach Spaces. Monotone Skew-Product Semiflows, *is an almost automorphic extension of the base flow.*

Proof. (i) We work in the case that a is a super-equilibrium, the proof being completely analogous in the case of a sub-equilibrium. Denote $K = \mathcal{O}(\omega_0, a(\omega_0))$ and let $(\omega, x) \in K$, i.e., for some $s_n \uparrow \infty$, $\omega_0 \cdot s_n \to \omega$ and $u(s_n, \omega_0, a(\omega_0)) \to x$. Since $C_a - \{(\widetilde{\omega}, \widetilde{x}) \mid \widetilde{x} \leq a(\widetilde{\omega})\}$ is closed and $u(s_n, \omega_0, a(\omega_0)) \leq a(\omega_0 \cdot s_n)$, we deduce that $x \leq a(\omega)$.

The proof of the result is done in Proposition 3.4, part II in [101], for a strongly monotone skew-product semiflow in a Banach space. A slight modification valid for our case is included here for completeness. From Lemma 3.2 there exists a residual set $\Omega_0 \subseteq \Omega$ of continuity points for the section map (3.17) associated to K. Let $\omega \in \Omega_0$ and take $(\omega, x), (\omega, y) \in K$. Thus, $\omega_0 \cdot s_n \to \omega$, $u(s_n, \omega_0, a(\omega_0)) \to x$ for some $s_n \uparrow \infty$. Besides, since $\lim_{n\to\infty} K_{\omega_0 \cdot s_n} = K_\omega$ and $y \in K_\omega$ there are points $(\omega_0 \cdot s_n, x_n) \in K$ such that $\lim_{n\to\infty} x_n = y$. In addition, since each point of K admits a backward orbit, $x_n = u(s_n, \omega_0, y_n)$ for some $(\omega_0, y_n) \in K$. Therefore, $y_n \leq a(\omega_0)$ and the monotone character of the semiflow yields to

$$x_n = u(s_n, \omega_0, y_n) \leq u(s_n, \omega_0, a(\omega_0)),$$

which as $n \to \infty$ provides $y \leq x$. Analogously, we show that $x \leq y$, that is, $y = x$ and $\mathrm{card}(K_\omega) = 1$ for each $\omega \in \Omega_0$. Note that the same argument implies that there can only be one minimal set inside $\mathscr{O}(\omega_0, a(\omega_0))$. Note that only (2) of Definition 3.9 has been used.

(ii) We have to check that there is a point $\omega \in \Omega$ such that card $\widehat{M}_\omega = 1$ where $\widehat{M}_\omega = \{\phi \in \widehat{M} \mid \phi(0) = (\omega, x) \in M\}$. In fact we prove that this is true for a residual set. As in Lemma 3.2, for each rational number $q \in \mathbb{Q}$ there exists a positively invariant residual set $\Omega_q \subset \Omega$ of continuity points for the map $\omega \mapsto M_{\omega \cdot q}$. The set $\Omega_0 = \cap_{q \in \mathbb{Q}} \Omega_q$ is also a residual set. We claim that it is an invariant set. It is positively invariant because Ω_q is positively invariant for each $q \in \mathbb{Q}$. In addition, if $s < 0$ and $\omega \in \Omega_0$, we have $\omega \in \Omega_q$ for each $q \in \mathbb{Q}$ and hence if we take $p \in \mathbb{Q}$ with $p + s > 0$ we deduce that $\omega \cdot (p + s) \in \Omega_q$, i.e., $\omega \cdot s \in \Omega_{p+q}$ for each $q \in \mathbb{Q}$, which implies that $\omega \cdot s \in \Omega_0$. Therefore, for each $\omega \in \Omega_0$ and each $t \in \mathbb{R}$ we deduce that card $M_{\omega \cdot t} = 1$, the backward extension is unique and card $\widehat{M}_\omega = 1$, as stated. □

Arnold and Chueshov [5, 6] show that, in the measurable case, the existence of a semi-equilibrium a with some additional compactness properties ensures the existence of an equilibrium for the semiflow. In our topological framework, also under the supplementary and somehow natural compactness conditions assumed on Proposition 3.11, the semicontinuous semi-equilibrium provides a semicontinuous equilibrium. We state the equivalent statements to this assumption. The next results hold in a more general case, provided that each invariant bounded set is relatively compact.

Proposition 3.12. *Let $a : \Omega \to X$ be a semicontinuous semi-equilibrium. The following statements are equivalent:*

  (i) *$\Gamma_a = \mathrm{closure}_X\{u(t, \omega, a(\omega)) \mid t \geq 0, \, \omega \in \Omega\}$ is a compact subset.*
  (ii) *For each $\omega \in \Omega$, $\mathrm{closure}_X\{u(t, \omega, a(\omega)) \mid t \geq 0\}$ is a compact subset.*
  (iii) *There is $\omega_0 \in \Omega$ such that $\mathrm{closure}_X\{u(t, \omega_0, a(\omega_0)) \mid t \geq 0\}$ is a compact subset.*

Theorem 3.13. *Let us assume the existence of a semicontinuous semi-equilibrium $a : \Omega \to X$ satisfying one of the equivalent statements of Proposition 3.12. Then,*

  (i) *There exists a semicontinuous equilibrium $c : \Omega \to X$ with $c(\omega) \in \Gamma_a$ for any $\omega \in \Omega$.*
  (ii) *Let ω_1 be a continuity point for c. Then, the restriction of the semiflow τ to*

$$K = \mathrm{closure}_{\Omega \times X}\{(\omega_1 \cdot t, c(\omega_1 \cdot t)) \mid t \geq 0\} \subset C_a$$

  is an almost automorphic extension of the base flow $(\Omega, \sigma, \mathbb{R})$.
  (iii) *K is the only minimal set contained in the omega-limit set $\mathscr{O}(\widehat{\omega}, a(\widehat{\omega}))$ for each point $\widehat{\omega} \in \Omega$.*
  (iv) *If $(K, \tau, \mathbb{R}^+)$ is uniformly stable then it is a copy of the base $(\Omega, \sigma, \mathbb{R})$.*

The result stated in Theorem 3.13 is optimum in the following sense: even the simultaneous existence of continuous super and sub-equilibria for an equation does not guarantee the occurrence of a minimal set given by a copy of the base, but just of an almost automorphic extension. As a first example, we mention the one constructed by Ortega and Tarallo in [87]. By taking the example $z' + a(t)z = b(t)$ due to Johnson [46] and explained in subsection 2.3.1 as starting point, they consider a first order scalar ordinary differential equation

$$z' + a(t)z = b(t) + D(z),$$

where D is smooth, vanishes in a previously fixed interval $[m_1, m_2]$, and satisfies $z\,D(z) < 0$ outside the interval and $\liminf_{|z|\to\infty} D(z)/z < -\|a\|_\infty$.

It is easy to check that the last property guarantees the existence of a large enough positive constant k such that k and $-k$ are super and sub-equilibria for the equation (and, in addition, they are ordered); whereas, as shown in [87], the remaining conditions preclude the occurrence of almost periodic solutions and hence the existence of an almost periodic minimal set.

The same phenomenon occurs for the equation analyzed by Fink and Frederickson in [27]: taking Opial's example [85] as starting point, they construct an equation which is almost periodic in time and for which all the solutions are ultimately uniformly bounded but no one is almost periodic. Again, this equation admits continuous super and sub-equilibria, which allows us to conclude the existence of an almost automorphic and not almost periodic extension of the base flow.

In despite of this optimum character of the result, the method we have just described has a strong limitation: it does not detect recurrent solutions which do not determine almost automorphic extensions of the base flow. To clarify this comment, we recall the system of equations constructed by Takáç [109] explained in subsection 3.1. It is easy to check that (k, k, k, k) and $(-k, -k, -k, -k)$ define super and sub-equilibria for the skew-product flow induced by the system (3.27) if k is large enough. Hence we can assert the existence of solutions determining almost automorphic extensions of the base flow (in fact they are two distinct 1-copies of the base, determined by periodic solutions: they are the upper and lower boundaries of the global attractor). But this tool gives no evidence of the presence of the 2-copy of the base.

We finish this subsection with the definition and some properties of the so called *strong semi-equilibria* needed through the rest of the section.

Definition 3.14. A continuous sub-equilibrium (resp. super-equilibrium) is *strong* if there exists an $s_* > 0$ such that $a(\omega \cdot s_*) \ll u(s_*, \omega, a(\omega))$ (resp. $\gg$) for every $\omega \in \Omega$.

According to Proposition 4.2 (i) in Novo et al. [72], it suffices for a continuous sub-equilibrium (resp. super-equilibrium) to be strong that there exist both an $s_* > 0$ and an $\omega_0 \in \Omega$ such that $a(\omega_0 \cdot s_*) \ll u(s_*, \omega_0, a(\omega_0))$ (resp. $\gg$).

On the other hand, a continuous function $\widetilde{a} : \Omega \to \mathbb{R}^m$ is said to be $\mathscr{C}^1$ *along the σ-orbits* if for every $\omega \in \Omega$ the function

$$\mathbb{R} \to \mathbb{R}^m, \ s \mapsto \widetilde{a}'(\omega \cdot s) = (d/dt)\widetilde{a}(\omega \cdot (s + t))|_{t=0}$$

exists and is continuous. Then, $\widetilde{a}$ is said to be a *lower solution* for the family of systems (3.15) if it is $\mathscr{C}^1$ along the σ-orbits and the function $a: \Omega \to X$ given by $a(\omega)(s) = \widetilde{a}(\omega \cdot s)$ for $s \in [-r, 0]$ satisfies that $u(t, \omega, a(\omega))$ is defined for any $\omega \in \Omega$ and any $t \geq 0$ and that $\widetilde{a}'(\omega) \leq F(\omega, a(\omega))$ for every $\omega \in \Omega$. In this situation the map $a : \Omega \to X$ turns out to be a continuous sub-equilibrium: see Novo et al. [72] and Núñez et al. [81] for further details. For the sake of completeness, we include a detailed proof of the following result (see Lemma 2.11(i) in [81]).

Lemma 3.15. *Let $\widetilde{a}$ be a lower solution for the family of systems* (3.15)*. If at a certain $\omega_0 \in \Omega$ it holds $\widetilde{a}'(\omega_0) \ll F(\omega_0, a(\omega_0))$, then a is a strong sub-equilibrium.*

Proof. As mentioned before, if suffices to show the existence of an $s_* > 0$ such that $a(\omega_0 \cdot s_*) \ll u(s_*, \omega_0, a(\omega_0))$. By the continuity of the maps involved and the fact that $\widetilde{a}'(\omega_0) \ll F(\omega_0, a(\omega_0))$, we find $\varepsilon > 0$ such that $\widetilde{a}'(\omega_0 \cdot t) \ll F(\omega_0 \cdot t, a(\omega_0 \cdot t))$ for any $t \in [0, \varepsilon)$. This can be rewritten as $\widetilde{a}'_{\omega_0}(t) \ll F(\omega_0 \cdot t, (\widetilde{a}_{\omega_0})_t)$ for any $t \in [0, \varepsilon)$, by denoting $\widetilde{a}_{\omega_0}(s) = \widetilde{a}(\omega_0 \cdot s)$ for $s \in \mathbb{R}$. Then, a standard comparison argument for delay differential equations leads to $\widetilde{a}_{\omega_0}(t) \ll z(t, \omega_0, a(\omega_0))$ for any $t \in (0, \varepsilon)$. Now define the function $y(t) = z(t, \omega_0, a(\omega_0)) - \widetilde{a}_{\omega_0}(t)$, fix $t_0 \in (0, \varepsilon)$, and note that for every component $i = 1, \ldots, m$ it holds that $y_i(t_0) > 0$. As $\widetilde{a}$ is a lower solution, it follows that $y(t)$ satisfies the linear delay inequality $y'(t) \geq L(t) y_t$ for

$$L(t) = \int_0^1 F_x(\omega_0 \cdot t, \lambda \, u(t, \omega_0, a(\omega_0)) + (1 - \lambda) \, a(\omega_0 \cdot t)) \, d\lambda \, .$$

Using again a comparison argument, we know that $y(t)$ remains above the solution of the linear delay system $z'(t) = L(t) z_t$ with initial data $z_{t_0} = y_{t_0}$, which we denote as $z(t, t_0, y_{t_0})$. At this point, as $z_i(t_0, t_0, y_{t_0}) = y_i(t_0) > 0$ for every component, we can apply Lemma 5.1.3 in Smith [105] to deduce that $y_i(t) \geq z_i(t, t_0, y_{t_0}) > 0$ for any $t \geq t_0$ and any $i = 1, \ldots, m$, that is, $y(t) \gg 0$ for any $t \geq t_0$. Therefore , we can take $s_* = t_0 + r > 0$ to complete the proof. $\square$

3.3   Monotone and Concave or Sublinear Cases

In this subsection, we consider the monotone skew-product semiflow (3.16) induced by (3.14) satisfying Hypotheses (C1)–(C4) and we will assume additional conditions providing the concave or sublinear character of the induced semiflow, important in applications, which will allow us to prove the existence of a globally asymptotically stable minimal set which is a copy of the base flow.

3.3.1   Monotone and Concave Semiflows

Theorem 3.13 shows how the existence of a semicontinuous sub-equilibrium a allows us to construct a minimal set $K \subset \Omega \times X$ which is an almost automorphic extension of the base Ω. The example given by Riccati equations (2.11) associated to Vinograd example (2.7) shows that even the existence of a strong super-equilibrium and concavity properties does not imply that K is a copy of the base. Now we will prove that, when the sub-equilibrium is strong and under an additional concavity assumption for the semiflow, K turns out to be a copy of the base Ω attracting all the solutions $u(t, \omega, x)$ with initial condition above a; i.e., with $a(\omega) \leq x$. In particular, the flow on this set is almost periodic when the base flow is, which is the case when the initial function f is uniformly almost periodic.

(C5) The function $f(t, x)$ is *concave* in x; i.e., for each $t \in \mathbb{R}$ and $\lambda \in [0, 1]$
$$f(t, \lambda x + (1 - \lambda) y) \geq \lambda f(t, x) + (1 - \lambda) f(t, y) \quad \text{whenever } x \leq y.$$

This property is satisfied simultaneously for all the functions of the hull Ω. Conditions (C4), (C5) and standard arguments of comparison of solutions (see [74]) ensure the *concave* character of the monotone semiflow $(\Omega \times X, \tau, \mathbb{R}^+)$. In other words, u is an *order concave map* in x:

$$u(t, \omega, \lambda x + (1 - \lambda)y) \geq \lambda u(t, \omega, x) + (1 - \lambda) u(t, \omega, y) \qquad (3.28)$$

for $x \leq y$ and for each $t \geq 0$, $\lambda \in [0, 1]$ and $\omega \in \Omega$.

The definition of *convex* semiflow is obtained by substituting the inequality $\geq$ by $\leq$ in (3.28). Finally, it is easy to check that if we take the set $X_- = \{-x \mid x \in X_+\}$ as positive cone, inverting in this way the order relation, a monotone and convex skew-product semiflow becomes monotone and concave. So that the results obtained for concave semiflows are easily adapted to the case of convexity.

Theorem 3.16. *Assume conditions* (C1)–(C5) *for the semiflow* (3.16). *Let* $a : \Omega \to X$ *be a strong semicontinuous sub-equilibrium satisfying one of the equivalent statements of Proposition 3.12, and let* c *and* K *be the equilibrium and the almost automorphic extension of the base flow provided by Theorem 3.13. Then,*

(i) *K defines a copy of the base flow; more precisely, $c : \Omega \to X$ is continuous and $K = \{(\omega, c(\omega)) \mid \omega \in \Omega\}$.*
(ii) *All the semiorbits corresponding to initial data (ω, x) with $a(\omega) \leq x$ are globally defined and approach asymptotically K, i.e.,*

$$\lim_{t \to \infty} \|u(t, \omega, x) - c(\omega \cdot t)\| = 0. \qquad (3.29)$$

Next, we can weaken the assumption of the sub-equilibrium to be strong by strength the assumption on concavity:

(C6) there is an $\omega_1 \in \Omega$ and a $t_1 > 0$ such that for each $\lambda \in (0, 1)$

$$u(t_1, \omega_1, \lambda x + (1 - \lambda)y) \gg \lambda u(t_1, \omega_1, x) + (1 - \lambda) u(t_1, \omega_1, y)$$

whenever $x \gg y$.

One way, but not the only one, of obtaining this is to assume that $f(t, x)$ is *strongly concave* in x for t in an interval I of length $r' > r$, i.e., for each $\lambda \in (0, 1)$ and $t \in I$

$$f(t, \lambda x + (1 - \lambda) y) \gg \lambda f(t, x) + (1 - \lambda) f(t, y) \quad \text{whenever } x \ll y.$$

As shown often in applications, a combination of strong monotonicity for the semiflow and strictly concavity for f also provides condition (C6) (see [73]).

Theorem 3.17. *Assume conditions* (C1)–(C6) *for the semiflow* (3.16). *Let* $a : \Omega \to X$ *be a semicontinuous sub-equilibrium and let* $(\omega_0, x_0) \in \Omega \times X$ *with* $a(\omega_0) \ll x_0$ *be such that its semitrajectory* $\{(\omega_0 \cdot t, u(t, \omega_0, x_0)) \mid t \geq 0\}$ *is bounded and there is a* $y_0 \gg 0$ *with* $u(t, \omega_0, x_0) - a(\omega_0 \cdot t) \geq y_0$ *for each* $t \geq 0$. *Then,*

(i) $K = \mathcal{O}(\omega_0, x_0)$ *defines a copy of the base flow, i.e.* $K = \{(\omega, c(\omega)) \mid \omega \in \Omega\}$.
(ii) *All the semiorbits corresponding to initial data* (ω, x) *with* $a(\omega) \ll x$ *are globally defined and approach asymptotically* K, *i.e.,*

$$\lim_{t \to \infty} \|u(t, \omega, x) - c(\omega \cdot t)\| = 0.$$

Remark 3.18. Núñez et al. [82] show that the convergence in Theorems 3.16 and 3.17 is exponential in the sense that for any $v \gg 0$ there exists $k_v > 1$ and $\rho > 0$ such that if $x \geq a(\omega) + v$ then $\|c(\omega \cdot t) - u(t, \omega, x)\| \leq k_v e^{-\rho t} \|c(\omega) - x\|$.

We refer the reader to Núñez et al. [82, 83] for an exhaustive study of the different dynamical situations in the monotone and concave case.

3.3.2 Monotone and Sublinear Semiflows

When dealing with sublinear systems, the natural space for solutions is the positive cone. For that reason in this subsection we restrict the study to systems given by functions $f : \mathbb{R} \times X_+ \to \mathbb{R}^m$. The purpose is to analyze the conditions ensuring the existence of a unique and asymptotically stable copy of the base when the concavity hypotheses are replaced by some sublinearity properties. Apart from (C1)–(C4) we will assume

(C7) The function $f : \mathbb{R} \times X^+ \to \mathbb{R}^m$ is *sublinear* in x; that is, for each $t \in \mathbb{R}$
$f(t, \lambda x) \geq \lambda f(t, x)$ whenever $x \in X^+$ and $\lambda \in [0, 1]$.

Again, this property is satisfied simultaneously for all the functions of the hull Ω, and together with condition (C4) provides the *sublinear* character of the monotone semiflow $(\Omega \times X, \tau, \mathbb{R}^+)$, that is, for each $t \geq 0$, $\lambda \in [0, 1]$ and $\omega \in \Omega$

$$u(t, \omega, \lambda x) \geq \lambda u(t, \omega, x) \quad \text{whenever } x \in X_+. \tag{3.30}$$

In addition, we will assume any condition on the family implying the following strong sublinearity property for the semiflow

(C8) there is an $\omega_1 \in \Omega$ and a $t_1 > 0$ such that for each $\lambda \in (0, 1)$

$$u(t_1, \omega_1, \lambda x) \gg \lambda u(t_1, \omega_1, x) \quad \text{whenever } x \gg 0. \tag{3.31}$$

One way is to assume that $f : \mathbb{R} \times X^+ \to \mathbb{R}^m$ is *strongly sublinear* in x when t belongs to an interval of length r, that is, for each $\lambda \in (0, 1)$ and $t \in [t_0, t_0 + r]$

$$f(t, \lambda x) \gg \lambda f(t, x) \quad \text{whenever } x \gg 0.$$

As we will check later in the example provided in Theorem 3.21, (C8) can also be obtained if the semiflow is strongly monotone and f is strictly sublinear (see [121]).

We include a proof of the main result of this subsection (see [80]).

Theorem 3.19. *Let us assume conditions* (C1)–(C4) *and* (C7)–(C8) *for the semiflow* (3.16). *Let* $(\omega_0, x_0) \in \Omega \times X$ *with* $0 \ll x_0$ *be such that its semitrajectory* $\{(\omega_0 \cdot t, u(t, \omega_0, x_0)) \mid t \geq 0\}$ *is bounded and there is a* $y_0 \gg 0$ *with* $u(t, \omega_0, x_0) \geq y_0$ *for each* $t \geq 0$. *Then,*

(i) $\mathcal{O}(\omega_0, x_0)$ *is a copy of the base flow, i.e.* $\mathcal{O}(\omega_0, x_0) = \{(\omega, e(\omega)) \mid \omega \in \Omega\}$.
(ii) *All the semiorbits corresponding to initial data* $(\omega, x) \in \Omega \times X_+$ *with* $0 \ll x$ *are globally defined and approach asymptotically* $\mathcal{O}(\omega_0, x_0)$, *i.e.,*

$$\lim_{t \to \infty} \|u(t, \omega, x) - e(\omega \cdot t)\| = 0.$$

Proof. First notice that the omega-limit set $\mathcal{O}(\omega_0, x_0)$ is strongly positive, that is, $y \gg 0$ for each $(\omega, y) \in \mathcal{O}(\omega_0, x_0)$ and we can fix e_1 and $e_2 \in X$ such that $0 \ll e_1 \leq y \leq e_2$ for each $(\omega, y) \in \mathcal{O}(\omega_0, x_0)$. Next, we check that all the semiorbits with initial data $x \geq 0$ are bounded and hence relatively compact. We take $(\omega, y) \in \mathcal{O}(\omega_0, x_0)$ and $\mu > 1$ such that $x \leq \mu y$. Monotonicity and sublinearity ensure that $0 \leq u(t, \omega, x) \leq \mu u(t, \omega, y)$ for $t \geq 0$, and the boundedness follows from the semimonotonicity of the norm. If, in addition, $x \gg 0$ we check that the semiorbit is uniformly stable. We choose $\lambda \in (0, 1)$ such that $\lambda y \leq x \leq (1/\lambda) y$. Again (3.19), (3.30) and $e_1 \leq u(t, \omega, y) \leq e_2$ lead us to

$$0 \ll \lambda e_1 \leq \lambda u(t, \omega, y) \leq u(t, \omega, x) \leq (1/\lambda) u(t, \omega, y) \leq (1/\lambda) e_2$$

for each $t \geq 0$. Let us now fix $\alpha \in (0, 1)$. It is easy to deduce the existence of $\delta = \delta(\alpha) > 0$ such that, if $z \in X_+$ satisfies $\|u(s, \omega, x) - z\| < \delta$ for certain $s \geq 0$, then $\alpha u(s, \omega, x) \leq z \leq (1/\alpha) u(s, \omega, x)$, and hence, again the monotonicity and sublinearity properties, $\alpha u(s + t, \omega, x) \leq u(t, \omega \cdot s, z) \leq (1/\alpha) u(s + t, \omega, x)$ for each $t \geq 0$. Therefore,

$$(1 - 1/\alpha)\lambda e_1 \leq (1 - 1/\alpha)u(s + t, \omega, x) \leq u(s + t, \omega, x) - u(t, \omega \cdot s, z)$$
$$\leq (1 - \alpha)u(s + t, \omega, x) \leq (1 - \alpha)(1/\lambda)e_2,$$

and the uniform stability follows easily. In particular, each $(\omega, x) \in \Omega \times X^+$ with $x \gg 0$ satisfies the same conditions of initial point (ω_0, x_0).

Let M be a strongly positive minimal set. It is clear that there is at least one because each minimal set $M \subset \mathcal{O}(\omega_0, x_0)$ satisfies $y \geq e_1 \gg 0$ for each $(\omega, y) \in M$. Then, let $\omega_1 \in \Omega$ be the point of condition (C8) and $(\omega_1, x) \in \Omega \times X_+$ with $0 \ll x \ll y$ for any $(\omega_1, y) \in M$. We fix $(\omega_1, y) \in M$ and define

$$\lambda(t) = \sup\{\lambda \in [0, 1] \mid u(t, \omega_1, x) \geq \lambda u(t, \omega_1, y)\}.$$

The cocycle property $u(t + s, \omega_1, x) = u(s, \omega_1 \cdot t, u(t, \omega_1, x))$, the monotonicity and the sublinearity imply that $u(t + s, \omega_1, x) \geq \lambda(t)u(t + s, \omega_1, y)$ if $s > 0$, which means that $\lambda(t + s) \geq \lambda(t)$, and $\lambda : [0, \infty) \to [0, 1]$ is an increasing function. Let $\lambda^* = \lim_{t \to \infty} \lambda(t)$. We claim that $\lambda^* = 1$. Assume on the contrary that $\lambda^* < 1$. Then, for an adequate sequence $\{t_n\} \uparrow \infty$

$$\lim_{n \to \infty} (\omega \cdot t_n, u(t_n, \omega_1, x)) = (\omega_1, x^*) \in \Omega \times X_+,$$

$$\lim_{n \to \infty} (\omega \cdot t_n, u(t_n, \omega_1, y)) = (\omega_1, y^*) \in M,$$

and together with $u(t_n, \omega_1, x) \geq \lambda(t_n)u(t_n, \omega_1, y)$ we deduce that $x^* \geq \lambda^* y^*$. Moreover, from $\lambda^* < 1$ the strong sublinearity (3.31) leads us to

$$u(t_1, \omega_1, x^*) \geq u(t_1, \omega_1, \lambda^* y^*) \gg \lambda^* u(t_1, \omega_1, y^*).$$

The above limits imply that there is n_0 with $u(t_n + t_1, \omega_1, x) \gg \lambda^* u(t_n + t_1, \omega_1, y)$ for each $n \geq n_0$, and hence $\lambda(t_n + t_1) > \lambda^*$, contradicting the definition of λ^*, and showing that $\lambda^* = 1$, as claimed.

In addition, from $u(t, \omega_1, y) \geq u(t, \omega_1, x) \geq \lambda(t)u(t, \omega_1, y)$ and $u(t, \omega_1, y) \leq e_2$ we have $0 \leq u(t, \omega_1, y) - u(t, \omega_1, x) \leq (1 - \lambda(t))e_2$, and we conclude that

$$\lim_{t \to \infty} \|u(t, \omega_1, x) - u(t, \omega_1, y)\| = 0,$$

$\mathcal{O}(\omega_1, x) = M$. Since this is true for each $x \gg 0$ sufficiently small, M is the only strongly positive minimal set and $M \subset \mathcal{O}(\omega_0, x_0)$.

The same proof of Proposition 3.11, taking now into account that $x \leq y$ for each $(\omega_1, y) \in M$, shows that M is an almost automorphic extension of the base flow, and we take $\hat{\omega} \in \Omega$ such that $M_{\hat{\omega}}$ reduces to a point $(\hat{\omega}, \hat{y})$. We claim that M is a copy of the base. Assume, on the contrary, that there are $(\widetilde{\omega}, \widetilde{y})$, $(\widetilde{\omega}, \widetilde{z}) \in M$ with $\|\widetilde{y} - \widetilde{z}\| > \varepsilon$, and let $\{s_n\} \downarrow -\infty$ be such that $\widetilde{\omega} \cdot s_n \to \omega_1$ as $n \to \infty$ and, hence for an adequate subsequence, let us take the whole sequence,

$$\lim_{n\to\infty} (\widetilde{\omega}\cdot s_n, u(s_n, \widetilde{\omega}, \overline{y})) = \lim_{n\to\infty} (\widetilde{\omega}\cdot s_n, u(s_n, \widetilde{\omega}, \overline{z})) = (\hat{\omega}, \hat{y}). \qquad (3.32)$$

As shown above, all the semiorbits are uniformly stable and, it is immediate from the proof that all the semiorbits in M have the same modulus of uniform stability $\delta(\varepsilon)$. From (3.32) we take n_0 such that $\|u(s_{n_0}, \widetilde{\omega}, \overline{y}) - u(s_{n_0}, \widetilde{\omega}, \overline{z})\| \le \delta(\varepsilon)$, and the uniform stability provides $\|\overline{y} - \overline{z}\| \le \varepsilon$, a contradiction.

We denote $M = \{(\omega, e(\omega)) \mid \omega \in \Omega\}$ for a continuous map $e: \Omega \to X$ and, in order to finish the proof we check that $M = \mathcal{O}(\omega, x)$ for each $\omega \in \Omega$ and $x \gg 0$, in particular, $M = \mathcal{O}(\omega_0, x_0)$. We take again $\delta(\varepsilon)$ the modulus of uniform stability of the trajectories in M. Since $(\omega, e(\omega)) \in M \subset \mathcal{O}(\omega, x)$,

$$\lim_{n\to\infty} (\omega\cdot t_n, u(t_n, \omega, x)) = (\omega, e(\omega)) = \lim_{n\to\infty} (\omega\cdot t_n, e(\omega\cdot t_n)),$$

and there is an n_0 such that $\|u(t_{n_0}, \omega, x) - e(\omega\cdot t_{n_0})\| \le \delta(\varepsilon)$. Hence, the uniform stability yields to $\|u(t + t_{n_0}, \omega, x) - e(\omega\cdot(t_{n_0} + t))\| \le \varepsilon$ for each $t > 0$, and the proof is finished. $\qquad\square$

We refer the reader to Núñez et al. [80, 81] for an exhaustive study of the different dynamical situations in the monotone and sublinear case.

3.4 A Non-autonomous Cyclic Feedback System

Next we show some application of the previous results, related to the mathematical model of biochemical feedback in protein synthesis represented by the non-autonomous system of finite-delay functional differential equations

$$\begin{aligned}
z_1'(t) &= g(t, z_m(t - r_m)) - \alpha_1(t) z_1(t), \\
z_j'(t) &= z_{j-1}(t - r_{j-1}) - \alpha_j(t) z_j(t), \quad \text{for } j = 2, \dots, m.
\end{aligned} \qquad (3.33)$$

Here, $\alpha_j(t)$ are positive almost periodic functions for $j = 1, \dots, m$, $g : \mathbb{R} \times \mathbb{R} \to \mathbb{R}^m$ is a C^1-admissible uniformly almost periodic function, $r_j \ge 0$ for $j = 1, \dots, m$ and $\max\{r_j \mid j = 1, \dots, m\} > 0$.

The system (3.33) expresses a model for a biochemical control circuit in which each of the z_j represents the concentration of an enzyme; hence $z_j \ge 0$ for $j = 1, \dots, m$. The autonomous ordinary case was firstly introduced by Selgrade [94]. Different extensions to the periodic and the autonomous functional cases are explored in Smith [104, 105], Krause and Ranft [55], Smith and Thieme [107], and references therein. Chueshov [18] analyzes the random case and Novo et al. [73] the ordinary non-autonomous case. The present case of finite-delay was considered, for the concave case, in Novo et al. [72].

3.4.1 The Concave Case

The next result provides conditions which ensure the existence of a global attractor for (3.33) in the interior of the positive cone X_+. In contrast to previous works, the approach we present here just requires the semiflow to be monotone and concave. Similar conclusions apply when we change almost periodicity for almost automorphy or, in general, recurrence.

Define $r = \max\{r_j \mid j = 1,\ldots,m\} > 0$ and consider, as in the previous subsections, the strongly ordered Banach space $X = C([-r,0],\mathbb{R}^m)$. As usual, given any element $x_0 \in X$, $z(t,x_0)$ represents the solution of (3.33) satisfying $z(t,x_0) = x_0(t)$ for $t \in [-r,0]$.

Theorem 3.20. *Let us assume that*

- $g(t,y) > 0$ *for each* $t \in \mathbb{R}$ *and* $y > 0$, *and* $g_y(t,y) \geq 0$ *for each* $t \in \mathbb{R}$ *and* $y \geq 0$;
- $g(t,y)$ *is concave in* y.

Moreover, there are positive constants β_j, α_j *with*

$$0 < \alpha_j \leq \alpha_j(t) < \beta_j \quad \text{for each } j = 1,\ldots,m \text{ and } t \in \mathbb{R}, \tag{3.34}$$

and a real function $g_0 \in C(\mathbb{R})$ *such that*

- $g_0(y) > 0$ *for each* $y > 0$;
- $\prod_{j=1}^m \beta_j < \limsup_{y\to 0+} g_0(y)/y \leq +\infty$;
- $g_0(y) \leq g(t,y) \leq \alpha\,y + \delta$ *for each* $t \in \mathbb{R}$, $y \geq 0$ *and some positive constants* α, $\delta > 0$ *with* $0 < \alpha < \prod_{j=1}^m \alpha_j$.

Then, there is a unique almost periodic solution $z^*(t) \gg 0$ *of (3.33) such that*

$$\lim_{t\to\infty} \|z(t,x_0) - z^*(t)\| = 0 \quad \text{for each } x_0 \in C([-r,0],\mathbb{R}^m) \text{ with } x_0 \gg 0.$$

Proof. We can write (3.33) as $z' = f(t,z_t)$ where $f : \mathbb{R} \times X \to \mathbb{R}^m$ is the C^1-admissible function defined for each $t \in \mathbb{R}$ and $x \in X$ by

$$\begin{aligned} f_1(t,x) &= g(t, x_m(-r_m)) - \alpha_1(t)\,x_1(0)\,, \\ f_j(t,x) &= x_{j-1}(-r_{j-1}) - \alpha_j(t)\,x_j(0) \quad \text{for } j = 2,\ldots,m\,. \end{aligned} \tag{3.35}$$

As explained in the introduction of section 3, system (3.33) is included in the family of systems $z'(t) = F(\omega{\cdot}t, z_t)$ for $\omega \in \Omega$, where Ω is the hull of f. The almost periodicity of the coefficients ensures that $(\Omega, \sigma, \mathbb{R})$ is minimal and almost periodic. Consequently, hypotheses (C1), (C2) and (C3) are satisfied.

It is also easy to check that condition (C4) and (C5) hold. Moreover, from $\prod_{j=1}^m \beta_j < \limsup_{y\to 0+} g_0(y)/y \leq +\infty$ we can find a sequence $\varepsilon_n > 0$ tending to 0 such that, for each $n \in \mathbb{N}$,

$$g_0(\varepsilon_n) - \varepsilon_n \prod_{j=1}^{m} \beta_j > 0 \tag{3.36}$$

Now we consider the constant function $a_n = (a_{n,1}, \dots, a_{n,m}) \gg 0$, with $a_{n,m} = \varepsilon_n$ and $a_{n,l} = \varepsilon_n \prod_{j=l+1}^{m} \beta_j$ for $l = 1, \dots, m-1$. From inequality $g_0(y) \le g(t,y)$ and relations (3.34) and (3.36),

$$f_1(t, a_n) = g(t, a_{n,m}) - \alpha_1(t) a_{n,1} > g_0(\varepsilon_n) - \varepsilon_n \prod_{j=1}^{m} \beta_j > 0$$

$$f_l(t, a_n) = a_{n,l-1} - \alpha_l(t) a_{n,l} = \varepsilon_n \prod_{j=l}^{m} \beta_j - \alpha_l(t) \varepsilon_n \prod_{j=l+1}^{m} \beta_j > 0, \quad l = 2, \dots, m.$$

which implies that $f(t, a_n) \gg 0$ for each $n \in \mathbb{N}$ and each $t \in \mathbb{R}$. Hence $F(\omega, a_n) \ge 0$ for each $n \in \mathbb{N}$ and $\omega \in \Omega$, and Lemma 3.15 allows us to assure that a_n is a strong sub-equilibrium for each $n \in \mathbb{N}$.

We define $\delta_1 = \delta/(\prod_{j=1}^{m} \alpha_j - \alpha) > 0$ and take $b = (b_1, \dots, b_m) \gg 0$, with $b_m = \delta_1$ and $b_l = \delta_1 \prod_{j=l+1}^{m} \alpha_j$ for $l = 1, \dots, m-1$. As before, from $g(t, y) \le \alpha y + \delta$ and (3.34) we check that $f(t, b) \le 0$ for each $t \in \mathbb{R}$, thus, $F(\omega, b) \le 0$, and hence b is a super-equilibrium. In addition, it is not hard to check that $a_n \ll b$ for each $n \in \mathbb{N}$. Consequently, since b is a super-equilibrium and the flow is monotone, we deduce that

$$a_n \le u(t, \omega, a_n) \ll u(t, \omega, b) \le b$$

for $t \ge 0$ and $n \in \mathbb{N}$. In particular, the equivalent statements of Proposition 3.12 are satisfied for each one of the sub-equilibria a_n, as deduced from Arzelà-Ascoli theorem.

Finally, Theorem 3.16 applied to each a_n yields to the existence of a copy of the base K_n which is a global attractor in the set $A_n = \{(\omega, x) \in \Omega \times X \mid x \ge a_n\}$. Consequently, $K_n = \{(\omega, e(\omega)) \mid \omega \in \Omega\}$ is the same for each $n \in \mathbb{N}$ and, since a_n tends to 0, it is a global attractor in $A = \{(\omega, x) \in \Omega \times X \mid x \gg 0\}$. The almost periodic function $z^*(t) = e(\omega_0 \cdot t)$ with $\omega_0 = f$ satisfies the statement. $\square$

3.4.2 The Sublinear Case

Now we increase the range of applications by changing concavity by sublinearity but we strength the monotonicity assumptions to provide new conditions which ensure the existence of a global attractor for (3.33) in the interior of the positive cone X_+. Again, similar conclusions apply when we change almost periodicity for almost automorphy or, in general, recurrence.

Theorem 3.21. *Let us assume that*

- $g(t, y) > 0$ *for each* $t \in \mathbb{R}$ *and* $y > 0$, *and* $g_y(t, y) > 0$ *for each* $t \in \mathbb{R}$ *and* $y > 0$;
- $g(t, y)$ *is sublinear in* y *and* $y\, g_y(t, y) < g(t, y)$ *for each* $t \in \mathbb{R}$ *and* $y > 0$.

Moreover, there are positive constants β_j, α_j *with*

$$0 < \alpha_j \leq \alpha_j(t) < \beta_j \quad \text{for each} \ \ j = 1, \ldots, m \ \ \text{and} \ t \in \mathbb{R},$$

and a real function $g_0 \in C(\mathbb{R})$ *such that*

- $g_0(y) > 0$ *for each* $y > 0$;
- $\prod_{j=1}^{n} \beta_j < \limsup_{y \to 0^+} g_0(y)/y \leq +\infty$;
- $g_0(y) \leq g(t, y) \leq \alpha\, y + \delta$ *for each* $t \in \mathbb{R}$, $y \geq 0$ *and some positive constants* $\alpha,\ \delta > 0$ *with* $0 < \alpha < \prod_{j=1}^{m} \alpha_j$.

Then, there is a unique almost periodic solution $z^*(t) \gg 0$ *of* (3.33) *such that*

$$\lim_{t \to \infty} \|z(t, x_0) - z^*(t)\| = 0 \quad \text{for each} \ x_0 \in C([-r, 0], \mathbb{R}^m) \ \text{with} \ x_0 \gg 0.$$

Proof. In this case, for simplicity in the notation, we will consider all the delays to be equal to r. If the delays are different, the proof is completely similar changing the phase space to $X = \prod_{j=1}^{m} C([-r_j, 0], \mathbb{R})$.

As in the proof of Theorem 3.20, it is easy to prove that conditions (C1)–(C4) and (C7) hold, the semitrajectory $\{(\omega \cdot t, u(t, \omega, b)) \mid t \geq 0\}$ is bounded and $u(t, \omega, b) \geq a_1 \gg 0$ for each $t \geq 0$. Therefore, if we check condition (C8), the result follows from Theorem 3.19.

We take $\omega_0 = f$, defined by (3.35), and note that $u(t, \omega_0, x)(s) = z(t + s, x)$ for $s \in [-r, 0]$. We check that there is a $t_0 > 0$ such that

$$u(t, \omega_0, x) \ll u(t, \omega_0, y) \quad \text{if } 0 \ll x < y \ \text{ and } t > t_0. \tag{3.37}$$

For this it is enough to check that if, as usual, we denote by $u_x(t, \omega, x)\colon X \to X$ the linear differential operator with respect to the third variable, we have

$$u_x(t, \omega_0, x)\, v \gg 0 \quad \text{if } v > 0,\ x \gg 0 \ \text{ and } t > t_0.$$

Note that $(u_x(t, \omega_0, x)\, v)(s) = h(t + s)$ where $h(t)$ satisfies the variational problem

$$
\begin{aligned}
h_1'(t) &= g_y(t, z_m(t - r, x))\, h_m(t - r) - \alpha_1(t)\, h_1(t), \\
h_j'(t) &= h_{j-1}(t - r) - \alpha_j(t)\, h_j(t), \quad \text{for } j = 2, \ldots, m.
\end{aligned}
\tag{3.38}
$$

with $h(s) = v(s)$ for $s \in [-r, 0]$. Therefore the assertion holds if there is $\widetilde{t} > r$ such that $h(t) \gg 0$ for each $t \geq \widetilde{t}$. Moreover, from Lemma 5.1.3 in [105], if $h_i(t_1) > 0$ for some $t_1 \geq 0$, then $h_i(t) > 0$ for each $t > t_1$. Then, we contradict the assertion if we assume that there is one $j \in \{1, \ldots, m\}$ such that $h_j(t) = 0$ for each $t \geq 0$. If $j \geq 2$, from (3.38) we would deduce that $h_{j-1}(t - r) = 0$ for each $t \geq 0$. In a recursive way we will obtain the same result for $j - 2, \ldots, 1$. Next, from (3.38) we obtain

$$g_y(t, z_m(t - r, x)) h_m(t - r) = 0 \quad \text{for each } t \geq 0.$$

Since $x \gg 0$ and the semiflow is monotone $u(t, \omega_0, x) = z(t + \cdot, x) \gg 0$ for each $t \geq 0$ and hence, $z_m(t - r, x) > 0$ for $t \geq 0$, and $g_y(t, z_m(t - r, x)) > 0$, which lead to $h_m(t - r) = 0$ for each $t \geq 0$. The result for the rest of the indices from $m - 1$ to $j + 1$ follows analogously, and we conclude that $h(t - r) = 0$ for each $t \geq 0$, which contradicts that $v > 0$. Note that relation (3.37), i.e, the eventually strongly monotone character of the semiflow, could have also been obtained by checking that conditions of Proposition 3.6 (iii) hold, and then $t_0 = m + 1$.

Next, we fix $\lambda \in (0, 1)$ and we consider $y(t) = \lambda z(t, x)$. Then $y'(t) = \lambda f(t, z_t)$, with f defined by (3.35). From $y g_y(t, y) < g(t, y)$ we deduce that g is strongly sublinear in y, that is, $g(t, \lambda y) > \lambda g(t, y)$ for $y > 0$, and consequently, $f(t, \lambda z_t) > \lambda f(t, z_t)$. Therefore $y' < f(t, y_t)$ and comparison theorems for this kind of ordinary differential equations [105] lead to $y(t) < z(t, \lambda x)$ for each $t > 0$, or equivalently $\lambda u(t, \omega_0, x) < u(t, \omega_0, \lambda x)$ for each $t > 0$. In addition, from relations (3.37) and (3.30), if we take $t^* > t_0$ we deduce that

$$u(t^* + t^*, \omega_0 \cdot (-t^*), \lambda x) = u(t^*, \omega_0, u(t^*, \omega_0, \lambda x)) \gg u(t^*, \omega_0, \lambda u(t^*, \omega_0, x))$$
$$\geq \lambda u(t^*, \omega_0, u(t^*, \omega_0, x)) = \lambda u(t^* + t^*, \omega_0 \cdot (-t^*), x),$$

and (C8) holds for $t_1 = 2 t^*$ and $\omega_1 = \omega_0(-t^*)$, which finishes the proof. □

3.5 Cellular Neural Networks

We finish this section by explaining some applications to neural networks. There has recently been increasing interest in the potential applications of the dynamics of artificial neural networks in signal and image processing, as well as in biological modelling, cognitive simulation or numerical computation. We refer the reader to Wu [117] and the references therein for a complete introduction to the subject.

The models we consider are among the so-called delayed cellular neural networks, which in particular include the Hopfield-type models (see the early works by Hopfield [36, 37] and Marcus and Westervelt [60] where the delay is introduced). The autonomous case, with finite or distributed delays, has been intensively investigated (see Gopalsamy and He [28], Van den Driessche and Zou [110], Van den Driessche et al. [111] and Zhao [120], among many others). The main interest has focused on determining sufficient conditions to guarantee the existence and uniqueness of a globally asymptotically stable equilibrium point for the system.

Nevertheless, the non-autonomous periodic or almost periodic case with constant or time-varying delays has just recently been considered in some papers (see for instance Liang and Cao [58], Mohamad [66], Fan-Ye [24], Jiang et al. [43], Chen and Cao [15] and Huang et al. [40]). Also recently, Novo et al. [73, 74] have considered quite general non-autonomous finite-delay monotone and concave models of Hopfield neural networks, giving conditions for the existence of attracting solutions with the same recurrence properties as the model coefficients. Besides, in Novo et al. [75] the existence of a global exponentially attracting solution of finite-delay cellular neural networks is deduced from the uniform asymptotical stability of the null solution of an associated non-autonomous linear system.

Let us consider a non-autonomous system of finite-delay FDEs

$$z_i'(t) = -\widetilde{a}_i(t)\, z_i(t) + \sum_{j=1}^{m} \widetilde{b}_{ij}(t)\, f_j(z_j(t))$$

$$+ \sum_{j=1}^{m} \widetilde{c}_{ij}(t) \int_{-r}^{0} g_j(z_j(t+s))\, d\mu_{ij}(s) + \widetilde{I}_i(t), \quad t \geq 0, \ i = 1,\ldots,m,$$

$$(3.39)$$

which describes the dynamics of a network of m neurons (or amplifiers) with delayed interconnections. The variable $z_i(t)$ represents the state of the i-th neuron in the network at time t. The coupling coefficients can be arranged in two interconnection matrices $[\widetilde{b}_{ij}(t)]$ and $[\widetilde{c}_{ij}(t)]$, whose entries, as well as the coefficients $\widetilde{a}_i$ and the external input functions $\widetilde{I}_i$ are bounded, uniformly continuous and present some recurrent behavior in time, such as almost periodicity. The functions f_j and g_j are the so-called real signal propagation functions or activation functions. More precisely, we make the following assumptions on system (3.39):

(A1) The coefficient functions $\widetilde{a}_i(t)$, $\widetilde{b}_{ij}(t)$, $\widetilde{c}_{ij}(t)$ and $\widetilde{I}_i(t)$ are all bounded and uniformly continuous on $\mathbb{R}$, and recurrent, that is, the *hull* is minimal.

(A2) $\widetilde{a}_i(t) \geq a_0 > 0$ for every $t \in \mathbb{R}$ and $i \in \{1,\ldots,m\}$.

(A3) $\widetilde{b}_{ij}(t)$, $\widetilde{c}_{ij}(t) \geq 0$ for every $t \in \mathbb{R}$ and $i, j \in \{1,\ldots,m\}$.

(A4) The activation functions f_j, $g_j : \mathbb{R} \to \mathbb{R}$ satisfy:

    (i) They are $\mathscr{C}^1$-functions on $\mathbb{R}$.
    (ii) $\lim_{|s|\to\infty} |f_j(s)|/|s| = \lim_{|s|\to\infty} |g_j(s)|/|s| = 0$ for each $j = 1,\ldots,m$.
    (iii) $0 \leq f_j'(s)$ and $0 \leq g_j'(s)$ for each $j = 1,\ldots,m$ and any $s \in \mathbb{R}$.

(A5) Each μ_{ij} is a normalized positive regular Borel measure on $(-r,0]$, i.e., $\mu_{ij}((-r,0]) = 1$, and $\mu_{ij}\{0\} = 0$.

(A6) f_j, g_j are convex for $s < 0$ and concave for $s > 0$, for each $j \in \{1,\ldots,m\}$.

Under condition (A1) we can build the so-called *hull* of the system, by considering the set of functions $\{(\widetilde{a}_i(t+s), \widetilde{b}_{ij}(t+s), \widetilde{c}_{ij}(t+s), \widetilde{I}_i(t+s)),\ (t \in \mathbb{R})\ |\ s \in \mathbb{R}\}$ formed by the time-translations of the coefficient functions determining system (3.39), together with their limit points for the compact-open topology. In this way, the hull turns out to be a compact metric space of functions, denoted by Ω, where a continuous flow $\sigma(t, \omega) = \omega \cdot t$ can be defined just by translation. Then we can consider the family of systems over the hull, which can be written for short as

$$z'(t) = F(\omega \cdot t, z_t), \quad t \geq 0, \ \omega \in \Omega, \tag{3.40}$$

where $z_t : [-r, 0] \rightarrow \mathbb{R}^m$ is defined by $z_t(s) = z(t + s)$ for $s \leq 0$ and $t \geq 0$ (whenever it makes sense) and the function F is given componentwise by

$$F_i(\omega, x) = -a_i(\omega) x_i(0) + \sum_{j=1}^m b_{ij}(\omega) f_j(x_j(0))$$

$$+ \sum_{j=1}^m c_{ij}(\omega) \int_{-r}^0 g_j(x_j(s)) \, d\mu_{ij}(s) + I_i(\omega), \quad i = 1, \ldots, m \tag{3.41}$$

for $\omega \in \Omega$ and a continuous map $x : [-r, 0] \rightarrow \mathbb{R}^m$, with $(a_i, b_{ij}, c_{ij}, I_i)$ defined on Ω by evaluation at 0, i.e., $(a_i, b_{ij}, c_{ij}, I_i)(\omega) = \omega(0)$, so that $(a_i, b_{ij}, c_{ij}, I_i)(\omega \cdot t) = \omega(t)$ for $t \in \mathbb{R}$. Therefore, taking $\omega = (\widetilde{a}_i, \widetilde{b}_{ij}, \widetilde{c}_{ij}, \widetilde{I}_i)$ we recover in (3.40) the initial system (3.39). Note that by the construction, the functions a_i are positive on Ω, and the functions b_{ij} and c_{ij} are nonnegative on Ω.

The recurrence hypothesis in (A1) means that the translation flow on Ω is minimal, that is, orbits are dense; this is a usual requirement in the non-autonomous setting and it happens for example if the coefficient functions are all almost periodic or almost automorphic. Consequently, hypothesis (C3) is satisfied. As in the introduction of the chapter, we consider the skew-product semiflow (3.16) induced by (3.40).

The condition stated in hypothesis (A2) is commonly imposed in the bibliography, representing the passive decay rates. Hypothesis (A3) together with (A4)-(iii) and (A5) will imply the quasimonotone condition (C4). With regard to hypothesis (A5), note also that the measures μ_{ij} are not necessarily absolutely continuous with respect to the Lebesgue measure. The concavity condition (C5) is deduced from Hypotheses (A5) and (A6).

Remark 3.22. We want to note that the previous formulation, in which the interconnection matrices have nonnegative entries, is very general in the sense that each system of cellular neural networks can be included in a bigger one with this property (see Van den Driessche et al. [111] for a different application of this fact).

For doing that we decompose b_{ij} and c_{ij} as the usual difference of two nonnegative functions, that is, $b_{ij} = b_{ij}^+ - b_{ij}^-$ and $c_{ij} = c_{ij}^+ - c_{ij}^-$; we denote $\widehat{f}(x) = -f(-x)$, $\widehat{g}(x) = -g(-x)$, and we introduce the new variables $\widehat{z}_i = -z_i$, $i = 1, \ldots, m$. It is easy to check that

$$z_i'(t) = - a_i(\omega \cdot t) z_i(t) + \sum_{j=1}^{m} b_{ij}^+(\omega \cdot t) f_j(z_j(t)) + \sum_{j=1}^{m} b_{ij}^-(\omega \cdot t) \widehat{f}_j(\widehat{z}_j(t))$$

$$+ \sum_{j=1}^{m} c_{ij}^+(\omega \cdot t) \int_{-r}^{0} g_j(z_j(t+s)) \, d\mu_{ij}(s)$$

$$+ \sum_{j=1}^{m} c_{ij}^-(\omega \cdot t) \int_{-r}^{0} \widehat{g}_j(\widehat{z}_j(t+s)) \, d\mu_{ij}(s) + I_i(\omega),$$

$$\widehat{z}_i'(t) = - a_i(\omega \cdot t)\widehat{z}_i(t) + \sum_{j=1}^{m} b_{ij}^+(\omega \cdot t) \widehat{f}_j(\widehat{z}_j(t)) + \sum_{j=1}^{m} b_{ij}^-(\omega \cdot t) f_j(z_j(t))$$

$$+ \sum_{j=1}^{m} c_{ij}^+(\omega \cdot t) \int_{-r}^{0} \widehat{g}_j(\widehat{z}_j(t+s)) \, d\mu_{ij}(s)$$

$$+ \sum_{j=1}^{m} c_{ij}^-(\omega \cdot t) \int_{-r}^{0} g_j(z_j(t+s)) \, d\mu_{ij}(s) - I_i(\omega),$$

whose interconnections matrices have nonnegative entries and properties (A1)–(A6) remain valid.

We now show the boundedness of solutions for the family of systems (3.40).

Proposition 3.23. *Assume that the initial system* (3.39) *satisfies hypotheses* (A1)–(A5). *Then, all solutions of the family of systems* (3.40) *are bounded and consequently the induced semiflow* (3.16) *is globally defined.*

Proof. Let $k \in \mathbb{R}$, and denote by **k** the constant map $(-r, 0] \to \mathbb{R}^m$, $s \mapsto (k, \dots, k)$ of $X = C([-r, 0], \mathbb{R}^m)$. As the induced semiflow (3.16) is monotone, we know that if $-\mathbf{k} \le x \le \mathbf{k}$ for some $k > 0$ and $x \in X$, then $u(t, \omega, -\mathbf{k}) \le u(t, \omega, x) \le u(t, \omega, \mathbf{k})$ for any $t \ge 0$ for which all the terms exist. Therefore, it suffices to make sure that the solutions $u(t, \omega, -\mathbf{k})$ and $u(t, \omega, \mathbf{k})$ remain bounded for sufficiently large $k > 0$. For this, it is enough to check that $-\mathbf{k}$ defines a lower solution and $\mathbf{k}$ defines an upper solution (see Remark 2.5 in [79]), that is, $F(\omega, -\mathbf{k}) \ge 0$ and $F(\omega, \mathbf{k}) \le 0$ for $k > 0$ big enough. In order to show this assertion, just note first that by hypothesis (A2), $0 < a_0 \le a_i(\omega)$ for any $\omega \in \Omega$, and that all the functions a_i, b_{ij}, c_{ij} and I_i are bounded on the compact set Ω. Note also that by hypothesis (A4)-(ii) given any $\varepsilon > 0$ there is k_0 such that for any $k \ge k_0$, $-\varepsilon k \le f_j(k)$, $g_j(k) \le \varepsilon k$ for any $j = 1, \dots, m$. The proof is complete. $\square$

First, we prove a global attractivity result for system (3.39), which is independent on the external input functions $\widetilde{I}_i$, $i = 1, \dots, m$. We define the $m \times m$ diagonal matrices $\mathsf{F} = \mathrm{diag}(f_1'(0), \dots, f_m'(0))$, $\mathsf{G} = \mathrm{diag}(g_1'(0), \dots, g_m'(0))$ and $A(\omega) = \mathrm{diag}(a_1(\omega), \dots, a_m(\omega))$, $\omega \in \Omega$, and the matrices $B(\omega) = [b_{ij}(\omega)]$, $C(\omega) = [c_{ij}(\omega)]$, for each $\omega \in \Omega$.

Theorem 3.24. *Under assumptions* (A1)–(A6), *if there exists a vector* $v \gg 0$ *with*

$$\left(A(\omega) - B(\omega)\,F - C(\omega)\,G\right) v \geq 0\,, \quad \omega \in \Omega\,, \tag{3.42}$$

and there is $\omega_0 \in \Omega$ *such that*

$$\left(A(\omega_0) - B(\omega_0)\,F - C(\omega_0)\,G\right) v \gg 0\,, \tag{3.43}$$

then there are positive constants k *and* α *such that any solution* $z(t, x_0)$ *of* (3.39) *satisfies*

$$\lim_{t \to \infty} \|z(t, x_0) - z(t, 0)\| \leq \lim_{t \to \infty} k\, e^{-\alpha t} \|x_0\| = 0\,.$$

In particular, if there are not external input functions, all the solutions tend to 0.

Proof. First of all, note that if $\omega \in \Omega$ and $x_0 \in X_+$,

$$0 \leq u(t, \omega, x_0) - u(t, \omega, 0) = \int_0^1 u_x(t, \omega, \lambda\, x_0)\, x_0\, d\lambda$$

where $u_x(t, \omega, \lambda\, x_0)\, x_0$ is the solution in X of the linearized equations along the orbit of $(\omega, \lambda\, x_0)$ with initial value x_0; the linearized equations being

$$y_i'(t) = -\, a_i(\omega{\cdot}t)\, y_i(t) + \sum_{j=1}^m b_{ij}(\omega{\cdot}t)\, f_j'(z_j(t, \omega, \lambda\, x_0))\, y_j(t)$$

$$+ \sum_{j=1}^m c_{ij}(\omega{\cdot}t) \int_{-r}^0 g_j'(z_j(t+s, \omega, \lambda\, x_0))\, y_j(t+s)\, d\mu_{ij}(s)\,, \quad i = 1, \ldots, m\,.$$

Now, consider the following family of monotone and linear systems for $\omega \in \Omega$,

$$y_i'(t) = -\, a_i(\omega{\cdot}t)\, y_i(t) + \sum_{j=1}^m b_{ij}(\omega{\cdot}t)\, f_j'(0)\, y_j(t)$$

$$+ \sum_{j=1}^m c_{ij}(\omega{\cdot}t)\, g_j'(0) \int_{-r}^0 y_j(t+s)\, d\mu_{ij}(s)\,, \quad i = 1, \ldots, m\,, \tag{3.44}$$

whose solution for each $\omega \in \Omega$ and each initial condition $x_0 \in X_+$ we denote by $\Phi(t, \omega)\, x_0 \in X_+$. Under the concavity and convexity assumptions, we have $f_j'(z_j(t, \omega, \lambda\, x_0)) \leq f_j'(0)$ and $g_j'(z_j(t+s, \omega, \lambda\, x_0)) \leq g_j'(0)$ for any $t \geq 0$ and $s \leq 0$. Therefore, a standard argument of comparison of solutions leads to the upper bound $u_x(t, \omega, \lambda\, x_0)\, x_0 \leq \Phi(t, \omega,)\, x_0$ for each λ, and consequently

$$0 \leq u(t, \omega, x_0) - u(t, \omega, 0) \leq \Phi(t, \omega)\, x_0\,.$$

Next we prove the asymptotic convergence of $\Phi(t,\omega)x_0$ to 0. For that, note that conditions (3.42), (3.43) and Lemma 3.15 means that $-k\,v$ defines a strong lower solution for the family of monotone and linear (in particular concave) systems (3.44) for each $k > 0$, or equivalently, a strong sub-equilibrium. Thus, given $x_0 \in X_+$ there exists $k > 0$ so that $-k\,v \le -x_0$ and from Theorem 3.16 we deduce that the semiorbit with initial data $(\omega, -x_0)$ approach exponentially the unique minimal set K which, in this linear case, coincide with $K = \{(\omega,0) \mid \omega \in \Omega\}$. Hence, $\Phi(t,\omega)x_0 \to 0$ as $t \to \infty$, as claimed, and $\lim_{t\to\infty} \|u(t,\omega,x_0) - u(t,\omega,0)\| = 0$. Analogously, we check that $\lim_{t\to\infty}\|u(t,\omega,x_0) - u(t,\omega,0)\| = 0$ for each $x_0 \in X_-$.

Finally, if $x_0 \in X$, there are $x_0^- \in X_-$ and $x_0^+ \in X_+$ such that $x_0^- \le x_0 \le x_0^+$, and the monotonicity of the semiflow yields to

$$u(t,\omega,x_0^-) \le u(t,\omega,x_0) \le u(t,\omega,x_0^+)$$

for each $t \ge 0$ to finish the proof. $\square$

Next we provide a global attractivity result in X_+ for the neural networks under consideration, assuming the external input functions to be nonnegative, and a similar result in X_- with nonpositive external input functions. We denote

$$\widetilde{I}^- = \inf_{\substack{i=1,\dots,m \\ t\in\mathbb{R}}} \widetilde{I}_i(t) \quad \text{and} \quad \widetilde{I}^+ = \sup_{\substack{i=1,\dots,m \\ t\in\mathbb{R}}} \widetilde{I}_i(t).$$

Theorem 3.25. *Assume that (A1)–(A6) hold, that $f_j(0) = g_j(0) = 0$, $j = 1,\dots,m$, and $\widetilde{I}^- \ge 0$ in (3.39). If, in addition, one of the following conditions is satisfied*

(i) *For each $i = 1,\dots,m$, there is a $t_i \ge 0$ such that $I_i(t_i) > 0$.*
(ii) *$f_i'(0) > 0$, $g_i'(0) > 0$, $i = 1,\dots,m$, there is a $j \in \{1,\dots m\}$ and $t_j \ge 0$ such that $I_j(t_j) > 0$, and $B(\omega) + C(\omega)$ is an irreducible matrix for each $\omega \in \Omega$.*
(iii) *There exists a vector $v \gg 0$ such that*

$$\big(A(\omega) - B(\omega)\,F - C(\omega)\,G\big)v \ll 0, \quad \omega \in \Omega; \tag{3.45}$$

then, there is a unique solution $z^*(t) \gg 0$ of (3.39) such that

$$\lim_{t\to\infty}\|z(t,x_0) - z^*(t)\| = 0 \quad \text{for each } x_0 \in X \text{ with } x_0 \gg 0.$$

Furthermore, the solution $z^*(t)$ has the same recurrence property as that of the coefficients, meaning for instance that if the coefficients of (3.39) are all almost periodic, then this solution is almost periodic too.

Proof. (i) The proof relies on an application of Theorem 3.16. For that, since $\widetilde{I}^- \ge 0$ and $f_j(0) = g_j(0) = 0$ for $j = 1,\dots,m$, we can simply take $a \equiv 0$ as a semicontinuous sub-equilibrium. In particular this implies that the set $\Omega \times X_+$ is invariant for the dynamics. We claim that $a \equiv 0$ is strong sub-equilibrium. From $I_1(t_1) > 0$ we deduce that there is $\omega_0 \in \Omega$ with $I_1(\omega_0) > 0$ and hence $z_1'(0, \omega_0, 0) > 0$, which implies that $z_1(t, \omega_0, 0) > 0$ for t near 0. Since $F(\omega, 0) \geq 0$ for each $\omega \in \Omega$, as in Lemma 3.15, denoting by $y(t) = z(t, \omega_0, 0)$ we have $y'(t) \geq L(t) y_t$ with

$$L(t) = \int_0^1 F_x(\omega_0 \cdot t, \lambda u(t, \omega_0, 0)) \, d\lambda \, ,$$

and consequently, from a comparison argument and Lemma 5.1.3 in [105], we deduce that $z_1(t, \omega_0, 0) > 0$ for each $t > 0$. Analogously, if $i > 1$, from $I_i(t_i) > 0$ we deduce that there is $t_i^* \geq 0$ such that $I_i(\omega_0 \cdot t_i^*) > 0$ and $z_i(t, \omega_0 \cdot t_i^*, 0) > 0$ for each $t > 0$. We take $t_0 > r + \max\{t_2^*, \ldots, t_m^*\}$ and we check that $u(t_0, \omega_0, 0) \gg 0$, that is, $u_i(t_0, \omega_0, 0)(s) = z_i(t_0 + s, \omega_0, 0) > 0$ for each $s \in [-r, 0]$, and $i = 1, \ldots, m$. It is already checked for $i = 1$. For $i > 2$, from the monotonicity and the cocycle property of the semiflow we deduce that

$$u(t_0, \omega_0, 0) = u(t_0 - t_i^*, \omega_0 \cdot t_i^*, u(t_i^*, \omega_0, 0)) \geq u(t_0 - t_i^*, \omega_0 \cdot t_i^*, 0) \, ,$$

and hence $u_i(t_0, \omega_0, 0)(s) \geq z_i(t_0 - t_i^* + s, \omega_0 \cdot t_i^*, 0) > 0$ because $t_0 - t_i^* + s > 0$. Consequently, $u(t_0, \omega_0, 0) \gg 0$, and $a \equiv 0$ is a strong sub-equilibrium. The rest of the proof follows from Theorem 3.16. Note that in this case the result holds for each $x_0 \geq 0$.

(ii) As in (i), there is $\omega_0 \in \Omega$ such that $z_j(t, \omega_0, 0) > 0$ for each $t > 0$. We fix $t_1 > r$. Since the matrix $B(\omega_0 \cdot t_1) + C(\omega_0 \cdot t_1)$ is irreducible, there is $i \neq j$ such that $b_{ij}(\omega_0 \cdot t_1) + c_{ij}(\omega_0 \cdot t_1) \succ 0$. If $z_i(t_1, \omega_0, 0) > 0$ we already know that $z_i(t, \omega_0, 0) > 0$ for each $t > t_1$. If $z_i(t_1, \omega_0, 0) = 0$, from $f'(0) > 0, g'(0) > 0$ we deduce that

$$z_i'(t_1, \omega_0, 0) \geq -a_i(\omega_0 \cdot t) z_i(t_1, \omega_0, 0) + b_{ij}(\omega_0 \cdot t_1) f_j(z_j(t_1, \omega_0, 0))$$

$$+ c_{ij}(\omega_0 \cdot t_1) \int_{-r}^0 g_j(z_j(t_1 + s, \omega_0, 0)) \, d\mu_{ij}(s) > 0 \, ;$$

hence, $z_i(t, \omega_0, 0) > 0$ near t_1, and as before, $z_i(t, \omega_0, 0) > 0$ for each $t > t_1$.

Analogously, in a second step we fix $t_2 > t_1 + r$ and we find $k \in \{1, \ldots, m\} - \{i, j\}$ such that $z_k(t, \omega_0, 0) > 0$ for each $t > t_2$. In a finite number of steps, we conclude that there is a $t_0 > mr$ such that $u(t_0, \omega_0, 0) \gg 0$ and $a \equiv 0$ is a strong sub-equilibrium. Finally, Theorem 3.16 finishes the proof, and as in (i), for each $x_0 \geq 0$.

(iii) We claim that δv, as a constant function from Ω to $\mathbb{R}^m$, defines a strong lower solution for the family of systems (3.40) for sufficiently small $\delta > 0$, or equivalently, δv, as a constant function from Ω to X is a strong sub-equilibrium.

It suffices to check that $F(\omega, \delta v) \gg 0$ where here δv is to be understood as an element in X. By Taylor's approximation theorem for the $\mathscr{C}^1$ activation functions, we can write $f_j(s) = f_j'(0)s + \varepsilon_j^1(s)s$ and $g_j(s) = g_j'(0)s + \varepsilon_j^2(s)s$ for certain functions ε_j^1, ε_j^2 with $\lim_{s \to 0} \varepsilon_j^i(s) = 0$ for any $j = 1, \ldots, m$ and $i = 1, 2$. Then, writing

$$F_i(\omega, \delta v) = \delta \left(-a_i(\omega) v_i + \sum_{j=1}^m b_{ij}(\omega) f_j'(0) v_j + \sum_{j=1}^m c_{ij}(\omega) g_j'(0) v_j \right.$$

$$\left. + \sum_{j=1}^m b_{ij}(\omega) \varepsilon_j^1(\delta v_j) v_j + \sum_{j=1}^m c_{ij}(\omega) \varepsilon_j^2(\delta v_j) v_j \right) + I_i(\omega),$$

and taking (3.45) into account, it is easy to check that $F_i(\omega, \delta v) > 0$, $i = 1, \ldots, m$, provided that $\delta > 0$ is taken small enough. An application of Theorem 3.16 for each δ proves the result. □

Remark 3.26. Different conditions on the initial system of neural networks (3.39) imply conditions (3.45) in Theorem 3.25. For example, with the notation $\widetilde{b}_{ij}^- = \inf_{t \in \mathbb{R}} \widetilde{b}_{ij}(t)$, $\widetilde{c}_{ij}^- = \inf_{t \in \mathbb{R}} \widetilde{c}_{ij}(t)$, and $\widetilde{a}_i^+ = \sup_{t \in \mathbb{R}} \widetilde{a}_i(t)$, it is sufficient for condition (3.45) to hold that for $i = 1, \ldots, m$,

$$\widetilde{a}_i^+ < \sum_{j=1}^m \widetilde{b}_{ij}^- f_j'(0) + \sum_{j=1}^m \widetilde{c}_{ij}^- g_j'(0).$$

The proof of the following result is analogous to the proof of Theorem 3.25, just making the obvious modifications because of convexity instead of concavity.

Theorem 3.27. *Assume that* (A1)–(A6) *hold, that* $f_j(0) = g_j(0) = 0$, $j = 1, \ldots, m$, *and* $\widetilde{I}^+ \leq 0$ *in* (3.39). *If, in addition, one of the following conditions is satisfied*

(i) *For each* $i = 1, \ldots, m$, *there is a* $t_i \leq 0$ *such that* $I_i(t_i) < 0$.
(ii) $f_i'(0) > 0$, $g_i'(0) > 0$, $i = 1, \ldots, m$, *there is a* $j \in \{1, \ldots m\}$ *and* $t_j \leq 0$ *such that* $I_j(t_j) < 0$, *and* $B(\omega) + C(\omega)$ *is an irreducible matrix for each* $\omega \in \Omega$.
(iii) *There exists a vector* $v \gg 0$ *such that condition* (3.45) *holds;*

then, there is a unique solution $z^*(t) \ll 0$ *of* (3.39) *such that*

$$\lim_{t \to \infty} \|z(t, x_0) - z^*(t)\| = 0 \quad \text{for each } x_0 \in X \text{ with } x_0 \ll 0.$$

Furthermore, the solution $z^*(t)$ *has the same recurrence property as that of the coefficients, meaning for instance that if the coefficients of* (3.39) *are all almost periodic, then this solution is almost periodic too.*

4 Non-autonomous Cyclic FDEs with Infinite Delay

This section is devoted to the study of the dynamical properties of a monotone skew-product semiflow determined by a family of functional differential equations with infinite delay. Many essential results in the theory of monotone dynamical systems deduced in the last decades require strong monotonicity. This condition never holds when we consider infinite delay differential equations with the usual order. Although the definition of an alternative order is possible in some particular cases (see for instance Wu [116]), this explains the reason why monotone methods have not been systematically applied to this kind of problems. We extend to this context recent results with significative dynamical meaning which only require the monotonicity of the semiflow.

Before considering this particular skew-product semiflow, we describe the structure of the compact invariant sets obtained in Novo et al. [76], which becomes essential to understand the global picture of the dynamics, for an abstract skew-product semiflow $(\Omega \times X, \tau, \mathbb{R}^+)$ where $(\Omega, \sigma, \mathbb{R})$ stands for a minimal flow on a compact metric space and X is a complete metric space.

4.1 Stability and Extensibility Results for Omega-Limit Sets

As explained above, in this subsection we give some new results in the area of topological dynamics for a continuous skew-product semiflow $(\Omega \times X, \tau, \mathbb{R}^+)$

$$\tau: \mathbb{R}^+ \times \Omega \times X \longrightarrow \Omega \times X, \quad (t, \omega, x) \mapsto (\omega{\cdot}t, u(t, \omega, x)), \qquad (4.46)$$

over a minimal base flow $(\Omega, \sigma, \mathbb{R})$ and a complete metric space (X, d). In particular, we extend classical stability and extensibility results to the case of a non-distal base flow, which allow us to generalize in a straightforward way known results for monotone semiflows induced by non-autonomous differential equations when the flow in the base is only minimal.

To begin with, we state the definitions of uniform stability and uniform asymptotic stability for a compact τ-invariant set $K \subset \Omega \times X$.

Definition 4.1. Let C be a positively invariant and closed set in $\Omega \times X$. A compact positively invariant set $K \subseteq C$ is *uniformly stable* (with respect to C) if for any $\varepsilon > 0$ there exists a $\delta(\varepsilon) > 0$, called the *modulus of uniform stability*, such that, if $(\omega, x) \in K$, $(\omega, y) \in C$ are such that $\mathsf{d}(x, y) < \delta(\varepsilon)$, then $\mathsf{d}(u(t, \omega, x), u(t, \omega, y)) \leq \varepsilon$ for all $t \geq 0$. K is *uniformly asymptotically stable* if it is uniformly stable and besides, there exists a $\delta_0 > 0$ such that, if $(\omega, x) \in K$, $(\omega, y) \in C$ satisfy $\mathsf{d}(x, y) < \delta_0$, then, uniformly in $(\omega, x) \in K$, $\lim_{t \to \infty} \mathsf{d}(u(t, \omega, x), u(t, \omega, y)) = 0$.

Very often one deals with either $C = \Omega \times X$ or $C = K$. If no mention to C is made, we assume that it is the whole space, whereas if the restricted semiflow $(K, \tau, \mathbb{R}^+)$

is said to be uniformly stable, we mean that $C = K$. Besides, as it is to be expected, if $C = \Omega \times X$, all the trajectories in a uniformly (asymptotically) stable set are uniformly (asymptotically) stable.

Conversely, if a trajectory has some stability properties, its omega-limit set inherits them: it is not difficult to prove that, if the semiorbit of certain (ω, x) is relatively compact and uniformly (asymptotically) stable, then the omega-limit set of (ω, x) is a uniformly (asymptotically) stable set with the same modulus of uniform stability as that of the semiorbit (see [96]).

Definition 4.2. We say that a compact τ-invariant set $K \subset \Omega \times X$ which admits a flow extension is *positively* (resp. *negatively*) *fiber distal* if for any $\omega \in \Omega$, any two distinct points (ω, x_1), $(\omega, x_2) \in K$ are positively (resp. negatively) distal, that is, $\inf_{t \geq 0} \mathsf{d}(u(t, \omega, x_1), u(t, \omega, x_2)) > 0$ (resp. $\inf_{t \leq 0} \mathsf{d}(u(t, \omega, x_1), u(t, \omega, x_2)) > 0$). The set K is *fiber distal* if it is both positively and negatively fiber distal, that is, $\inf_{t \in \mathbb{R}} \mathsf{d}(u(t, \omega, x_1), u(t, \omega, x_2)) > 0$.

The next result, proved in [76], relates the property of uniform stability to that of fiber distallity, provided that there exists a flow extension. The important topological tool, known as the section map of a positively invariant set $K \subset \Omega \times X$, was already recalled in subsection 3.2.

Theorem 4.3. *Let* $K \subset \Omega \times X$ *be a compact* τ-*invariant set admitting a flow extension. If* $(K, \tau, \mathbb{R})$ *is uniformly stable as* $t \to \infty$, *then it is a fiber distal flow which is also uniformly stable as* $t \to -\infty$. *Furthermore, the section map for* K, $\omega \in \Omega \mapsto K_\omega = \{x \in X \mid (\omega, x) \in K\} \in \mathscr{P}_c(K_X)$, *is continuous at every* $\omega \in \Omega$.

The next step is to prove the same result without assuming that K has a flow extension but considering the existence of backward extensions of semiorbits (see also [76]).

Theorem 4.4. *Let* $K \subset \Omega \times X$ *be a compact positively invariant set such that every point of K admits a backward orbit. If the semiflow* $(K, \tau, \mathbb{R}^+)$ *is uniformly stable, then it admits a flow extension which is fiber distal and uniformly stable as* $t \to -\infty$. *Besides, the section map for* K, $\omega \in \Omega \mapsto K_\omega \in \mathscr{P}_c(K_X)$, *is continuous at every* $\omega \in \Omega$.

We can now easily state the theorem on the structure of uniformly asymptotically stable sets admitting backward semiorbits. We prove that these sets K are N-covers of the base flow, that is, maintaining the notation introduced for the section map (3.17), $\operatorname{card}(K_\omega) = N$ for every $\omega \in \Omega$. Without distallity on the base flow, we combine Theorem 4.4 with previous ideas by Sacker and Sell [91].

Theorem 4.5. *Consider a compact positively invariant set* $K \subset \Omega \times X$ *for the skew-product semiflow* (4.46) *and assume that every semiorbit in K admits a backward extension. If* $(K, \tau, \mathbb{R}^+)$ *is uniformly asymptotically stable, then it is an N-cover of the base flow* $(\Omega, \sigma, \mathbb{R})$.

Proof. By Theorem 4.4 we know that K admits a flow extension which is fiber distal. Let us fix any $\omega \in \Omega$ and let us check that card(K_ω) must be finite. Suppose for contradiction that it is infinite. Then, we can take a sequence of pairwise distinct elements $\{x_n\} \subset K_\omega$ such that $\lim_{n \to \infty} x_n = x_0 \in K_\omega$. Let $\delta_0 > 0$ be the positive radius of attraction for K given in Definition 4.1. Choosing n sufficiently large, we have that $0 < \mathsf{d}(x_n, x_0) < \delta_0$, so that $\lim_{t \to \infty} \mathsf{d}(u(t, \omega, x_n), u(t, \omega, x_0)) = 0$, in contradiction with the fiber distallity of K. Therefore, there is a finite N such that card$(K_\omega) = N$.

Finally, it suffices to apply a classical result by Sacker and Sell (see Theorem 3 in [91]) or just the continuity of the section map proved in Theorem 4.4 to conclude that it must be card$(K_\omega) = N$ for all $\omega \in \Omega$, as we claimed. $\qquad \square$

As a consequence, we extend old results by Miller [65] and Sacker and Sell [91] on the structure of omega-limit sets with an almost periodic minimal base flow, to the case of a non-distal base flow.

Proposition 4.6. *Let $\{\tau(t, \widetilde{\omega}, \widetilde{x}) \mid t \geq 0\}$ be a forward orbit of the skew-product semiflow (4.46) which is relatively compact and let $\overline{K}$ denote the omega-limit set of $(\widetilde{\omega}, \widetilde{x})$. The following statements hold:*

  (i) *If $\overline{K}$ contains a minimal set K which is uniformly stable, then $\overline{K} = K$ and it admits a fiber distal flow extension.*
 (ii) *If the semiorbit is uniformly stable, then the omega-limit set $\overline{K}$ is a uniformly stable minimal set which admits a fiber distal flow extension.*
(iii) *If the semiorbit is uniformly asymptotically stable, then the omega-limit set $\overline{K}$ is a uniformly asymptotically stable minimal set which is an N-cover of the base flow.*

Proof. (i) We just need to show that $\overline{K} \subseteq K$. So, take an element $(\omega, x) \in \overline{K}$ and let us prove that $(\omega, x) \in K$. As K is in particular closed, it suffices to see that for any fixed $\varepsilon > 0$ there exists $(\omega, x^*) \in K$ such that $\mathsf{d}(x, x^*) \leq \varepsilon$. Let $\delta(\varepsilon) > 0$ be the modulus of uniform stability for K.

First of all, there exists $s_n \uparrow \infty$ such that $\lim_{n \to \infty} (\widetilde{\omega} \cdot s_n, u(s_n, \widetilde{\omega}, \widetilde{x})) = (\omega, x)$. Now, take a pair $(\omega, x_0) \in K \subseteq \overline{K}$. Then, there exists a sequence $t_n \uparrow \infty$ such that
$$(\omega, x_0) = \lim_{t \to \infty} (\widetilde{\omega} \cdot t_n, u(t_n, \widetilde{\omega}, \widetilde{x})).$$

As it is well-known, in omega-limit sets and in minimal sets there always exist backward continuations of semiorbits. Then, we can apply Theorem 4.4 to K so that the section map (3.17) turns out to be continuous at any point. As $\widetilde{\omega} \cdot t_n \to \omega$, we deduce that $K_{\widetilde{\omega} \cdot t_n} \to K_\omega$ in the Hausdorff metric. Then, for $x_0 \in K_\omega$ there exists a sequence $x_n \in K_{\widetilde{\omega} \cdot t_n}$, $n \geq 1$, such that $x_n \to x_0$ as $n \to \infty$. Altogether, there exists $n_0 \in \mathbb{N}$ such that $\mathsf{d}(u(t_{n_0}, \widetilde{\omega}, \widetilde{x}), x_{n_0}) < \delta(\varepsilon)$. By the uniform stability,

$$\mathsf{d}(u(t + t_{n_0}, \widetilde{\omega}, \widetilde{x}), u(t, \widetilde{\omega} \cdot t_{n_0}, x_{n_0})) \leq \varepsilon \quad \text{for all } t \geq 0.$$

In particular, if n_1 is such that $s_n - t_{n_0} \geq 0$ for $n \geq n_1$, we obtain that

$$d(u(s_n, \widetilde{\omega}, \widetilde{x}), u(s_n - t_{n_0}, \widetilde{\omega} \cdot t_{n_0}, x_{n_0})) \leq \varepsilon \quad \text{for all } n \geq n_1. \tag{4.47}$$

Now, it remains to note that, as $(\widetilde{\omega} \cdot t_{n_0}, x_{n_0}) \in K$, also $\tau(s_n - t_{n_0}, \widetilde{\omega} \cdot t_{n_0}, x_{n_0}) = (\widetilde{\omega} \cdot s_n, u(s_n - t_{n_0}, \widetilde{\omega} \cdot t_{n_0}, x_{n_0})) \in K$ for all $n \geq n_1$. Therefore, there is a convergent subsequence towards a pair $(\omega, x^*) \in K$, and taking limits in (4.47), we deduce that $d(x, x^*) \leq \varepsilon$, as we wanted.

(ii) We already remarked that in this case $\widetilde{K}$ is uniformly stable. The fact that it is minimal is a straightforward consequence of (i). For the fiber distal flow extension one just needs to apply Theorem 4.4.

(iii) It follows from previous comments as well as from Theorem 4.5. $\qquad \square$

Remark 4.7. Note that the stability and extensibility results obtained in this subsection allow us to extend many of the results of Shen and Yi [101] and Jiang and Zhao [44], proved with distallity on the base, to the case of just a minimal base flow.

4.2 FDEs with Infinite Delay

Let $(\Omega, \sigma, \mathbb{R})$ be a minimal flow over a compact metric space (Ω, d) and denote $\sigma(t, \omega) = \omega \cdot t$ for each $\omega \in \Omega$ and $t \in \mathbb{R}$. In $\mathbb{R}^m$ we take the maximum norm $\|v\| = \max_{j=1,\dots,m} |v_j|$ and the usual partial order relation, already introduced in section 3. We consider the Fréchet space $X = C((-\infty, 0], \mathbb{R}^m)$ endowed with the compact-open topology, i.e., the topology of uniform convergence over compact subsets, which is a metric space for the distance

$$d(x, y) = \sum_{n=1}^{\infty} \frac{1}{2^n} \frac{\|x - y\|_n}{1 + \|x - y\|_n}, \quad x, y \in X,$$

where $\|x\|_n = \sup_{s \in [-n, 0]} \|x(s)\|$. The subset

$$X_+ = \{x \in X \mid x(s) \geq 0 \text{ for each } s \in (-\infty, 0]\}$$

is a normal positive cone in X and has an empty interior. As usual, a partial order relation in X is induced, given by

$$x \leq y \quad \Longleftrightarrow \quad x(s) \leq y(s) \quad \text{for each } s \in (-\infty, 0],$$

$$x < y \quad \Longleftrightarrow \quad x \leq y \quad \text{and} \quad x \neq y.$$

Let $BU \subset X$ be the Banach space

$$BU = \{x \in X \mid x \text{ is bounded and uniformly continuous}\}$$

with the supremum norm $\|x\|_\infty = \sup_{s \in (-\infty, 0]} \|x(s)\|$. Given $r > 0$ we will denote

$$B_r = \{x \in BU \mid \|x\|_\infty \leq r\}.$$

As usual, given $I = (-\infty, a] \subset \mathbb{R}$, $t \in I$, and a continuous function $z : I \to \mathbb{R}^m$, z_t will denote the element of X defined by $z_t(s) = z(t + s)$ for $s \in (-\infty, 0]$.

We consider the family of non-autonomous infinite delay FDEs

$$z'(t) = F(\omega \cdot t, z_t), \quad t \geq 0, \ \omega \in \Omega, \tag{48}_\omega$$

defined by a function $F : \Omega \times BU \to \mathbb{R}^m$, $(\omega, x) \mapsto F(\omega, x)$ satisfying the following conditions:

(H1) The functions $F_x : \Omega \times BU \to L(BU, \mathbb{R}^m)$, $(\omega, x) \mapsto F_x(\omega, x)$ and F are continuous on $\Omega \times BU$.

(H2) For each $r > 0$, $F(\Omega \times B_r)$ is a bounded subset of $\mathbb{R}^m$.

(H3) For each $r > 0$, $F : \Omega \times B_r \to \mathbb{R}^m$ is continuous when we take the restriction of the compact-open topology to B_r, i.e., if $\omega_n \to \omega$ and $x_n \overset{d}{\to} x$ as $n \to \infty$ with $x \in B_r$, then $\lim_{n \to \infty} F(\omega_n, x_n) = F(\omega, x)$.

From hypothesis (H1), the standard theory of infinite delay functional differential equations (see [33]) assures that for each $x \in BU$ and each $\omega \in \Omega$ the system $(48)_\omega$ locally admits a unique solution $z(t, \omega, x)$ with initial value x, i.e., $z(s, \omega, x) = x(s)$ for each $s \in (-\infty, 0]$. Therefore, the family $(48)_\omega$ induces a local skew-product semiflow

$$\tau : \mathbb{R}^+ \times \Omega \times BU \longrightarrow \Omega \times BU, \quad (t, \omega, x) \mapsto (\omega \cdot t, u(t, \omega, x)), \tag{4.49}$$

where $u(t, \omega, x) \in BU$ and $u(t, \omega, x)(s) = z(t + s, \omega, x)$ for $s \in (-\infty, 0]$.

As shown in [76], from hypotheses (H1) and (H2), each bounded solution $z(t, \omega_0, x_0)$ provides a relatively compact trajectory for the compact-open topology. We include the proof for completeness.

Proposition 4.8. *Let* $(\omega_0, x_0) \in \Omega \times BU$. *The following statements are equivalent:*

(i) $z(t, \omega_0, x_0)$ *is a bounded solution, i.e.,* $r = \sup_{t \in \mathbb{R}} \|z(t, \omega_0, x_0)\| < \infty$.

(ii) *The closure* $_x\{u(t, \omega_0, x_0) \mid t \geq 0\}$ *is a compact subset of* X *for the compact-open topology.*

Moreover, the closure is a subset of BU.

Proof. (i) $\Rightarrow$ (ii). Consider the set $\mathscr{F} = \{u(t, \omega_0, x_0) \mid t \geq 0\} \subset BU \subset X$, with the compact-open topology. According to Theorem 8.1.4. in Hino et al. [33], $\mathscr{F}$ is relatively compact in X if, and only if, for every $s \in (-\infty, 0]$ $\mathscr{F}$ is equicontinuous at s and $\mathscr{F}(s) = \{u(t, \omega_0, x_0)(s) \mid t \geq 0\}$ is relatively compact in $\mathbb{R}^m$.

The second condition holds, as $\|u(t, \omega_0, x_0)(s)\| = \|z(t + s, \omega_0, x_0)\| \leq r$ for any $t \geq 0$ and $s \leq 0$, i.e., $\mathscr{F} \subset B_r$. As for the equicontinuity, fix $\varepsilon > 0$. Let $\delta_1 > 0$ be such that, if $s, s' \in (-\infty, 0]$ with $|s - s'| < \delta_1$, $\|x_0(s) - x_0(s')\| \leq \varepsilon/2$, and let $\delta_2 = \varepsilon/(2c)$, for the constant $c = \sup\{\|F(\omega, x)\| \mid (\omega, x) \in \Omega \times B_r\} < \infty$, thanks to (H2). Then, if we take $s, s_1 \in (-\infty, 0]$ with $|s - s_1| < \delta = \min(\delta_1, \delta_2)$ and $s \leq s_1$ (the case $s_1 \leq s$ is analogous), we have that

$$\|u(t, \omega_0, x_0)(s) - u(t, \omega_0, x_0)(s_1)\| = \|z(t + s, \omega_0, x_0) - z(t + s_1, \omega_0, x_0)\|$$

$$\leq \begin{cases} \|x_0(t + s) - x_0(t + s_1)\| \leq \dfrac{\varepsilon}{2} & \text{if } t + s,\ t + s_1 \leq 0; \\[2mm] c\,|s - s_1| \leq \dfrac{\varepsilon}{2} & \text{if } t + s,\ t + s_1 \geq 0; \\[2mm] \|x_0(t + s) - x_0(0)\| + \|x_0(0) - z(t + s_1, \omega_0, x_0)\| \leq \varepsilon & \text{if } s \leq -t \leq s_1. \end{cases}$$

Note that the second case holds from the mean value theorem and (H2), and in the last case we have combined the application of the mean value theorem and the uniform continuity of the initial function x_0. With this, we have actually proved that $\mathscr{F}$ is uniformly equicontinuous on $(-\infty, 0]$.

(ii) $\Rightarrow$ (i). Since the initial data $x_0 \in BU$ is bounded, it follows from the compactness of $K = \text{closure}_X\{u(t, \omega_0, x_0) \mid t \geq 0\}$ and the continuity of the map $K \to \mathbb{R}^m$, $x \mapsto x(0)$.

To finish, we have to prove that the limit points of $\mathscr{F}$ remain inside BU. Obviously for any limit point v, $\|v\|_\infty \leq r$, and we only have to check uniform continuity. So, assume that for some sequence $\{t_n\} \subset \mathbb{R}^+$, $u(t_n, \omega_0, x_0) \overset{d}{\to} v$. If $\{t_n\}$ is bounded and we suppose without loss of generality that $t_n \to t_0$ as $n \to \infty$, then, by continuity of the solution, it must be $v = u(t_0, \omega_0, x_0) \in BU$. If $\{t_n\}$ is not bounded and again without loss of generality we put that $t_n \to \infty$ as $n \to \infty$, we easily get that v is Lipschitzian with the former Lipschitz constant c, and we are done. $\qquad\square$

In addition, from hypotheses (H1), (H2) and (H3) and the next result, we can deduce the continuity of the semiflow restricted to some compact subsets $K \subset \Omega \times BU$ when the compact-open topology is considered in BU.

Proposition 4.9. *Let $\{(\omega_n, x_n)\} \subset \Omega \times B_R$ for some $R > 0$ be such that $\omega_n \to \omega$ and $x_n \overset{d}{\to} x$ with $(\omega, x) \in \Omega \times B_R$. If there is $t > 0$ such that $u(t, \omega, x)$ is defined, then there is n_0 such that $u(t, \omega_n, x_n)$ is defined for each $n \geq n_0$ and $u(t, \omega_n, x_n) \overset{d}{\to} u(t, \omega, x)$.*

Proof. If $s \leq -t$, $u(t, \omega_n, x_n)(s) - u(t, \omega, x)(s) = x_n(t + s) - x(t + s)$, and $x_n \overset{d}{\to} x$. Thus, it suffices to show that $u(t, \omega_n, x_n)(s) \to u(t, \omega, x)(s)$ uniformly for $s \in [-t, 0]$ or, equivalently, $z(s, \omega_n, x_n) \to z(s, \omega, x)$ uniformly for $s \in [0, t]$.

First, we claim that the result holds if $z(t, \omega_n, x_n)$ is defined for each $n \geq 1$ and $\sup\{\|z(s, \omega_n, x_n)\| \mid s \in [0, t], \, n \geq 1\} \leq r_0$ for some $r_0 > 0$. In fact, the set $\mathscr{F} = \{z(\cdot, \omega_n, x_n)|_{[0,t]} \mid n \geq 1\} \subset (C([0, t], \mathbb{R}^m), \|\cdot\|_\infty)$ is uniformly bounded, and it is uniformly equicontinuous, because of the mean value theorem and (H2). Then, by Arzelà-Ascoli theorem, $\mathscr{F}$ is relatively compact. We just need to prove that $z(\cdot, \omega, x)|_{[0,t]}$ is its only limit point. So, assume for simplicity that $z(s, \omega_n, x_n) \to v(s)$ uniformly on $[0, t]$. We extend the function v with continuity to all $(-\infty, t]$ by defining $v(s) = x(s)$ for any $s \leq 0$. Then, it trivially holds that $u(s, \omega_n, x_n) \overset{d}{\to} v_s$ and $v_s \in B_{r_0}$ for every $s \in [0, t]$. Now, for each $n \geq 1$, integrating in the equation it satisfies, we have that for any $s \in [0, t]$,

$$z(s, \omega_n, x_n) = x_n(0) + \int_0^s F(\omega_n \cdot r, u(r, \omega_n, x_n)) \, dr \, .$$

Because of (H2) we can apply Lebesgue convergence theorem, and because of the continuity of the flow on Ω and (H3), when we take limits we obtain that

$$v(s) = x(0) + \int_0^s F(\omega \cdot r, v_r) \, dr \quad \text{for every } s \in [0, t] \, .$$

As we have uniqueness of solutions for the initial value problem, it must be $v(s) = z(s, \omega, x)$ for every $s \in [0, t]$, as we wanted to see.

Next, we take $r_1 \geq R + 1$ such that $\sup\{\|z(s, \omega, x)\| \mid s \in [0, t]\} < r_1 - 1$ and we define a C^∞ function $\varphi \colon \mathbb{R}^m \to \mathbb{R}$ such that $\varphi(y) = 1$ if $\|y\| \leq r_1 - 1$ and $\varphi(y) = 0$ if $\|y\| \geq r_1$. We consider the new family

$$y'(s) = F(\omega \cdot s, y_s)\varphi(y(s)), \quad \omega \in \Omega \, . \tag{4.50}$$

It is clear that $y(s) = z(s, \omega, x)$ satisfies the equation for $s \in [0, t]$. Moreover, denoting as usual, by $y(s, \omega_n, x_n)$ the solution of (4.50) for ω_n with initial data x_n, $y(t, \omega_n, x_n)$ is defined for $n \geq 1$ and $\sup\{\|y(s, \omega_n, x_n)\| \mid s \in [0, t], \, n \geq 1\} \leq r_1$. Then, we can apply the first part of the proof to deduce that $y(s, \omega_n, x_n) \to y(s) = z(s, \omega, x)$ uniformly in $[0, t]$. Finally from $\|z(s, \omega, x)\| < r_1 - 1$, there is n_0 such that $\|y(s, \omega_n, x_n)\| < r_1 - 1$ for each $n \geq n_0$ and $s \in [0, t]$. Therefore, $z(s, \omega_n, x_n) = y(s, \omega_n, x_n)$ for $s \in [0, t]$, and the proof is finished. $\qquad\square$

From Proposition 4.8, a bounded solution $z(t, \omega_0, x_0)$ provides a relatively compact trajectory and we can define the omega-limit set as

$$\mathscr{O}(\omega_0, x_0) = \{(\omega, x) \in \Omega \times BU \mid \exists t_n \uparrow \infty \text{ with } \omega_0 \cdot t_n \to \omega, \, u(t_n, \omega_0, x_0) \overset{d}{\to} x\} \, .$$

It is a positively invariant compact set, the restriction of the semiflow to $\mathscr{O}(\omega_0, x_0)$ is continuous when the compact open topology is considered in BU, as shown in Proposition 4.9, and it admits a flow extension because each trajectory has a unique backward orbit (see [76]).

Remark 4.10. A FDE with finite delay could be considered as a FDE with infinite delay, extending the initial data to BU. The main difference is that, as explained above, in infinite delay, omega-limit sets, and hence minimal sets always admit a flow extension. Now we relate this fact with the minimal lifting skew-product flow, introduced in subsection 2.4. Let $(M, \tau, \mathbb{R}^+)$ be a minimal semiflow for the skew-product semiflow (3.16) induced by a FDE with finite delay, and the minimal lifting skew-product flow $(\widehat{M}, \widehat{\tau}, \mathbb{R})$. For $(\omega, \phi) \in \widehat{M}$ we consider $(\omega, x) \in \Omega \times BU$, defined as $x(s) = \phi(s)$, $s \leq 0$, and its omega-limit set $\widetilde{M} = \mathcal{O}(\omega, x) \subset \Omega \times BU$ for the corresponding induced semiflow with infinite delay. It is easy to check that $\widehat{M}$ and $\widetilde{M}$ are isomorphic. That is, the infinite delay formulation extends to flows those minimal semiflows of $\Omega \times C([-r, 0], \mathbb{R}^m)$ appearing in finite delay.

Next we characterize the omega-limit sets when a Lyapunov function exists.

Definition 4.11. Let $V : \Omega \times BU \times BU \to \mathbb{R}^+$, $(\omega, x, y) \mapsto V(\omega, x, y)$ be a continuous map for the norm in BU such that for each $r > 0$, $V : \Omega \times B_r \times B_r \to \mathbb{R}^+$ is continuous when we take the restriction of the compact-open topology to B_r, i.e., if $\omega_n \to \omega$ and $x_n \overset{d}{\to} x$, $y_n \overset{d}{\to} y$, as $n \to \infty$ with $x, y \in B_r$, then $\lim_{n \to \infty} V(\omega_n, x_n, y_n) = V(\omega, x, y)$. It is said that V is a *Lyapunov function* for the skew-product semiflow (4.49) if it satisfies the following properties:

(1) $V(\omega, x, y) = 0 \Rightarrow x(0) = y(0)$;
(2) $V(\omega \cdot t, u(t, \omega, x), u(t, \omega, y)) \leq V(\omega, x, y)$ for each $t \geq 0$.

It is said that V is a *strict Lyapunov function* if, in addition:

(3) For each $(\omega, x, y) \in \Omega \times BU \times BU$ with $V(\omega, x, y) \neq 0$, there is a t_0 (depending on ω, x and y) such that $V(\omega \cdot t_0, u(t_0, \omega, x), u(t_0, \omega, y)) < V(\omega, x, y)$.

Lemma 4.12. *Let V be a strict Lyapunov function for the skew-product semiflow (4.49). Then for each (ω_0, x_0), $(\omega_0, y_0) \in \Omega \times BU$ with bounded trajectories*

$$\lim_{t \to \infty} V(\omega_0 \cdot t, u(t, \omega_0, x_0), u(t, \omega_0, y_0)) = 0.$$

Proof. Note that the limit exists because of (2). Assume on the contrary that it does not vanish and let $t_n \uparrow \infty$ be a sequence such that $\lim_{n \to \infty} \omega_0 \cdot t_n = \omega_0$, $\lim_{n \to \infty} (\omega_0 \cdot t_n, u(t_n, \omega_0, x_0)) = (\omega_0, x_0^*)$ and $\lim_{n \to \infty} (\omega_0 \cdot t_n, u(t_n, \omega_0, y_0)) = (\omega_0, y_0^*)$. Hence, $V(\omega_0, x_0^*, y_0^*) \neq 0$ and since V is a strict Lyapunov function, there is a $t_0 > 0$ such that

$$V(\omega_0 \cdot t_0, u(t_0, \omega_0, x_0^*), u(t_0, \omega_0, y_0^*)) < V(\omega, x_0^*, y_0^*).$$

Moreover, from $V(\omega, x_0^*, y_0^*) = \lim_{n \to \infty} V(\omega_0 \cdot t_n, u(t_n, \omega_0, x_0), u(t_n, \omega_0, y_0))$

$$V(\omega_0, x_0^*, y_0^*) > V(\omega_0 \cdot t_0, u(t_0, \omega_0, x_0^*), u(t_0, \omega_0, y_0^*))$$

$$= \lim_{n \to \infty} V(\omega_0 \cdot (t_0 + t_n), u(t_0 + t_n, \omega_0, x_0), u(t_0 + t_n, \omega_0, y_0)),$$

and if we choose a subsequence $\{t_{m_n}\}$ with $t_{m_n} > t_n + t_0$ we conclude that

$$V(\omega_0, x_0^*, y_0^*) > \lim_{n \to \infty} V(\omega_0 \cdot t_{m_n}, u(t_{m_n}, \omega_0, x_0), u(t_{m_n}, \omega_0, y_0)) = V(\omega, x_0^*, y_0^*),$$

a contradiction. $\qquad\square$

Theorem 4.13. *Let V be a Lyapunov function for the skew-product semiflow* (4.49) *and a point $(\omega_0, x_0) \in \Omega \times BU$ such that its trajectory is bounded, and in addition, $\lim_{t \to \infty} V(\omega_0 \cdot t, u(t, \omega_0, x_0), u(t, \omega_0, x_0)) = 0$. Then, its omega-limit set $K = \mathcal{O}(\omega_0, x_0)$ is minimal and $(K, \tau, \mathbb{R}^+)$ is uniformly stable.*

If V is a strict Lyapunov function, then

(i) *K is a copy of the base, that is, $K = \{(\omega, c(\omega)) \mid \omega \in \Omega\}$;*

(ii) *K is the only compact positively invariant set, and it is the global attractor, i.e., for each $(\omega, x) \in \Omega \times BU$ with relatively compact trajectory in X*

$$\lim_{t \to \infty} d(u(t, \omega, x), c(\omega \cdot t)) = 0.$$

Proof. First of all, note that from $\lim_{t \to \infty} V(\omega_0 \cdot t, u(t, \omega_0, x_0), u(t, \omega_0, x_0)) = 0$ we deduce that $K \subset \{(\omega, x) \mid V(\omega, x, x) = 0\}$. Since K is a compact set, V is uniformly continuous on K, and we deduce that given $\varepsilon > 0$ there is a $\delta_1(\varepsilon) > 0$ such that, if $(\omega, x), (\omega, y) \in K$ and $d(x, y) \leq \delta_1(\varepsilon)$ then $V(\omega, x, y) < \varepsilon$. Next, we claim that given $\varepsilon > 0$ there is a $\delta_2(\varepsilon) > 0$ such that if $(\omega, x), (\omega, y) \in K$ and $V(\omega, x, y) \leq \delta_2(\varepsilon)$ then $\|x(0) - y(0)\| \leq \varepsilon$. Assume on the contrary that there is $\varepsilon_0 > 0$ and a sequence $(\omega_n, x_n), (\omega_n, y_n) \in K$ with $V(\omega_n, x_n, y_n) \leq 1/n$ and $\|x_n(0) - y_n(0)\| > \varepsilon_0$. For an adequate subsequence $\lim_{j \to \infty} (\omega_{n_j}, x_{n_j}) = (\omega^*, x^*)$, $\lim_{j \to \infty} (\omega_{n_j}, y_{n_j}) = (\omega^*, y^*)$, and hence $\lim_{j \to \infty} V(\omega_{n_j}, x_{n_j}, y_{n_j}) = V(\omega^*, x^*, y^*) = 0$, which implies that $x^*(0) = y^*(0)$ and contradicts that $\|x^*(0) - y^*(0)\| \geq \varepsilon_0$.

Now, given $\varepsilon > 0$ we take $\delta = \delta_1(\delta_2(\varepsilon))$. Therefore, if $(\omega, x), (\omega, y) \in K$ with $d(x, y) < \delta$ we deduce that $V(\omega \cdot t, u(t, \omega, x), u(t, \omega, y)) \leq V(\omega, x, y) \leq \delta_2(\varepsilon)$ for each $t \geq 0$. Hence $\|z(t, \omega, x) - z(t, \omega, y)\| \leq \epsilon$ for $t \geq 0$, from which it is easily deduced that $(K, \tau, \mathbb{R}^+)$ is uniformly stable.

In addition, we prove that $(\omega, x), (\omega, y) \in K$ and $V(\omega, x, y) = 0$ imply $x = y$. We take a sequence $t_n \uparrow \infty$ such that $\lim_{n \to \infty} \omega \cdot t_n = \omega$. Since $(K, \tau, \mathbb{R}^+)$ is uniformly stable, as shown in Theorem 3.3 in [76], there is a subsequence $\{t_{n_j}\}$ such that the map

$$U : K_\omega \to K_\omega, \quad z \mapsto U(z) = \lim_{j \to \infty} u(t_{n_j}, \omega, z)$$

is a homeomorphism. Since $V(\omega{\cdot}t, u(t, \omega, x), u(t, \omega, y)) \leq V(\omega, x, y) = 0$ for each $t \geq 0$, we deduce that $V(\omega{\cdot}t, u(t, \omega, x), u(t, \omega, y)) = 0$ and therefore $u(t, \omega, x)(0) = u(t, \omega, y)(0)$ for each $t \geq 0$. Consequently, $u(t, \omega, x)(s) = u(t, \omega, y)(s)$ for each $s \leq 0$ with $t + s \geq 0$, and from the definition of U we conclude that $U(x)(s) = U(y)(s)$, i.e., $U(x) = U(y)$, which yields to $x = y$. Then, as above, by contradiction we check that given $\varepsilon > 0$ there is a $\delta_3(\varepsilon) > 0$ such that if $(\omega, x), (\omega, y) \in K$ and $V(\omega, x, y) \leq \delta_3(\varepsilon)$ then $\mathsf{d}(x, y) \leq \varepsilon$.

Now we prove that K is a minimal set. Let M be a minimal subset $M \subset K$. We just need to show that $K \subseteq M$. So, take an element $(\omega, x) \in K$ and let us prove that $(\omega, x) \in M$. As M is in particular closed, it suffices to see that for any fixed $\varepsilon > 0$ there exists $(\omega, x^*) \in M$ such that $\mathsf{d}(x, x^*) \leq \varepsilon$. First of all, there exists $s_n \uparrow \infty$ such that $\lim_{n\to\infty}(\omega_0{\cdot}s_n, u(s_n, \omega_0, x_0)) = (\omega, x)$. Now, take a pair $(\omega, y) \in M$ and a sequence $t_n \uparrow \infty$ such that

$$\lim_{n\to\infty} (\omega_0{\cdot}t_n, u(t_n, \omega_0, x_0)) = (\omega, y). \tag{4.51}$$

From Theorem 4.3 the section map for M is continuous at every $\omega \in \Omega$, hence $M_{\omega_0{\cdot}t_n} \to M_\omega$, and there are $(\omega_0{\cdot}t_n, y_n) \in M$ with $\lim_{n\to\infty}(\omega_0{\cdot}t_n, y_n) = (\omega, y)$, which together with (4.51) yields to the existence of an n_0 such that $\mathsf{d}(u(t_{n_0}, \omega_0, x_0), y_{n_0}) < \delta_1(\delta_3(\varepsilon))$ and $V(\omega_0{\cdot}t_{n_0}, u(t_{n_0}, \omega_0, x_0), y_{n_0}) \leq \delta_3(\varepsilon)$. Therefore

$$V(\omega_0{\cdot}(t + t_{n_0}), u(t + t_{n_0}, \omega_0, x_0), u(t, \omega_0{\cdot}t_{n_0}, y_{n_0})) \leq \delta_3(\varepsilon).$$

for each $t \geq 0$. In particular, if n_1 is such that $s_n - t_{n_0} \geq 0$ for $n \geq n_1$, we obtain that

$$V(\omega_0{\cdot}s_n, u(s_n, \omega_0, x_0), u(s_n - t_{n_0}, \omega_0{\cdot}t_{n_0}, y_{n_0})) \leq \delta_3(\varepsilon) \quad \text{for all } n \geq n_1. \tag{4.52}$$

Now, it remains to note that, as $(\omega_0{\cdot}t_{n_0}, y_{n_0}) \in M$, also $\tau(s_n - t_{n_0}, \omega_0{\cdot}t_{n_0}, y_{n_0}) = (\omega_0{\cdot}s_n, u(s_n - t_{n_0}, \omega_0{\cdot}t_{n_0}, y_{n_0})) \in M$ for all $n \geq n_1$. Therefore, there is a convergent subsequence towards a pair $(\omega, x^*) \in M$, and taking limits in (4.52), we deduce that $V(\omega, x, x^*) \leq \delta_3(\varepsilon)$ and hence, $\mathsf{d}(x, x^*) \leq \varepsilon$, as we wanted.

(i) We assume that V is a strict Lyapunov function and we check that K is a copy of the base, that is card$K_\omega = 1$ for each $\omega \in \Omega$. We fix $\omega \in \Omega$. Assume on the contrary that there is a pair $(\omega, x_1), (\omega, x_2) \in K$ with $x_1 \neq x_2$. Let $t_n \uparrow \infty$ be such that $\lim_{n\to\infty} \omega{\cdot}t_n = \omega$ and $\lim_{n\to\infty}(\omega{\cdot}t_n, u(t_n, \omega, x_1)) = (\omega, x_1^*) \in K$, $\lim_{n\to\infty}(\omega{\cdot}t_n, u(t_n, \omega, x_2)) = (\omega, x_2^*) \in K$. From Theorem 4.3, K is fiber distal; hence $x_1^* \neq x_2^*$ and $V(\omega, x_1^*, x_2^*) \neq 0$. However, since V is a strict Lyapunov function we deduce from Lemma 4.12 that $V(\omega, x_1^*, x_2^*) = 0$, a contradiction. Therefore, $x_1 = x_2$ and K is a copy of the base, as claimed.

(ii) On the contrary assume the existence of two minimal subsets M_1 and M_2. From (i) they are copies of the base, that is, $M_i = \{(\omega, x_i(\omega)) \mid \omega \in \Omega\}, i = 1, 2$. We fix $\omega \in \Omega$. Since $x_1(\omega) \neq x_2(\omega)$, there is $s_0 \leq 0$ such that $x_1(\omega)(s_0) \neq x_2(\omega)(s_0)$, and hence $V(\omega{\cdot}s_0, x_1(\omega{\cdot}s_0), x_2(\omega{\cdot}s_0)) \neq 0$. As before, let $t_n \uparrow \infty$

be such that $\lim_{n\to\infty} \omega \cdot t_n = \omega$, from Lemma 4.12 we deduce that

$$\lim_{n\to\infty} V(\omega \cdot (t_n + s_0), u(t_n, \omega \cdot s_0, x_1(\omega \cdot s_0)), u(t_n, \omega \cdot s_0, x_2(\omega \cdot s_0))) = 0,$$

i.e., $V(\omega \cdot s_0, x_1(\omega \cdot s_0), x_2(\omega \cdot s_0)) = 0$, a contradiction. Thus $M_1 = M_2$ and there is a unique minimal subset. Finally, for each $(\omega, x) \in \Omega \times BU$ with relatively compact trajectory, $\mathcal{O}(\omega, x) = K$ and $\lim_{t\to\infty} \mathsf{d}(u(t, \omega, x), c(\omega \cdot t)) = 0$, as stated.

$\square$

Remark 4.14. If the base flow is almost periodic and a strict Lyapunov function exists, the previous theorem says that every relatively forward orbit is asymptotically almost periodic.

The main problem for the construction of a Lyapunov function in the non-autonomous case is the structure of the minimal subsets, as well as the continuous variation of its section map with respect to time. This difficulty can be avoided in the study of the stability of the null solution in the non-autonomous linear case.

4.2.1 Example: Linear Cellular Neural Networks with Infinite Delay

We apply the previous results on Lyapunov functions to study the behavior of the solutions of the family of linear FDEs with infinite delay

$$z'(t) = -a(\omega \cdot t) z(t) + b(\omega \cdot t) \int_{-\infty}^{0} z(t+s) \alpha(s) \, ds, \quad \omega \in \Omega, \qquad (4.53)$$

satisfying:

(a1) $a, b \in C(\Omega, \mathbb{R}^+)$.

(a2) $\alpha(s) \geq 0$ for each $s \in (-\infty, 0]$, $\int_{-\infty}^{0} \alpha(s) \, ds = 1$ and $\int_{-\infty}^{0} s \, \alpha(s) \, ds < \infty$.

(a3) $a(\omega) \geq \int_{-\infty}^{0} b(\omega \cdot (-s)) \alpha(s) \, ds$ for each $\omega \in \Omega$.

(a4) There is $\widetilde{\omega} \in \Omega$ such that $a(\widetilde{\omega}) > \int_{-\infty}^{0} b(\widetilde{\omega} \cdot (-s)) \alpha(s) \, ds$.

As usual, this family may have been constructed from a single almost periodic or recurrent equation by a Bebutov process.

It is easy to check that if $x \in BU_+$ then $z(t, \omega, x) \geq 0$ for each $t \geq 0$, which implies the monotonicity of the induced semiflow. In particular, this can be shown, as explained in subsection 4.3, by checking the quasimonotone condition (H4). If in addition, $x(0) > 0$, it is shown that $z(t, \omega, x) > 0$ for each $t > 0$.

We define $V : \Omega \times BU_+ \times BU_+ \rightarrow \mathbb{R}^+$ by $V(\omega, x, y) = W(\omega, x) + W(\omega, y)$ with

$$W(\omega, x) = x(0) + \int_{-\infty}^{0} \left[\int_{s}^{0} b(\omega \cdot (r - s)) \, x(r) \, dr \right] \alpha(s) \, ds .$$

It is easy to check that V satisfies the assumptions of continuity in Definition 4.11. Moreover, $V(\omega, x, y) = 0$ implies $x(0) = y(0) = 0$ and, since

$$W(\omega \cdot t, x_t) = x(t) + \int_{-\infty}^{0} \left[\int_{s}^{0} b(\omega \cdot (t + r - s)) \, x(t + r) \, dr \right] \alpha(s) \, ds$$

$$= x(t) + \int_{-\infty}^{0} \left[\int_{t+s}^{t} b(\omega \cdot (\tau - s)) \, x(\tau) \, d\tau \right] \alpha(s) \, ds ,$$

from (a3) we deduce that

$$\frac{d}{dt} W(\omega \cdot t, u(t, \omega, x)) = - a(\omega \cdot t) \, z(t, \omega, x) + b(\omega \cdot t) \int_{-\infty}^{0} z(t + s, \omega, x) \, \alpha(s) \, ds$$

$$+ \int_{-\infty}^{0} [b(\omega \cdot (t - s)) \, z(t, \omega, x) - b(\omega \cdot t) \, z(t + s, \omega, x)] \, \alpha(s) \, ds$$

$$= \left[-a(\omega \cdot t) + \int_{-\infty}^{0} b(\omega \cdot (t - s) \, \alpha(s) \, ds \right] z(t, \omega, x) \leq 0 ,$$

and, consequently $V(\omega \cdot t, u(t, \omega, x), u(t, \omega, y)) \leq V(\omega, x, y)$ for each $t \geq 0$. In particular, all the solutions with initial data $x \in BU_+$ are bounded.

Moreover, from (a4) and the density of the trajectories in Ω, there is a $t_0 \geq 0$ depending on ω such that

$$a(\omega \cdot t_0) > \int_{-\infty}^{0} b(\omega \cdot (t_0 - s)) \, \alpha(s) \, ds ,$$

and hence, if we take $x, y \in BU_+$ with $V(\omega, x, y) \neq 0$, i.e., $x(0) > 0$ or $y(0) > 0$, we have $V(\omega \cdot t_0, u(t_0, \omega, x), u(t_0, \omega, y)) < V(\omega, x, y)$, i.e., V is a strict Lyapunov function on $\Omega \times BU_+ \times BU_+$.

Next, for each $(\omega_0, x_0) \in \Omega \times BU_+$, we know that

$$K = \mathcal{O}(\omega_0, x_0) \subset \{(\omega, x) \in \Omega \times BU_+ \mid V(\omega, x, x) = 0\} \subset \{(\omega, x) \mid x(0) = 0\} .$$

In addition, if $(\omega, x) \in K$, also $(\omega \cdot s, x_s) \in K$ for each $s \leq 0$, and we deduce that $x(s) = 0$ for each $s \leq 0$, i.e, $x = 0$ and $K = \Omega \times \{0\}$. We claim that it is a global atractor, i.e, for each $(\omega, x) \in BU$

$$\lim_{t \to \infty} u(t, \omega, x) = 0$$

for the compact open-topology, which implies that the null solution is uniformly asymptotically stable. From Theorem 4.13 the result is true for each $(\omega, x) \in \Omega \times BU_+$. Since the family is linear, the same may be said for $(\omega, x) \in BU_-$. Finally, if $(\omega, x) \in \Omega \times BU$, there are $x_- \in BU_-$ and $x_+ \in BU_+$ such that $x_- \leq x \leq x_+$, which implies

$$u(t, \omega, x_-) \leq u(t, \omega, x) \leq u(t, \omega, x_+)$$

for each $t \geq 0$ to finish the proof.

4.3 Monotone FDEs with Infinite Delay

In this subsection, in addition to Hypotheses (H1)–(H3) we consider the quasimonotone condition

(H4) If $x, y \in BU$ with $x \leq y$ and $x_j(0) = y_j(0)$ holds for some $j \in \{1, \ldots, m\}$, then $F_j(\omega, x) \leq F_j(\omega, y)$ for each $\omega \in \Omega$.

From hypothesis (H4) the monotone character of the semiflow is deduced, that is, for each $\omega \in \Omega$ and $x, y \in BU$ such that $x \leq y$ it holds that $u(t, \omega, x) \leq u(t, \omega, y)$ whenever they are defined. The proof is completely analogous to the one given in Theorem 2.6 of Wu [116] or Theorem 5.1.1 of Smith [105].

The techniques and conclusions derived in the in subsection 4.1 allow us to prove results concerning the existence of minimal sets which are almost automorphic extensions of the flow on the base. These minimal sets are copies of the base flow assuming additional hypotheses of stability. More precisely, in [76] we extend previous results of Novo et al. [72], explained and stated in subsection 3.2, deducing the presence of almost automorphic dynamics from the existence of a semicontinuous semi-equilibrium which satisfies additional compactness conditions. In the present situation it is natural to assume that the range of a semi-equilibrium is the set BU. If the base $(\Omega, \sigma, \mathbb{R})$ is almost periodic these methods ensure the existence of almost automorphic minimal sets, which in many cases become exact copies of the base and hence are almost periodic. We refer the reader to [76] but we include the definitions and statements of the results for completeness.

Definition 4.15. A measurable map $a : \Omega \rightarrow BU$ such that $u(t, \omega, a(\omega))$ is defined for any $\omega \in \Omega, t \geq 0$ is

(a) An *equilibrium* if $a(\omega \cdot t) = u(t, \omega, a(\omega))$ for any $\omega \in \Omega$ and $t \geq 0$,
(b) A *super-equilibrium* if $a(\omega \cdot t) \geq u(t, \omega, a(\omega))$ for any $\omega \in \Omega$ and $t \geq 0$, and
(c) A *sub-equilibrium* if $a(\omega \cdot t) \leq u(t, \omega, a(\omega))$ for any $\omega \in \Omega$ and $t \geq 0$.

We will call *semi-equilibrium* to either a super or a sub-equilibrium.

Definition 4.16. A super-equilibrium (resp. sub-equilibrium) $a : \Omega \to BU$ is *semicontinuous* if the following properties hold:

(1) $\Gamma_a = \text{closure}_X\{a(\omega) \mid \omega \in \Omega\}$ is a compact subset of X for the compact-open topology, and
(2) $C_a = \{(\omega, x) \mid x \leq a(\omega)\}$ (resp. $C_a = \{(\omega, x) \mid x \geq a(\omega)\}$) is a closed subset of $\Omega \times X$ for the product metric topology.

An equilibrium is *semicontinuous* in any of these cases.

Note that considering a as a map from Ω to X, it satisfies conditions (1) and (2) of Definition 3.9. These topological properties are now enough to obtain almost automorphic or almost periodic dynamics.

Proposition 4.17. *Let $a : \Omega \to BU$ be a semicontinuous semi-equilibrium and assume that there is an $\omega_0 \in \Omega$ such that the solution $z(t, \omega_0, a(\omega_0))$ is bounded, i.e., $\text{closure}_X\{u(t, \omega_0, a(\omega_0)) \mid t \geq 0\}$ is a compact subset of X for the compact-open topology. Then:*

(i) *The omega-limit set $\mathcal{O}(\omega_0, a(\omega_0))$ contains a unique minimal set which is an almost automorphic extension of the base flow.*
(ii) *If the orbit $\{\tau(t, \omega_0, a(\omega_0)) \mid t \geq 0\}$ is uniformly stable, then $\mathcal{O}(\omega_0, a(\omega_0))$ is a copy of the base.*

Proposition 4.18. *Let $a : \Omega \to BU$ be a semicontinuous semi-equilibrium such that $\sup_{\omega \in \Omega} \|a(\omega)\|_\infty < \infty$ and $\Gamma_a \subset BU$. The following statements are equivalent:*

(i) *$\Gamma_a = \text{closure}_X\{u(t, \omega, a(\omega)) \mid t \geq 0, \omega \in \Omega\}$ is a compact subset of BU for the compact-open topology.*
(ii) *For each $\omega \in \Omega$, the $\text{closure}_X\{u(t, \omega, a(\omega)) \mid t \geq 0\}$ is a compact subset of BU for the compact-open topology.*
(iii) *There is an $\omega_0 \in \Omega$ such that the $\text{closure}_X\{u(t, \omega_0, a(\omega_0)) \mid t \geq 0\}$ is a compact subset of BU for the compact-open topology.*

Theorem 4.19. *Let us assume the existence of a semicontinuous semi-equilibrium $a : \Omega \to BU$ satisfying $\sup_{\omega \in \Omega} \|a(\omega)\|_\infty < \infty$, $\Gamma_a \subset BU$ and one of the equivalent statements of Proposition 4.18. Then,*

(i) *There exists a semicontinuous equilibrium $c : \Omega \to BU$ with $c(\omega) \in \Gamma_a$ for any $\omega \in \Omega$.*
(ii) *Let ω_1 be a continuity point for c. Then, the restriction of the semiflow τ to the minimal set*

$$K^* = \text{closure}_{\Omega \times X}\{(\omega_1 \cdot t, c(\omega_1 \cdot t)) \mid t \geq 0\} \subset C_a \qquad (4.54)$$

is an almost automorphic extension of the base flow $(\Omega, \sigma, \mathbb{R})$.
(iii) *K^* is the only minimal set contained in the omega-limit set $\mathcal{O}(\widehat{\omega}, a(\widehat{\omega}))$ for each point $\widehat{\omega} \in \Omega$.*

(iv) *If there is a point $\widetilde{\omega} \in \Omega$ such that the trajectory $\{\tau(t, \widetilde{\omega}, a(\widetilde{\omega})) \mid t \geq 0\}$ is uniformly stable, then for each $\widehat{\omega} \in \Omega$,*

$$\mathcal{O}(\widehat{\omega}, a(\widehat{\omega})) = K^* = \{(\omega, c(\omega)) \mid \omega \in \Omega\},$$

i.e., it is a copy of the base determined by the equilibrium c of (i), which is a continuous map.

In the study of cooperative and irreducible autonomous ODEs and FDEs with finite delay, it is well known that the generic convergence of the solutions to the set of equilibria is the essential result describing the global dynamics of the corresponding semiflows. The example of subsection 2.3.2 shows that this kind of results are not valid for non-autonomous and cooperative ordinary differential equations. Consequently, the study of the non-autonomous case require new and different arguments and techniques.

Next we obtain an infinite delay version of significant results proved by Jiang and Zhao [44], again without the assumption of distal flow on the base. They established the 1-covering property of omega-limit sets for monotone and uniformly stable skew-product semiflows with the componentwise separating property of bounded and ordered full orbits, where the ordered space is a product Banach space. As a consequence, our results also hold for the finite delay case.

A componentwise separation property has been frequently considered for ordinary and finite delayed cooperative differential equations (see [102, 105]). We show that this is also a natural condition for cooperative retarded differential equations with infinite delay; in fact, can be deduced from Hypotheses (H1)–(H4).

Proposition 4.20. *Under Hypotheses* (H1)–(H4), *if $x, y \in BU$ with $x \leq y$ and $x_i(0) < y_i(0)$ holds for some $i \in \{1, \ldots, m\}$, then $z_i(t, \omega, x) < z_i(t, \omega, y)$ for each $\omega \in \Omega$ and whenever they are defined.*

Proof. We fix $\omega \in \Omega$ and $x, y \in BU$ satisfying the assumptions in the statement. We take $\widetilde{x} = y + (x - y)g$ with $g : (-\infty, 0] \to \mathbb{R}$ continuous, $0 \leq g \leq 1$, $g(t) = 0$ if $t \leq -1$ and $g(0) = 1$. Note that $y - \widetilde{x} = (y - x)g$ has compact support and $x \leq \widetilde{x} \leq y$.

Let $J = [0, T]$ be an interval of definition of $z(t, \omega, y)$ and $z(t, \omega, x)$, and hence, also of $z(t, \omega, \widetilde{x})$. We denote $h(t) = z(t, \omega, y) - z(t, \omega, \widetilde{x})$ for $t \in J = [0, T]$. Then

$$h'(t) = F(\omega \cdot t, u(t, \omega, y)) - F(\omega \cdot t, u(t, \omega, \widetilde{x}))$$

$$= \int_0^1 F_x(\omega \cdot t, r\, u(t, \omega, y) + (1 - r)\, u(t, \omega, \widetilde{x}))\, h_t\, dr = L(t)\, h_t$$

where $L(t) : BU \to \mathbb{R}^m$ is linear and continuous for the norm.

From Riesz representation theorem we obtain that for each φ of compact support, i.e., i.e., $\varphi_i \in C_c(-\infty, 0]$, $i = 1, \ldots, m$,

$$L(t)\varphi = \int_{-\infty}^{0} [d\mu(t)(s)] \, \varphi(s) \,,$$

where $\mu(t) = [\mu_{ij}(t)]$ is a matrix of real regular Borel measures $\mu_{ij}(t)$ with finite total variation $|\mu_{ij}(t)|(-\infty, 0] < \infty$, for all $i, j \in \{1, \dots, m\}$.

Next, we can express $L(t)\varphi$ as

$$L(t)\varphi = D(t)\,\varphi(0) + \int_{-\infty}^{0} [d\nu(t)(s)] \, \varphi(s) = D(t)\,\varphi(0) + \widetilde{L}(t)\varphi \,,$$

where $D(t) = \text{diag}(a_1(t), \dots, a_m(t))$ with $a_i(t) = \mu_{ii}(t)(\{0\})$, $\nu(t) = [\nu_{ij}(t)]$ with $\nu_{ij}(t) = \mu_{ij}(t)$ if $i \neq j$, and $\nu_{ii}(t)(A) = \mu_{ii}(t)(A - \{0\})$ for each Borel set $A \subset (-\infty, 0]$.

Moreover, from (H4) we deduce that whenever $\varphi \geq 0$ and $\varphi_i(0) = 0$ then $L_i(t)\varphi \geq 0$, and from (H1) we know that $L: J \rightarrow L(BU, \mathbb{R}^m)$ is continuous. Hence, as in Lemma 5.1.2 in [105], it is shown that $\widetilde{L}(t)\varphi \geq 0$ whenever $\varphi \geq 0$, and both D and $\widetilde{L}$ vary continuously with t.

Fix $t > 0$ and note that $h_t = u(t, \omega, y) - u(t, \omega, \widetilde{x}) \geq 0$ has compact support because $h_t(s) = y(t + s) - \widetilde{x}(t + s)$ for each $s \leq -t$. Therefore,

$$h_i'(t) = L_i(t)\, h_t = a_i(t)\, h_i(t) + \widetilde{L}_i(t)\, h_t \geq a_i(t)\, h_i(t) \,,$$

which implies, since $h_i(0) = y_i(0) - \widetilde{x}_i(0) > 0$, that $h_i(t) = z_i(t, \omega, y) - z_i(t, \omega, \widetilde{x}) > 0$ for each $t \in J$. Finally, from $x \leq \widetilde{x}$ and the monotonicity we deduce that $z(t, \omega, x) \leq z(t, \omega, \widetilde{x})$ and hence, $z_i(t, \omega, x) < z_i(t, \omega, y)$ for each $t \in J$, as claimed.   □

Definition 4.21. A forward orbit $\{\tau(t, \omega_0, x_0) \mid t \geq 0\}$ is said to be *uniformly stable* in the ball $B_{r'}$, if for every $\varepsilon > 0$ there is a $\delta(\varepsilon) > 0$ such that, if $s \geq 0$ and $\mathsf{d}(u(s, \omega_0, x_0), x) \leq \delta(\varepsilon)$ for certain $x \in B_{r'}$ and $u(s, \omega_0, x_0) \in B_{r'}$, then for each $t \geq 0$,

$$\mathsf{d}(u(t + s, \omega_0, x_0), u(t, \omega_0 \cdot s, x)) = \mathsf{d}(u(t, \omega_0 \cdot s, u(s, \omega_0, x_0)), u(t, \omega_0 \cdot s, x)) \leq \varepsilon \,.$$

We establish the 1-covering property of omega-limit sets when, in addition to hypotheses (H1)–(H4), the uniform stability is assumed:

(H5) There is an $r > 0$ such that all the trajectories with initial data in B_r are uniformly stable in $B_{r'}$ for each $r' > r$, and relatively compact for the product metric topology.

Theorem 4.22. *Assume that Hypotheses* (H1)–(H5) *hold and let* $(\omega_0, x_0) \in \Omega \times B_r$ *be such that* $K = \mathcal{O}(\omega_0, x_0) \subset \Omega \times B_r$. *Then* $K = \overline{\mathcal{O}}(\omega_0, x_0) = \{(\omega, c(\omega)) \mid \omega \in \Omega\}$ *is a copy of the base and*

$$\lim_{t \to \infty} \mathsf{d}(u(t, \omega_0, x_0), c(\omega_0 \cdot t)) = 0 \,,$$

where $c : \Omega \rightarrow BU$ *is a continuous equilibrium.*

Proof. For each $\omega \in \Omega$ we define the map $a(\omega)$ on $(-\infty, 0]$ by

$$a(\omega)(s) = \inf\{x(s) \mid (\omega, x) \in K\} \quad \text{for each } s \leq 0. \tag{4.55}$$

Then, we claim that $a: \Omega \to BU, \omega \mapsto a(\omega)$ is well-defined, it is a continuous super-equilibrium with $\Gamma_a = \text{closure}_X \{a(\omega) \mid \omega \in \Omega\} \subset BU$, $\sup_{\omega \in \Omega} \|a(\omega)\|_\infty < \infty$, and it satisfies the equivalent statements of Proposition 4.18.

It is not hard to check that for any $(\widetilde{\omega}, \widetilde{x}) \in K, \widetilde{x}$ is Lipschitzian with Lipschitz constant $L = \sup\{\|F(\omega, x)\| \mid (\omega, x) \in \Omega \times B_r\}$. From this one can prove that each $a(\omega)$ is also Lipschitzian with the same constant L and so, $a(\omega) \in B_r$ for any $\omega \in \Omega$ (see Proposition 5.6 in [73] for more details). Then, it holds that Γ_a is a compact subset of X, and actually $\Gamma_a \subset BU$.

Let us check that a defines a super-equilibrium. Note that, as $a(\omega) \in B_r$, it follows from hypothesis (H5) that $u(t, \omega, a(\omega))$ exists for any $\omega \in \Omega$ and $t \geq 0$. Now, fix $\omega \in \Omega$ and $t \geq 0$ and consider any $(\omega \cdot t, y) \in K$. As we have a flow on K, $\tau(-t, \omega \cdot t, y) = (\omega, u(-t, \omega \cdot t, y)) \in K$ and therefore, $a(\omega) \leq u(-t, \omega \cdot t, y)$. Applying monotonicity, $u(t, \omega, a(\omega)) \leq y$. As this happens for any $(\omega \cdot t, y) \in K$, we get that $u(t, \omega, a(\omega)) \leq a(\omega \cdot t)$. Besides, as done in Proposition 5.6 in [73], we have that, if $\omega_n \to \omega$ and $a(\omega_n) \overset{d}{\to} x$, then $a(\omega) \leq x$.

Now let us prove that a is continuous on Ω. From hypothesis (H5) and Proposition 4.6 we know that K is uniformly stable, and then Theorem 4.3 asserts that the section map (3.17) for K, $\omega \in \Omega \mapsto K_\omega$, is continuous at every $\omega \in \Omega$. Fix $\omega \in \Omega$ and $\omega_n \to \omega$ such that $a(\omega_n) \overset{d}{\to} x$. As we have just noted, $a(\omega) \leq x$. On the other hand, as $K_{\omega_n} \to K_\omega$ in the Hausdorff metric, for any $y \in K_\omega$ there exist $x_n \in K_{\omega_n}, n \geq 1$, such that $x_n \overset{d}{\to} y$. Then, $(\omega_n, x_n) \in K$ implies that $a(\omega_n) \leq x_n$ and taking limits, $x \leq y$. As again this happens for any $y \in K_\omega$, we conclude that $x \leq a(\omega)$. In all, $a(\omega) = x$, as wanted.

Hence, from Theorem 4.19 we deduce that there is a continuous equilibrium $c : \Omega \to BU$ such that for each $\widehat{\omega} \in \Omega$,

$$\mathcal{O}(\widehat{\omega}, a(\widehat{\omega})) = K^* = \{(\omega, c(\omega)) \mid \omega \in \Omega\}. \tag{4.56}$$

The definition of a yields to $a(\omega) \leq x$ for each $(\omega, x) \in K$ and hence $c(\omega) \leq x$ by the construction of c. As in [44] we prove that there is a subset $J \subset \{1, \ldots, m\}$ such that

$$\begin{aligned} c_i(\omega) = x_i \quad & \text{for each } (\omega, x) \in K \text{ and } i \notin J, \\ c_i(\omega) < x_i \quad & \text{for each } (\omega, x) \in K \text{ and } i \in J. \end{aligned} \tag{4.57}$$

It is enough to check that if $c_i(\widetilde{\omega})(0) = \widetilde{x}_i(0)$ for some $i \in \{1, \ldots, m\}$ and $(\widetilde{\omega}, \widetilde{x}) \in K$, then $c_i(\omega) = x_i$ for any $(\omega, x) \in K$. We first note that $c_i(\widetilde{\omega}) = \widetilde{x}_i$. Otherwise, there would be $s \in (-\infty, 0]$ with $c_i(\widetilde{\omega})(s) < \widetilde{x}_i(s)$. Then, since $u_i(s, \widetilde{\omega}, \widetilde{x})(0) = \widetilde{x}_i(s)$ because K admits a flow extension, $u(t, \widetilde{\omega}, c(\widetilde{\omega})) = c(\widetilde{\omega} \cdot t)$ for each $t \in \mathbb{R}$ because c is an equilibrium, and Proposition 4.20, we would deduce that $c_i(\widetilde{\omega})(0) < \widetilde{x}_i(0)$, a contradiction. Next, as K is minimal from (H5) and

Proposition 4.6, we take $(\omega, x) \in K$ and a sequence $s_n \downarrow -\infty$ such that $\widetilde{\omega} \cdot s_n \to \omega$ and $u(s_n, \widetilde{\omega}, \widetilde{x})^{\mathrm{d}} \to x$. Then,

$$x_i(0) = \lim_{n \to \infty} u_i(s_n, \widetilde{\omega}, \widetilde{x})(0) = \lim_{n \to \infty} \widetilde{x}_i(s_n)$$

$$= \lim_{n \to \infty} c_i(\widetilde{\omega})(s_n) = \lim_{n \to \infty} c_i(\widetilde{\omega} \cdot s_n)(0) = c_i(\omega)(0) \,,$$

and as before this implies that $c_i(\omega) = x_i$, as wanted.

Let $(\omega, x) \in K$ and define $x_\alpha = (1 - \alpha) a(\omega) + \alpha x \in B_r \subset BU$ for $\alpha \in [0, 1]$, and

$$L = \{\alpha \in [0, 1] \mid \mathcal{O}(\omega, x_\alpha) = K^*\} \,.$$

If we prove that $L = [0, 1]$, then $K = K^*$, $J = \emptyset$ and the proof is finished. From the monotone character of the semiflow and since $\mathcal{O}(\omega, a(\omega)) = K^*$, it is immediate to check that if $0 < \alpha \in L$ then $[0, \alpha] \subset L$.

Next we show that L is closed, that is, if $[0, \alpha) \subset L$ then $\alpha \in L$. From Hypothesis (H5), $\{\tau(t, \omega, x_\alpha) \mid t \geq 0\}$ is uniformly stable; let $\delta(\varepsilon) > 0$ be the modulus of uniform stability for $\varepsilon > 0$. Thus, we take $\beta \in [0, \alpha)$ with $\mathsf{d}(x_\alpha, x_\beta) < \delta(\varepsilon)$ and we obtain $\mathsf{d}(u(t, \omega, x_\alpha), u(t, \omega, x_\beta)) < \varepsilon$ for each $t \geq 0$. Moreover, $\mathcal{O}(\omega, x_\beta) = K^*$ and hence, there is a t_0 such that $\mathsf{d}(u(t, \omega, x_\beta), c(\omega \cdot t)) < \varepsilon$ for each $t \geq t_0$. Then, we deduce that $\mathsf{d}(u(t, \omega, x_\alpha), c(\omega \cdot t)) < 2\varepsilon$ for each $t \geq t_0$ and $\mathcal{O}(\omega, x_\alpha) = K^*$, as claimed.

Finally, we prove that the case $L = [0, \alpha]$ with $0 \leq \alpha < 1$ is impossible. For each $i \in J$ we consider the continuous map

$$K \longrightarrow (0, \infty) \,, \quad (\widetilde{\omega}, \widetilde{x}) \mapsto \widetilde{x}_i(0) - c_i(\widetilde{\omega})(0) \,.$$

Hence, there is an $\varepsilon > 0$ such that $\widetilde{x}_i(0) - c_i(\widetilde{\omega})(0) \geq \varepsilon > 0$ for each $i \in J$ and $(\widetilde{\omega}, \widetilde{x}) \in K$. Moreover, since $(\widetilde{\omega} \cdot s, u(s, \widetilde{\omega}, \widetilde{x})) \in K$, $u_i(s, \widetilde{\omega}, \widetilde{x})(0) = \widetilde{x}_i(s)$ for each $s \leq 0$ because K admits a flow extension, and $c_i(\widetilde{\omega})(s) = c_i(\widetilde{\omega} \cdot s)(0)$, we deduce that $\widetilde{x}_i(s) - c_i(\widetilde{\omega})(s) \geq \varepsilon > 0$ for each $s \in (-\infty, 0]$ and $(\widetilde{\omega}, \widetilde{x}) \in K$.

As before, let $\delta(\varepsilon/4) > 0$ be the modulus of uniform stability for the trajectory $\{\tau(t, \omega, x_\alpha) \mid t \geq 0\}$ and take $\alpha < \gamma \leq 1$ with $\mathsf{d}(x_\alpha, x_\gamma) < \delta(\varepsilon/4)$. For each $t \geq 0$ we have $\|u(t, \omega, x_\alpha)(0) - u(t, \omega, x_\gamma)(0)\| < \varepsilon/4$ and, as above, from $\mathcal{O}(\omega, x_\alpha) = K^*$ we deduce that there is a $t_0 \geq 0$ such that $\|u(t, \omega, x_\alpha)(0) - c(\omega \cdot t)(0)\| < \varepsilon/4$ for each $t \geq t_0$. Consequently, for each $t \geq t_0$

$$\|u(t, \omega, x_\gamma)(0) - c(\omega \cdot t)(0)\| < \frac{\varepsilon}{2} \,. \tag{4.58}$$

Let $(\widetilde{\omega}, \widetilde{x}) \in \mathcal{O}(\omega, x_\gamma)$, i.e., $(\widetilde{\omega}, \widetilde{x}) = \lim_{n \to \infty}(\omega \cdot t_n, u(t_n, \omega, x_\gamma))$ for some $t_n \uparrow \infty$. The monotonicity and $c(\omega) \leq x_\gamma$ imply that $c(\omega \cdot t_n) \leq u(t_n, \omega, x_\gamma)$, which yields to $c(\widetilde{\omega}) \leq \widetilde{x}$. Moreover, from $c(\omega) \leq x_\gamma \leq x$ we have $c(\omega \cdot t_n) \leq u(t_n, \omega, x_\gamma) \leq u(t_n, \omega, x)$ and hence from (4.57) we deduce that $c_i(\omega \cdot t_n) = u_i(t_n, \omega, x_\gamma)$ for each $i \notin J$. This yields to $c_i(\widetilde{\omega}) = \widetilde{x}_i$ for $i \notin J$. Given any $(\widetilde{\omega}, z) \in K$, from (4.57) we know that $c_i(\widetilde{\omega}) = z_i$ for each $i \notin J$ and, as shown above,

$$z_i(s) - c_i(\widetilde{\omega})(s) \geq \varepsilon \quad \text{for each } s \in (-\infty, 0] \text{ and } i \in J. \tag{4.59}$$

From (4.58) there is an n_0 such that $0 \leq u_i(t_n, \omega, x_y)(0) - c_i(\omega \cdot t_n)(0) < \varepsilon/2$ for each $n \geq n_0$, and consequently, $0 \leq \widetilde{x}_i(0) - c_i(\widetilde{\omega})(0) \leq \varepsilon/2$. As before, since this is true for each $(\widetilde{\omega}, \widetilde{x}) \in \mathcal{O}(\omega, x_y)$ admitting a flow extension, we deduce that $0 \leq \widetilde{x}_i(s) - c_i(\widetilde{\omega})(s) \leq \varepsilon/2$ for each $s \in (-\infty, 0]$ and $i \in J$, which combined with (4.59) and $c_i(\widetilde{\omega}) = \widetilde{x}_i = z_i$ for $i \notin J$ show that $c(\widetilde{\omega}) \leq \widetilde{x} \leq z$. Since this holds for each $(\widetilde{\omega}, z) \in K$, the definition of a provides $c(\widetilde{\omega}) \leq \widetilde{x} \leq a(\widetilde{\omega})$. From (4.56) we know that $\mathcal{O}(\widetilde{\omega}, a(\widetilde{\omega})) = K^*$ and therefore $\mathcal{O}(\widetilde{\omega}, \widetilde{x}) = K^* \subseteq \mathcal{O}(\omega, x_y)$. Once more from (H5) and Proposition 4.6 we conclude that $\mathcal{O}(\widetilde{\omega}, \widetilde{x}) = \mathcal{O}(\omega, x_y) = K^*$, a contradiction. Therefore, $L = [0, 1]$, i.e., $J = \emptyset$ and $\mathcal{O}(\omega_0, x_0) = K^*$, as stated. $\qquad\square$

Jiang and Zhao [44] and Hu and Jiang [38] apply this theorem to the study of families of monotone ODEs and FDEs with finite delay. Another application of this result to comparable skew-product semiflows, which agree with eventually strongly monotone semiflows on minimal sets, can be found in Cao et al. [14].

4.4 Compartmental Systems

Compartmental systems have been widely used as a mathematical model for the study of the dynamical behavior of many processes in biological and physical sciences which depend on local mass balance conditions (see Jacquez and Simon [41, 42] for a review of compartmental systems with or without delay, Györi [29], Györi and Eller [30] and Wu and Freedman [118]).

In this subsection, we apply the previous result, that is, the 1-covering property of omega-limit sets, to show that the solutions of a compartmental system given by a monotone FDE with infinite delay are asymptotically of the same type as the transport functions.

Firstly, we introduce the model with which we are going to deal as well as some notation. Let us suppose that we have a system formed by m compartments $C_1, \ldots, C_m$, denote by C_0 the environment surrounding the system, and by $z_i(t)$ the amount of material within compartment C_i at time t for each $i \in \{1, \ldots, m\}$. Material flows from compartment C_j into compartment C_i through a pipe P_{ij} having a transit time distribution given by a positive regular Borel measure μ_{ij} with finite total variation $\mu_{ij}(-\infty, 0] = 1$, for each $i, j \in \{1, \ldots, m\}$. Let $\widetilde{g}_{ij} : \mathbb{R} \times \mathbb{R} \to \mathbb{R}^+$ be the so-called *transport function* determining the volume of material flowing from C_j to C_i given in terms of the time t and the value of z_j in t for $i \in \{0, \ldots, m\}, j \in \{1, \ldots, m\}$. For each $i \in \{1, \ldots, m\}$, we will assume that there exists an incoming flow of material $\widetilde{I}_i$ from the environment into the compartment C_i which only depends on time.

Thus, taking into account that the change of the amount of material of any compartment C_i, $1 \leq i \leq m$, equals to the difference between the amount of total influx into and total outflux out of C_i, we obtain a model governed by the the following system of infinite delay differential equations:

$$z_i'(t) = -\widetilde{g}_{0i}(t, z_i(t)) - \sum_{j=1}^n \widetilde{g}_{ji}(t, z_i(t)) + \sum_{j=1}^n \int_{-\infty}^0 \widetilde{g}_{ij}(t+s, z_j(t+s)) \, d\mu_{ij}(s) + \widetilde{I}_i(t),$$

(4.60)

$i \in \{1, \ldots, m\}$. For simplicity, we denote $\widetilde{g}_{i0} : \mathbb{R} \times \mathbb{R} \to \mathbb{R}^+$, $(t, v) \mapsto \widetilde{I}_i(t)$ for each $i \in \{1, \ldots, m\}$ and let $\widetilde{g} = (\widetilde{g}_{ij})_{i,j} : \mathbb{R} \times \mathbb{R} \to \mathbb{R}^{m(m+2)}$. We will assume that:

(C1) $\widetilde{g}$ is C^1-*admissible*, i.e., $\widetilde{g}$ is C^1 in its second variable and $\widetilde{g}$, $\frac{\partial}{\partial v}\widetilde{g}$ are uniformly continuous and bounded on $\mathbb{R} \times \{v_0\}$ for all $v_0 \in \mathbb{R}$.

(C2) All the component of $\widetilde{g}$ are monotone in the second variable, and $\widetilde{g}_{ij}(t, 0) = 0$ for each $t \in \mathbb{R}$, $i \in \{0, \ldots, m\}$ and $j \in \{1, \ldots, m\}$.

(C3) $\widetilde{g}$ is a recurrent function, i.e., its *hull* is minimal.

(C4) $\mu_{ij}(-\infty, 0] = 1$ and $\int_{-\infty}^0 |s| \, d\mu_{ij}(s) < \infty$.

As usual, we include the non-autonomous system (4.60) into a family of non-autonomous FDEs with infinite delay of the form $(48)_\omega$ as follows.

Let Ω be the *hull* of $\widetilde{g}$, namely, the closure of the set of mappings $\{\widetilde{g}_t \mid t \in \mathbb{R}\}$, with $\widetilde{g}_t(s, v) = \widetilde{g}(t + s, v)$, $(s, v) \in \mathbb{R}^2$, with the topology of uniform convergence on compact sets, which from (C1) is a compact metric space. Let $(\Omega, \sigma, \mathbb{R})$ be the continuous flow defined on Ω by translation, $\sigma : \mathbb{R} \times \Omega \to \Omega$, $(t, \omega) \mapsto \omega{\cdot}t$, with $\omega{\cdot}t(s, v) = \omega(t + s, v)$. By hypothesis (C3), the flow $(\Omega, \sigma, \mathbb{R})$ is minimal. In addition, if $\widetilde{g}$ is almost periodic (resp. almost automorphic) the flow will be almost periodic (resp. almost automorphic). Note that these two cases are included in our formulation.

Let $g : \Omega \times \mathbb{R} \to \mathbb{R}^{m(m+2)}$, $(\omega, v) \mapsto \omega(0, v)$, continuous on $\Omega \times \mathbb{R}$ and denote $g = (g_{ij})_{i,j}$. It is easy to check that, for all $\omega = (\omega_{ij})_{i,j} \in \Omega$ and all $i \in \{1, \ldots, m\}$, ω_{i0} is a function dependent only on t; thus, we can define $I_i = \omega_{i0}$, $i \in \{1, \ldots, m\}$. Let $F : \Omega \times BU \to \mathbb{R}^m$ be the map defined by

$$F_i(\omega, x) = -g_{0i}(\omega, x_i(0)) - \sum_{j=1}^m g_{ji}(\omega, x_i(0)) + \sum_{j=1}^m \int_{-\infty}^0 g_{ij}(\omega{\cdot}s, x_j(s)) \, d\mu_{ij}(s) + I_i(\omega),$$

for $(\omega, x) \in \Omega \times BU$ and $i \in \{1, \ldots, m\}$. Hence, the family

$$z'(t) = F(\omega{\cdot}t, z_t), \quad t \geq 0, \ \omega \in \Omega,$$

(61)$_\omega$

includes system (4.60) when $\omega = \widetilde{g}$.

It is easy to check that this family satisfies hypotheses (F1)–(F4). Next we will study some cases in which hypothesis (F5) is satisfied. In order to do this, we define $M : \Omega \times BU \to \mathbb{R}$, the *total mass* of the system $(61)_\omega$ as

$$M(\omega, x) = \sum_{i=1}^{m} x_i(0) + \sum_{i=1}^{m} \sum_{j=1}^{m} \int_{-\infty}^{0} \left(\int_{s}^{0} g_{ji}(\omega \cdot r, x_i(r)) \, dr \right) d\mu_{ji}(s), \quad (4.62)$$

for all $\omega \in \Omega$ and $x \in BU$, which is well defined from condition (C4). The next result shows the continuity properties of M and its variation along the flow.

Proposition 4.23. *The total mass M is a uniformly continuous function on all the sets of the form $\Omega \times B_r$ with $r > 0$ for the product metric topology. Moreover, for each $(\omega, x) \in \Omega \times BU$ and each $t \geq 0$*

$$M(\omega \cdot t, z_t(\omega, x)) = M(\omega, x) + \sum_{i=1}^{m} \int_{0}^{t} [I_i(\omega \cdot s) - g_{0i}(\omega \cdot s, z_i(s, \omega, x))] \, ds.$$

$$(4.63)$$

In the case of closed systems, that is, without incoming and outgoing flow from and to the environment, the total mass is constant and hence all the solutions are bounded. Of course, the existence of bounded solutions is possible in more general situations.

Theorem 4.24. *Under Assumptions (C1)–(C4), if there exists $\omega_0 \in \Omega$ such that $(61)_{\omega_0}$ has a bounded solution, then all solutions of $(61)_{\omega}$ are bounded as well, hypothesis (H5) holds, and all omega-limit sets are copies of the base.*

Proof. First of all, note that the existence of a bounded solution $z(t, \omega_0, x_0)$ implies, by considering the omega-limit of (ω_0, x_0), the existence of a bounded solution of $(61)_{\omega}$ for each $\omega \in \Omega$.

As explained before, from hypothesis (H4) the monotone character of the semiflow is deduced, that is, for each $\omega \in \Omega$ and $x, y \, l \in BU$ such that $x \leq y$ it holds that $u(t, \omega, x) \leq u(t, \omega, y)$ whenever they are defined. Therefore, $z_i(t, \omega, x) \leq z_i(t, \omega, y)$ for each $i = 1, \ldots, m$. In addition, the monotonicity of transport functions yields $g_{ij}(\omega, z_j(t, \omega, x)) \leq g_{ij}(\omega, z_j(t, \omega, y))$ for each $\omega \in \Omega$. From all these inequalities, (4.62) and (4.63) we deduce that

$$0 \leq z_i(t, \omega, y) - z_i(t, \omega, x) \leq M(\omega \cdot t, z_t(\omega, y)) - M(\omega \cdot t, z_t(\omega, x))$$
$$\leq M(\omega, y) - M(\omega, x),$$

for each $i = 1, \ldots, m$ and whenever $z(t, \omega, x)$ and $z(t, \omega, y)$ are defined. Hence, from the uniform continuity of the total mass M, given $\varepsilon > 0$ there is a $\delta > 0$ such that $\|z(t, \omega, y) - z(t, \omega, x)\| < \varepsilon$ provided that $x, y \in B_r$, $d(x, y) < \delta$ and $x \leq y$. The case in which x and y are not ordered follows easily from this one, by taking the supremum and the infimum of x and y. Hence, this fact, together with the existence of a bounded solution of $(61)_{\omega}$ for each $\omega \in \Omega$, yields to the boundedness of all solutions.

Let $(\omega, x) \in \Omega \times BU$ and $r' > 0$ such that $z_t(\omega, x) \in B_{r'}$ for all $t \geq 0$. Then, as above, we deduce that given $\varepsilon > 0$ there exists a $\delta > 0$ such that

$$\|z(t + s, \omega, x) - z(t, \omega \cdot s, y)\| = \|z(t, \omega \cdot s, z_s(\omega, x)) - z(t, \omega \cdot s, y)\| < \varepsilon$$

for all $t \geq 0$ whenever $y \in B_{r'}$ and $\mathsf{d}(z_s(\omega, x), y) < \delta$, which shows the uniform stability of the trajectories in $B_{r'}$ for each $r' > 0$, hypothesis (H5) holds for all $r > 0$, and Theorem 4.22 applies for all initial data, which finishes the proof.   □

The above statement is compatible with the presence of a unique minimal set which is globally asymptotically stable, and also with the existence of infinitely many minimal sets.

Concerning the solutions of the original compartmental system, we obtain the following result providing a non trivial generalization of the autonomous case, in which the asymptotically constancy of the solutions was shown (see [118]). Although the theorem is stated in the almost periodic case, similar conclusions are obtained changing almost periodicity for periodicity, almost automorphy or recurrence, that is, all solutions are asymptotically of the same type as the transport functions.

Theorem 4.25. *Under Assumptions* (C1)–(C4) *and in the almost periodic case, if there is a bounded solution of* (4.60), *then there is at least an almost periodic solution and all the solutions are asymptotically almost periodic. For closed systems, i.e.,* $\widetilde{I}_i \equiv 0$ *and* $\widetilde{g}_{0i} \equiv 0$ *for each* $i = 1, \ldots, m$, *there are infinitely many almost periodic solutions and the rest of them are asymptotically almost periodic.*

Proof. The first statement is an easy consequence of the previous theorem. We take $\omega_0 = \widetilde{g}$. The omega-limit of each solution $z(t, \omega_0, x_0)$ is a copy of the base $\mathcal{O}(\omega_0, x_0) = \{(\omega, x(\omega)) \mid \omega \in \Omega\}$ and hence, $z(t, \omega_0, x(\omega_0)) = x(\omega_0 \cdot t)(0)$ is an almost periodic solution of (4.60) and

$$\lim_{t \to \infty} \|z(t, \omega_0, x_0) - z(t, \omega_0, x(\omega_0))\| = 0.$$

The statement for closed systems follows in addition from (4.63), which implies that the mass is constant along the trajectories. Hence, there are infinitely many minimal subsets because from the definition of the mass, given $c > 0$ there is an $(\omega_0, x_0) \in \Omega \times BU^+$ such that $M(\omega_0, x_0) = c$ and hence $M(\omega, \dot{x}) = c$ for each $(\omega, x) \in \mathcal{O}(\omega_0, x_0)$.   □

It is important to mention that the papers by Múñoz-Villarragut et al. [67], Novo et al. [77], and Obaya and Villarragut [84] improve the theory stated here. They contain an study of the compartmental geometry for the standard ordering and they analyze compartmental systems described by neutral functional differential equations with infinite delay which are monotone for the exponential ordering.

Acknowledgements The authors were partly supported by Junta de Castilla y León under project VA060A09, and Ministerio de Ciencia e Innovación under project MTM2008-00700/MTM.

References

1. A.I. Alonso, R. Obaya, The structure of the bounded trajectories set of a scalar convex differential equation. Proc. Roy. Soc. Edinb. **133 A**, 237–263 (2003)
2. A.I. Alonso, R. Obaya, A.M. Sanz, A note on non-autonomous scalar functional differential equations with small delay. C. R. Acad. Sci. Paris, Ser. I **340**, 155–160 (2005)
3. H. Amann, Fixed point equations and nonlinear eigenvalue problems in ordered Banach spaces. SIAM Rev. **18**, 620–709 (1976)
4. L. Amerio, G. Prouse, in *Almost-Periodic Functions and Functional Equations*. The University Series in Higher Mathematics (Van Nostrand Reinhold Company, New York, 1971)
5. L. Arnold, I.D. Chueshov, Order-preserving random dynamical systems: equilibria, attractors, applications. Dynam. Stabil. Syst. **13**, 265–280 (1998)
6. L. Arnold, I.D. Chueshov, Cooperative random and stochastic differential equations. Discrete Contin. Dynam. Syst. **7**, 1–33 (2001)
7. A.S. Besicovitch, *Almost Periodic Functions* (Dover, New York, 1954)
8. K. Bjerklöv, Positive Lyapunov exponent and minimality for the continuous 1-d quasi-periodic Schrödinger equations with two basic frequencies. Ann. Henri Poincaré **8**(4), 687–730 (2007)
9. S. Bochner, A new approach to almost periodicity. Proc. Natl. Acad. Sci. USA **48**, 2039–2043 (1962)
10. H. Bohr, Zur Theorie der Fastperiodischen Funktionen. Acta Math. **46**(1), 29–127 (1925)
11. H. Bohr, *Almost Periodic Functions* (Chelsea Publishing Company, New York, 1947)
12. T. Caraballo, D. Cheban, Almost periodic and almost automorphic solutions of linear differential/difference equations without Favard's separation condition. I. J. Differ. Equat. **246**(1), 108–128 (2009)
13. T. Caraballo, D. Cheban, Almost periodic and almost automorphic solutions of linear differential/difference equations without Favard's separation condition. II. J. Differ. Equat. **246**(3), 1164–1186 (2009)
14. F. Cao, M. Gyllenberg, Y. Wang, Asymptotic behaviour of comparable skew-product semiflows with applications. Proc. Lond. Math. Soc. **103**, 271–293 (2011)
15. A. Chen, J. Cao, Existence and attractivity of almost periodic solutions for cellular neural networks with distributed delays and variable coefficients. Appl. Math. Comp. **134**, 125–140 (2003)
16. C. Chicone, Y. Latushkin, in *Evolution Semigroups in Dynamical Systems and Differential Equations*. Mathematical Surveys and Monographs, vol. 70 (American Mathematical Society, Providence, 2002)
17. G. Choquet, in *Lectures on Analysis. Integration and Topological Vector Spaces*. Mathematics Lecture Notes, vol. I (Benjamin, New York, 1969)
18. I.D. Chueshov, in *Monotone Random Systems. Theory and Applications*. Lecture Notes in Mathematics, vol. 1779 (Springer, Berlin, 2002)
19. C.C. Conley, R.A. Miller, Asymptotic stability without uniform stability: almost periodic coefficients. J. Differ. Equat. **1**, 333–336 (1965)
20. C. Corduneanu, *Almost Periodic Functions* (Chelsea Publishing Company, New York, 1968)
21. R.D. Driver, Linear differential systems with small delays. J. Differ. Equat. **21**, 149–167 (1976)
22. R. Ellis, *Lectures on Topological Dynamics* (Benjamin, New York, 1969)
23. R. Fabbri, R. Johnson, F. Mantellini, A nonautonomous saddle-node bifurcation pattern. Stochast. Dynam. **4**(3), 335–350 (2004)
24. M. Fan, D. Ye, Convergence dynamics and pseudo almost periodicity of a class of nonautonomous RFDEs with applications. J. Math. Anal. Appl. **309**, 598–625 (2005)
25. J. Favard, *Leçons sur les Fonctions Presque-périodiques* (Gauthier-Villars, Paris, 1933)
26. A.M. Fink, in *Almost Periodic Differential Equations*. Lecture Notes in Mathematics, vol. 377 (Springer, Berlin, 1974)

27. A.M. Fink, P.O. Frederickson, Ultimately boundedness does not imply almost periodicity. J. Differ. Equat. **9**, 280–284 (1971)
28. K. Gopalsamy, X.Z. He, Stability in asymmetric Hopfield nets with transmission delays. Phys. D **76**, 344–358 (1994)
29. I. Györi, Connections between compartmental systems with pipes and integro-differential equations. Math. Model. **7**, 1215–1238 (1986)
30. I. Györi, J. Eller, Compartmental systems with pipes. Math. Biosci. **53**, 223–247 (1981)
31. J.K. Hale, *Ordinary Differential Equations*, 2nd edn. (Kreiger Publ. Co., Mabar Florida, 1980)
32. J.K. Hale, S.M. Verduyn Lunel, in *Introduction to Functional Differential Equations*. Applied Mathematical Sciences, vol. 99 (Springer, Berlin, 1993)
33. Y. Hino, S. Murakami, T. Naiko, in *Functional Differential Equations with Infinite Delay*. Lecture Notes in Mathematics, vol. 1473 (Springer, Berlin, 1991)
34. M. Hirsch, Systems of differential equations which are competitive or cooperative I: limit sets. SIAM J. Appl. Math. **13**, 167–179 (1982)
35. M. Hirsch, Systems of differential equations which are competitive or cooperative II: convergence almost everywhere. SIAM J. Math. Anal. **16**, 423–439 (1985)
36. J.J. Hopfield, Neural networks and physical systems with emegernt collective computational abilities. Proc. Natl. Acad. Sci. USA **79**, 2554–2558 (1982)
37. J.J. Hopfield, Neurons with grade response have collective computational properties like those of two-state neurons. Proc. Natl. Acad. Sci. USA **81**, 3088–3092 (1984)
38. H. Hu, J. Jiang, Translation-invariant monotone systems II: Almost periodic/automorphic case. Proc. Am. Math. Soc. **138**(11), 3997–4007 (2010)
39. W. Huang, Y. Yi, Almost periodically forced circle flows. J. Funct. Anal. **257**(3), 832–902 (2009)
40. X. Huang, J. Cao, D.W.C. Ho, Existence and attractivity of almost periodic solutions for recurrent neural networks with unbounded delays and variable coefficients. Nonlinear Dynam. **45**, 337–351 (2006)
41. J.A. Jacquez, C.P. Simon, Qualitative theory of compartmental systems. SIAM Rev. **35**(1), 43–49 (1993)
42. J.A. Jacquez, C.P. Simon, Qualitative theory of compartmental systems with lags. Math. Biosci. **180**, 329–362 (2002)
43. H. Jiang, L. Zhang, Z. Teng, Existence and global exponential stability of almost periodic solution for cellular neural networks with variable coefficients and time-varying delays. IEEE Trans. Neural Networks **16**, 1340–1351 (2005)
44. J. Jiang, X.-Q. Zhao, Convergence in monotone and uniformly stable skew-product semiflows with applications. J. Reine Angew. Math. **589**, 21–55 (2005)
45. R. Johnson, On a Floquet theory for almost-periodic, two-dimensional linear systems. J. Differ. Equat. **37**, 184–205 (1980)
46. R. Johnson, A linear, almost periodic equation with an almost automorphic solution. Proc. Am. Math. Soc. **82**(2), 199–205 (1981)
47. R. Johnson, Bounded solutions of scalar, almost periodic linear equations. Illinois J. Math. **25**(4), 632–643 (1981)
48. R. Johnson, On almost-periodic linear differential systems of Millionščikov and Vinograd. J. Math. Anal. Appl. **85**, 452–460 (1982)
49. R. Johnson, Exponential dichotomy, rotation number, and linear differential operators with bounded coefficients. J. Differ. Equat. **61**(1), 54–78 (1986)
50. R. Johnson, F. Mantellini, Non-autonomous differential equations, in *Dynamical Systems*. Lecture Notes in Mathematics, vol. 1822 (Springer, Berlin, 2003), pp. 173–229
51. R. Johnson, S. Novo, R. Obaya, Ergodic properties and Weyl M-functions for linear Hamiltonian systems. Proc. Roy. Soc. Edinb. **130A**, 1045–1079 (2000)
52. R. Johnson, S. Novo, R. Obaya, An ergodic and topological approach to disconjugate linear Hamiltonian systems. Illinois J. Math. **45**, 1045–1079 (2001)
53. A. Jorba, C. Núñez, R. Obaya, J.C. Tatjer, Old and new results on strange nonchaotic attractors. Int. J. Bifur. Chaos Appl. Sci. Eng. **17**(11), 3895–3928 (2007)

54. M.A. Krasnoselskii, J.A. Lisfshits, A.V. Sobolev, *Positive Linear Systems: The Method of Positive Operators.* Sigma Series in Appl. Math., vol. 5 (Heldermann, Berlin, 1989)
55. U. Krause, P. Ranft, A limit set trichotomy for monotone nonlinear dynamical systems. Nonlinear Anal. **19**, 375–392 (1992)
56. Y. Kuang, *Delay Differential Equations with Applications in Population Dynamics* (Academic, New York, 1993)
57. B. Levitan, V. Zhikov, *Almost Periodic Functions and Differential Equations* (Cambridge University Press, Cambridge, 1982)
58. J. Liang, J. Cao, Boundedness and stability for recurrent neural networks with variable coefficients and time-varying delays. Phys. Lett. A **318**, 53–64 (2003)
59. R. Mañé, *Ergodic Theory and Differentiable Dynamics* (Springer, Berlin, 1987)
60. C.M. Marcus, R.M. Westervelt, Stability of analog neural networks with delay. Phys. Rev. A **39**, 347–359 (1989)
61. N.G. Markey, M.E., Paul, Almost automorphic symbolic minimal sets without unique ergodicity. Israel J. Math. **34**, 259–272 (1979)
62. J. Mawhin, First order ordinary differential equations with several periodic solutions. Z. Angew. Math. Phys. **38**, 257–265 (1987)
63. A. Miller, A relation between almost automorphic and Levitan almost periodic point in compact minimal flows. J. Dynam. Differ. Equat. **20**(2), 519–529 (2008)
64. R. Miller, Almost periodic differential equations as dynamical systems with applications to the existence of almost periodic solutions. J. Differ. Equat. **1**, 337–345 (1965)
65. V.M. Millionščikov, Proof of the existence of irregular systems of linear differential equations with almost periodic coefficients. Differ. Uravn. **4**(3), 391–396 (1968)
66. S. Mohamad, Convergence dynamics of delayed Hopfield-type neural networks under almost periodic stimuli. Acta Appl. Math. **76**, 117–135 (2003)
67. V. Muñoz-Villarragut, S. Novo, R. Obaya, Neutral functional differential equations with applications to compartmental systems. SIAM J. Math. Anal. **40**(3), 1003–1028 (2008)
68. V. Nemytskii, V. Stepanoff, *Qualitative Theory of Differential Equations* (Princeton University Press, Princeton, 1960)
69. T.Y. Nguyen, T.S. Doan, T. Jäger, S. Siegmund, Nonautonomous saddle-node bifurcations in the quasiperiodically forced logistic map. Int. J. Bifur. Chaos Appl. Sci. Eng. **21**(5), 1427–1438 (2011)
70. S. Novo, R. Obaya, Strictly ordered minimal subsets of a class of convex monotone skew-product semiflows. J. Differ. Equat. **196**, 249–288 (2004)
71. S. Novo, R. Obaya, A.M. Sanz, Almost periodic and almost automorphic dynamics for scalar convex differential equations. Israel J. Math. **144**, 157–189 (2004)
72. S. Novo, C. Núñez, R. Obaya, Almost automorphic and almost periodic dynamics for quasimonotone non-autonomous functional differential equations. J. Dynam. Differ. Equat. **17**(3), 589–619 (2005)
73. S. Novo, R. Obaya, A.M. Sanz, Attractor minimal sets for cooperative and strongly convex delay differential systems. J. Differ. Equat. **208**(1), 86–123 (2005)
74. S. Novo, R. Obaya, A.M. Sanz, Attractor minimal sets for non-autonomous delay functional differential equations with applications for neural networks. Proc. Roy. Soc. Lond. Ser. A Math. Phys. Eng. Sci. **461**(2061), 2767–2783 (2005)
75. S. Novo, R. Obaya, A.M. Sanz, Exponential stability in non-autonomous delayed equations with applications to neural networks. Discrete Contin. Dynam. Syst. **18**(2–3), 517–536 (2007)
76. S. Novo, R. Obaya, A.M. Sanz, Stability and extensibility results for abstract skew-product semiflows. J. Differ. Equat. **235**(2), 623–646 (2007)
77. S. Novo, R., Obaya, V.M. Villarragut, Exponential ordering for nonautonomous neutral functional differential equations. SIAM J. Math. Anal. **41**(3), 1025–1053 (2009)
78. C. Núñez, R. Obaya, A non-autonomous bifurcation theory for deterministic scalar differential equations. Discrete Contin. Dynam. Syst. B **9**(3–4), 701–730 (2008)
79. C. Núñez, R. Obaya, A.M. Sanz, Global attractivity in concave or sublinear monotone infinite delay differential equations. J. Differ. Equat. **246**, 3332–3360 (2009)

80. C. Núñez, R. Obaya, A.M. Sanz, Minimal sets in monotone and sublinear skew-product semiflows I: The general case. J. Differ. Equat. **248**, 1879–1897 (2010)
81. C. Núñez, R. Obaya, A.M. Sanz, Minimal sets in monotone and sublinear skew-product semiflows II: Two-dimensional systems of differential equations. J. Differ. Equat. **248**, 1899–1925 (2010)
82. C. Núñez, R. Obaya, A.M. Sanz, Minimal sets in monotone and concave skew-product semiflows I: A general theory. J. Differ. Equat. **252**(10), 5492–5517 (2012)
83. C. Núñez, R. Obaya, A.M. Sanz, Minimal sets in monotone and concave skew-product semiflows II: Two-dimensional systems of differential equations. J. Differ. Equat. **252**(5), 3575–3607 (2012)
84. R. Obaya, V.M. Villarragut, Exponential ordering for neutral functional differential equations with non-autonomous linear D-operator. J. Dynam. Differ. Equat. **23**(3), 695–725 (2011)
85. Z. Opial, Sur une équation différentielle presque-périodique sans solution presque-périodic. Bull. Acad. Polon. Sci. Ser. Sci. Math. Astron. Phys. **9**, 673–676 (1961)
86. R. Ortega, M. Tarallo, Almost periodic equations and conditions of Ambrosetti-Prodi type. Math. Proc. Camb. Philos. Soc. **135**(2), 239–254 (2003)
87. R. Ortega, M. Tarallo, Almost periodic upper and lower solutions. J. Differ. Equat. **193**, 343–358 (2003)
88. R. Ortega, M. Tarallo, Almost periodic linear differential equations with non-separated solutions. J. Funct. Anal. **237**(2), 402–426 (2006)
89. R. Phelps, in *Lectures on Choquet's Theory*. Van Nostrand Mathematical Studies (American Book Co., New York, 1966)
90. P. Poláčik, I. Tereščák, Exponential separation and invariant bundles for maps in ordered Banach spaces with applications to parabolic equations. J. Dynam. Differ. Equat. **5**(2), 279–303 (1993)
91. R.J. Sacker, G.R. Sell, in *Lifting Properties in Skew-Products Flows with Applications to Differential Equations*. Mem. Amer. Math. Soc., vol. 190 (American Mathematical Society, Providence, 1977)
92. R.J. Sacker, G.R. Sell, A spectral theory for linear differential systems. J. Differ. Equat. **27**, 320–358 (1978)
93. A.M. Sanz, Dinámicas casi periódica y casi automórfica en sistemas diferenciales monótonos y convexos. Ph.D. Dissertation, Universidad de Valladolid, 2004
94. J.F. Selgrade, Asymptotic behavior of solutions to single loop positive feedback systems. J. Differ. Equat. **38**, 80–103 (1980)
95. G.R. Sell, Non-autonomous differential equations and topological dynamics I, II. Trans. Am. Math. Soc. **127**, 241–283 (1967)
96. G.R. Sell, *Topological Dynamics and Ordinary Differential Equations* (Van Nostrand-Reinhold, London, 1971)
97. W. Shen, Y. Yi, Asymptotic almost periodicity of scalar parabolic equations with almost periodic time dependence. J. Differ. Equat. **122**, 373–397 (1995)
98. W. Shen, Y. Yi, Dynamics of almost periodic scalar parabolic equations. J. Differ. Equat. **122**, 114–136 (1995)
99. W. Shen, Y. Yi, On minimal sets of scalar parabolic equations with skew-product structures. Trans. Am. Math. Soc. **347**(11), 4413–4431 (1995)
100. W. Shen, Y. Yi, Ergodicity of minimal sets in scalar parabolic equations. J. Dynam. Differ. Equat. **8**(2), 299–323 (1996)
101. W. Shen, Y. Yi, in *Almost Automorphic and Almost Periodic Dynamics in Skew-Product Semiflows*. Mem. Amer. Math. Soc., vol. 647 (American Mathematical Society, Providence, 1998)
102. W. Shen, X.-Q. Zhao, Convergence in almost periodic cooperative systems with a first integral. Proc. Am. Math. Soc. **133**, 203–212 (2004)
103. H.L. Smith, Cooperative systems of differential equations with concave nonlinearities. Nonlinear Anal. **10**, 1037–1052 (1986)

104. H.L. Smith, Monotone semiflows generated by functional differential equations. J. Differ. Equat. **66**, 420–442 (1987)
105. H.L. Smith, *Monotone Dynamical Systems. An introduction to the Theory of Competitive and Cooperative Systems* (American Mathematical Society, Providence, 1995)
106. H.L. Smith, *An Introduction to Delay Differential Equations with Applications to the Life Sciences*. Texts in Applied Mathematics, 57 (Springer, New York, 2011)
107. H.L. Smith, H.R. Thieme, Strongly order preserving semiflows generated by functional-differential equations. J. Differ. Equat. **93**(2), 332–363 (1991)
108. P. Takáč, Asymptotic behavior of discrete-time semigroups of sublinear, strongly increasing mappings with applications to biology. Nonlinear Anal. **14**(1), 35–42 (1990)
109. P. Takáč, Linearly stable subharmonic orbits in strongly monotone time-periodic dynamical systems. Proc. Am. Math. Soc. **115**(3), 691–698 (1992)
110. P. Van den Driessche, X. Zou, Global attractivity in delayed Hopfield neural networks models. SIAM J. Appl. Math. **58**(6), 1878–1890 (1998)
111. P. Van den Driessche, J. Wu, X. Zou, Stabilization role of inhibitory self-connections in a delayed neural network. Phys. D **150**, 84–90 (2001)
112. W.A. Veech, Almost automorphic functions on groups. Am. J. Math. **87**, 719–751 (1965)
113. W.A. Veech, Properties of minimal functions on abelian groups. Am. J. Math. **91**, 415–441 (1969)
114. W.A. Veech, Topological dynamics. Bull. Am. Math. Soc. **83**, 775–830 (1977)
115. R.E. Vinograd, A problem suggested by N.P. Erugin. Differ. Uravn. **11**(4), 632–638 (1975)
116. J. Wu, Global dynamics of strongly monotone retarded equations with infinite delay. J. Integr. Equat. Appl. **4**(2), 273–307 (1992)
117. J. Wu, in *Introduction to Neural Dynamics and Signal Transmission Delay*. Nonlinear Analysis and Aplications, vol. 6 (Walter de Gruyter, Berlin, 2001)
118. J. Wu, H.I. Freedman, Monotone semiflows generated by neutral functional differential equations with application to compartmental systems. Can. J. Math. **43**(5), 1098–1120 (1991)
119. Y. Yi, On almost automorphic oscillations, in *Differences and Differential Equations*. Fields Inst. Commun., vol. 42 (American Mathematical Society, Providence, 2004), pp. 75–99
120. H. Zhao, Global asymptotic stability of Hopfield neural networks involving distributed delays. Neural Netw. **17**, 47–53 (2004)
121. X. Q. Zhao, Global attractivity in monotone and subhomogeneous almost periodic systems. J. Differ. Equat. **187**, 494–509 (2003)

Twist Mappings with Non-Periodic Angles

Markus Kunze and Rafael Ortega

Abstract Consider an annulus A with coordinates (θ, r), $\theta + 2\pi \equiv \theta$, $r \in [a,b]$. A map $(\theta, r) \mapsto (\theta_1, r_1)$ is twist if it satisfies $\frac{\partial \theta_1}{\partial r} > 0$. Twist maps have been extensively studied and they are useful to understand the dynamics of autonomous or periodic Hamiltonian systems. In this course we study twist maps without assuming periodicity in θ. In other words, the annulus A is replaced by a strip $S = \mathbb{R} \times [a,b]$. This new class of twist maps can be applied to the study of generalized standard maps or ping-pong models with a general non-autonomous time dependence.

1 Introduction

Consider the map

$$f : \quad \theta_1 = F(\theta, r), \quad r_1 = G(\theta, r).$$

The functions F and G are defined for $\theta \in \mathbb{R}$, $r \in]a, b[$, and satisfy the periodicity conditions

$$F(\theta + 2\pi, r) = F(\theta, r) + 2\pi, \quad G(\theta + 2\pi, r) = G(\theta, r). \tag{1}$$

After the identification $\theta + 2\pi \equiv \theta$, the domain of f can be interpreted as an annulus or a cylinder. Let us think that it is a cylinder with vertical coordinate r. We say that the map f has *twist* if

M. Kunze
Universität Duisburg-Essen, Fakultät für Mathematik, 45117 Essen, Germany
e-mail: markus.kunze@uni-due.de

R. Ortega (✉)
Departamento de Matemática Aplicada, Universidad de Granada, 18071 Granada, Spain
e-mail: rortega@ugr.es

A. Capietto et al., *Stability and Bifurcation Theory for Non-Autonomous Differential Equations*, Lecture Notes in Mathematics 2065, DOI 10.1007/978-3-642-32906-7_5,
© Springer-Verlag Berlin Heidelberg 2013

$$\frac{\partial F}{\partial r} > 0,$$

and it is *exact symplectic* if the differential form $r_1 d\theta_1 - r d\theta$ is exact. This means that there exists a smooth function $H = H(\theta, r)$ that is 2π-periodic in θ and such that

$$r_1 d\theta_1 - r d\theta = dH.$$

The above definitions have simple geometrical interpretations which will be discussed later. The reversed inequality $\frac{\partial F}{\partial r} < 0$ is also admissible as a twist condition.

Exact symplectic twist maps play an important role in the qualitative theory of Hamiltonian systems of low dimension. See [4, 7, 11, 26] for the general theory and [3, 9, 17, 27, 34, 36–38] for applications. Typically these maps appear in the study of systems of the type

$$\dot{q} = \frac{\partial \mathcal{H}}{\partial p}, \quad \dot{p} = -\frac{\partial \mathcal{H}}{\partial q}, \quad (q, p) \in \mathbb{R}^{2d},$$

in the cases

- 2 degrees of freedom and autonomous, $d = 2$ and $\mathcal{H} = \mathcal{H}(q, p)$
- 1 degree of freedom and time periodic, $d = 1$ and $\mathcal{H} = \mathcal{H}(t, q, p)$ with $\mathcal{H}(t + 2\pi, q, p) = \mathcal{H}(t, q, p)$.

The second case is sometimes referred to as the case of 1.5 degrees of freedom. The periodicity in time is usually employed to guarantee the periodicity of the angle θ in the associated twist map. In this course we will show that twist maps are also useful in the study of Hamiltonian systems with one degree of freedom but with general dependence on time. The key point will be to change the domain of the map f: Instead of a cylinder we will work on the horizontal strip $-\infty < \theta < \infty$, $a < r < b$, without any periodicity assumption on the angle θ. Before entering into the details we will discuss some results at an intuitive (non-rigorous) level. This will be useful to describe the contents of the course.

Let us start with the *integrable twist map*

$$T: \quad \theta_1 = \theta + \varphi(r), \quad r_1 = r,$$

where $\varphi : [a, b] \to \mathbb{R}$ is a smooth function such that $\varphi' > 0$. This map has twist ($\frac{\partial F}{\partial r} = \varphi'$) and it is exact symplectic on the cylinder. To check this last property we notice that $r_1 d\theta_1 - r d\theta = d\Phi$ where $\Phi = \Phi(r)$ is a primitive of the function $r\varphi'(r)$. The function $I(\theta, r) = r$ is a first integral, that is $I(\theta_1, r_1) = I(\theta, r)$, and each circle $r = r_*$ is invariant under T. The twist condition implies that the rotation number $\omega = \varphi(r_*)$ associated to each of these circles increases with r_*. When the rotation number ω is commensurable with 2π, say $\frac{\omega}{2\pi} = \frac{p}{q}$ in reduced form, then all orbits in the invariant circle are periodic and satisfy $\theta_{n+q} = \theta_n + 2\pi p$, $r_{n+q} = r_n$.

On the contrary, when $\frac{\omega}{2\pi}$ is irrational, orbits are quasi-periodic with frequencies 2π and ω. A key property of the map T is that many of its invariant sets persist under small perturbations in the class of exact symplectic twist mappings. This is a delicate theory because there are different cases depending on the arithmetic properties of ω. Given a compact interval $[\varphi_-, \varphi_+]$ with $\varphi(a+) < \varphi_- < \varphi_+ < \varphi(b-)$ and a small perturbation T_ϵ of T in the class of exact symplectic twist maps, then for each $\omega \in [\varphi_-, \varphi_+]$, ω commensurable with 2π, there are at least two periodic orbits with rotation number ω. This is a consequence of the Poincaré–Birkhoff theorem (see[2, 29]). In the case where ω is not commensurable with 2π there are two possibilities: Either the invariant curve associated to ω persists and all motions on this curve are quasi-periodic with frequencies 2π and ω, or the invariant circle breaks down and an invariant Cantor set appears. The dynamics of the Cantor set is of Denjoy type and has rotation number ω. These are consequences of KAM and Aubry–Mather theories (see [3, 22, 34]).

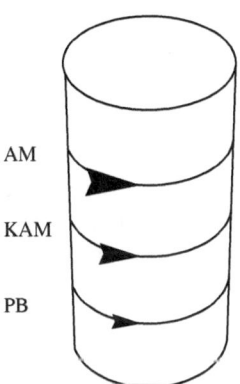

AM

KAM

PB

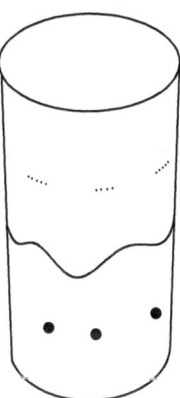

In the above discussions it is essential that the perturbation T_ϵ is exact symplectic. In the cylinder $r \in]-1, 1[$ and for $\epsilon > 0$ the map

$$T_{\epsilon,1}: \quad \theta_1 = \theta + \varphi(r), \quad r_1 = (1 - \epsilon)r,$$

has no invariant set with rotation number $\omega \neq 0$, and the map

$$T_{\epsilon,2}: \quad \theta_1 = \theta + \varphi(r), \quad r_1 = r + \epsilon,$$

has no invariant sets at all.

All the previously mentioned results can be derived from by now classical theorems in the theory of twist maps. Let us now go to a less standard situation and consider maps f on the strip $-\infty < \theta < +\infty$, $r \in]a, b[$. In particular the periodicity conditions on F and G as imposed before will in general be dropped. The twist condition still makes sense and we replace the concept of exact symplectic map in the cylinder with the following definition. The map f on the strip is *E-symplectic* if the differential form $r_1 d\theta_1 - r d\theta$ is exact and its primitive $H = H(\theta, r)$ is bounded on each region $\mathbb{R} \times [A, B]$ with $a < A < B < b$. Notice that now the function H is not periodic in θ and so this boundedness condition is not automatic. The integrable twist map T is E-symplectic because the function Φ is bounded on compact intervals. The sets $r = r_*$, invariant under T, are now straight lines where the orbits move with increasing velocities as r_* goes from a to b. The rotation number is recovered from the limit

$$\lim_{|n| \to \infty} \frac{\theta_n}{n}, \tag{2}$$

which exists for each orbit $(\theta_n, r_n)_{n \in \mathbb{Z}}$ and coincides with $\omega = \varphi(r_*)$. Assuming that the strip is wide enough, we will prove that there is still some persistence of invariant sets for small perturbations of T in the class of E-symplectic maps. In particular we will prove the existence of complete orbits which are bounded in the variable r. However, it does not seem possible to associate a rotation number to these sets. As an example consider for $\epsilon > 0$ the map

$$T_{\epsilon,3}: \quad \theta_1 = \theta + r, \quad r_1 = r + \frac{\epsilon}{1 + \theta_1^2},$$

where $r \in]a, b[$ for $0 < a < b$. All orbits of this map are strictly increasing in θ and so rotation numbers cannot exist, at least if they are understood in the sense of (2). On the other hand this map is an E-symplectic perturbation of T. Actually, $r_1 d\theta_1 - r d\theta = dH$ with

$$H(\theta, r) = \frac{1}{2} r^2 + \epsilon \arctan \theta_1.$$

These results on perturbations of the map T have many consequences. As an application to mechanical problems we can consider the following ping-pong game. Two players move their rackets ($\equiv$ parallel moving walls) according to known protocols, say $x = \rho_1(t)$ and $x = \rho_2(t)$ with $\rho_1(t) < \rho_2(t)$. The ball is hit alternatively by the players and all impacts are assumed to be perfectly elastic.

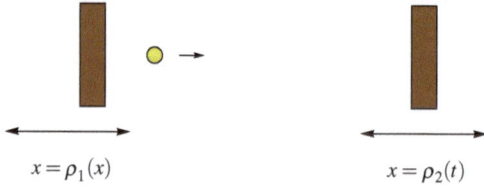

$$x = \rho_1(x) \qquad\qquad\qquad x = \rho_2(t)$$

In the absence of gravity the motion of the ball is described by a Hamiltonian system with one degree of freedom defined by

$$\mathscr{H}(t,q,p) = \frac{1}{2}p^2 + V(t,q) \quad \text{with} \quad V(t,q) = \begin{cases} 0 & : \quad \rho_1(t) \leq q \leq \rho_2(t) \\ +\infty & : \quad \text{otherwise} \end{cases}.$$

The question is to decide whether the velocity of the ball remains bounded or can become arbitrarily large. For simplicity we assume that one of the walls is fixed, say $\rho_1 \equiv 0$. Then we can consider the successor map $(t_0, v_0) \mapsto (t_1, v_1)$, associating consecutive impacts against the fixed wall. Times of impact are t_0, t_1 and velocities after impacts are $v_0 \geq 0$, $v_1 \geq 0$. Ideally we would like to determine if $\sup v_n < +\infty$ or $\sup v_n = +\infty$, for each sequence of iterates (t_n, v_n). When the function $\rho_2(t)$ is 2π-periodic, the successor map satisfies the periodicity conditions (1) and it seems reasonable to interpret t_0 as an angle. However, the successor map is not E-symplectic. A computation shows that the form $v_1^2 dt_1 - v_0^2 dt_0$ is closed and this fact suggests the use of new variables: time t_0 and kinetic energy $E_0 = \frac{1}{2}v_0^2$. With the identification $\theta = t_0$ and $r = E_0$, the map $(t_0, E_0) \mapsto (t_1, E_1)$ becomes exact symplectic when ρ_2 is periodic and E-symplectic in the non-periodic case. These properties hold for large energies, as well as the twist condition

$$\frac{\partial t_1}{\partial E_0} < 0 \quad \text{if } E_0 \gg 0.$$

The intuition behind this formula is that the time employed by the ball to go back to the fixed wall will decrease as the energy increases. We have reformulated our problem in terms of an exact symplectic twist map but in general this map is not a perturbation of the integrable twist map. The symplectic change of variables

$$\tau(t) = \int_0^t \frac{ds}{\rho_2(s)}, \quad W = \rho_2(t)^2 E,$$

leads to a new map $(\tau_0, W_0) \mapsto (\tau_1, W_1)$ which is close to T for $\varphi(W) = \sqrt{\frac{2}{W}}$ and W_0 large enough. The results on twist maps described previously are applicable and many consequences for the ping-pong model can be deduced. When $\rho_2(t)$ is a 2π-periodic and smooth function, say of class C^5, KAM theory implies that the map has invariant curves in $W_0 \gg 0$. These curves act as barriers for the orbits (τ_n, W_n) so that all motions have bounded velocity. A complete proof of this result can be found in [17] and some extensions to the quasi-periodic case with diophantine conditions can be found in [38]. The use of KAM theory forces the assumption on the smoothness of ρ_2. An ingenious example in [37] shows that motions with unbounded velocity can exist for certain functions $\rho_2(t)$ which are still periodic but only continuous. Without the periodicity assumption, motions with unbounded velocity can exist even if $\rho_2(t)$ is very smooth. Examples are constructed in [16]. Also in that paper it is proved that there exist infinitely many motions with bounded velocity when $\rho_2 \in C^3(\mathbb{R})$ satisfies $\|\rho_2^{(k)}\|_\infty < \infty$ for $k = 0, 1, 2, 3$. Moreover, unbounded motions remain close to some of these bounded motions for arbitrarily long periods of time.

The plan of these notes is as follows. First we will discuss the theory of twist maps in the plane and present a result on the persistence of bounded orbits. The notion of *generating function* will be crucial. This is a classical tool in mechanics which allows to represent E-symplectic maps in terms of one single function. Some connections between generating functions and the Calculus of Variations will be discussed. We will restrict ourselves to the variational framework associated to Newtonian equations. Finally the general theory will be applied to the study of a ping-pong model. The one mentioned before is technically difficult, so we will analyze a simpler variant which only involves one single racket and gravity. The notes are concluded with some bibliographical comments.

2   Symplectic Maps in the Plane and in the Cylinder

We will work on the plane $\mathbb{R}^2$ with cartesian coordinates (θ, r). Sometimes we will also work on the cylinder $\mathbb{T} \times \mathbb{R}$ with $\mathbb{T} = \mathbb{R}/2\pi\mathbb{Z}$. A generic point in the cylinder will be denoted by $(\overline{\theta}, r)$ with $\overline{\theta} = \theta + 2\pi\mathbb{Z}$. The covering map $p : \mathbb{R}^2 \to \mathbb{T} \times \mathbb{R}$, $(\theta, r) \mapsto (\overline{\theta}, r)$, is useful to lift maps from the cylinder to the universal covering $\mathbb{R}^2$.

Let us start with the plane. We work with C^k embeddings, $k \geq 1$, defined on a strip $\Sigma = \mathbb{R} \times]a, b[$, $-\infty \leq a < b \leq +\infty$. More precisely, consider a C^k map

$$f : \Sigma \subset \mathbb{R}^2 \to \mathbb{R}^2, \quad (\theta, r) \mapsto (\theta_1, r_1),$$

satisfying

(i) f is one-to-one
(ii) $\det f'(\theta, r) \neq 0 \quad \forall (\theta, r) \in \Sigma$.

The class of these maps will be denoted by $\mathscr{E}^1(\Sigma)$. A map $f \in \mathscr{E}^1(\Sigma)$ is called *symplectic* if it preserves the differential form $\omega = d\theta \wedge dr$. This means that, in the set Σ,

$$d\theta_1 \wedge dr_1 = d\theta \wedge dr. \tag{3}$$

If we express the map f in coordinates

$$\theta_1 = F(\theta, r), \quad r_1 = G(\theta, r),$$

then the condition (3) can be reformulated as

$$\det f' = \frac{\partial F}{\partial \theta}\frac{\partial G}{\partial r} - \frac{\partial F}{\partial r}\frac{\partial G}{\partial \theta} = 1 \quad \text{on } \Sigma.$$

This is the classical definition of area-preserving map.

Exercise 2.1. Prove that $f \in \mathscr{E}^1(\Sigma)$ is symplectic if and only if the two conditions below hold:

(a) f is orientation-preserving
(b) For each (Lebesgue) measurable set $\Omega \subset \Sigma$, the image $\Omega_1 = f(\Omega)$ is also measurable and $\mu(\Omega) = \mu(\Omega_1)$. Here μ is the Lebesgue measure in the plane.

Given $f \in \mathscr{E}^2(\Sigma)$ we consider the 1-form

$$\alpha = r_1 d\theta_1 - r d\theta.$$

Then $d\alpha = -d\theta_1 \wedge dr_1 + d\theta \wedge dr$ and so α is closed if and only if f is symplectic. The strip Σ is contractible and therefore closed and exact forms coincide. In particular, if f is symplectic there must exist a function H with $\alpha = dH$. After taking differentials in this identity we conclude that the converse is also true. Summing up, a map $f \in \mathscr{E}^2(\Sigma)$ is symplectic if and only if

$$dH = r_1 d\theta_1 - r d\theta \quad \text{for some } H \in C^2(\Sigma). \tag{4}$$

This identity can be expressed as

$$H_\theta = GF_\theta - r, \quad H_r = GF_r. \tag{5}$$

When f is only in $\mathscr{E}^1(\Sigma)$, the components of α given by (5) are only continuous and the operation of taking the differential of α becomes more delicate.

Exercise 2.2. Assume that $f \in \mathscr{E}^1(\Sigma)$. Prove that f is symplectic if and only if there exists a function $H \in C^1(\Sigma)$ such that $dH = \alpha$. Hint: If f is symplectic define the potential H by the standard integral of α and check directly that (5) holds. For the converse construct sequences $F^\epsilon, G^\epsilon \in C^2(\Sigma)$ with $\|F - F^\epsilon\|_{C^1(\Sigma)} + \|G - G^\epsilon\|_{C^1(\Sigma)} \to 0$ as $\epsilon \to 0$ and consider the integral

$$\int_\Sigma \{[G^\epsilon(F^\epsilon)_\theta - r]\phi_r - G^\epsilon(F^\epsilon)_r \phi_\theta\} \, d\theta \, dr$$

for each test function $\phi \in \mathscr{D}(\Sigma)$. Integrate by parts and pass to the limit.

The equivalence between closed and exact 1-forms is no longer true in the cylinder. Consider the strip immersed in the cylinder $\overline{\Sigma} = \mathbb{T} \times]a, b[$. Since Σ is its universal covering, all 1-forms on $\overline{\Sigma}$ of class C^1 can be expressed as

$$\beta = A(\theta, r)d\theta + B(\theta, r)dr$$

with $A, B \in C^1(\Sigma)$ and 2π-periodic in θ. When β is closed it is possible to find a function $H = H(\theta, r)$ with $dH = \beta$. The problem is that sometimes H is not periodic in θ and so it becomes a multi-valued function when regarded in the cylinder.

Exercise 2.3. Prove that exact 1-forms in the cylinder can be characterized as closed 1-forms satisfying

$$\int_0^{2\pi} A(\theta, r_*) d\theta = 0$$

for some $r_* \in]a, b[$.

This difference between the plane and the cylinder plays a role when one tries to extend the notion of symplectic map to $\mathbb{T} \times \mathbb{R}$. Let us start with a map

$$\overline{f} : \overline{\Sigma} \subset \mathbb{T} \times \mathbb{R} \to \mathbb{T} \times \mathbb{R}, \quad (\overline{\theta}, r) \mapsto (\overline{\theta}_1, r_1),$$

satisfying the same conditions as in the case of the plane. The class of these maps is $\mathscr{E}^1(\overline{\Sigma})$. Every $\overline{f} \in \mathscr{E}^1(\overline{\Sigma})$ has a lift $f = (F, G)$ in $\mathscr{E}^1(\Sigma)$. The coordinates satisfy

$$F(\theta + 2\pi, r) = F(\theta, r) + 2n\pi, \quad G(\theta + 2\pi, r) = G(\theta, r).$$

In principle n could be any integer but since our map is an orientation-preserving embedding it can only take the values $n = -1$ or $n = 1$. We say that $\overline{f}$ is symplectic if its lift is symplectic as a map of the plane. Notice that, up to an additive constant $2N\pi$, the lift is unique and so this definition is all right.

Exercise 2.4. Extend Exercise 2.1 to the cylinder using the measure transported from the plane via the covering map,

$$\mu_{\mathbb{T} \times \mathbb{R}}(\overline{A}) = \mu_{\mathbb{R}^2}(p^{-1}(\overline{A}) \cap ([0, 2\pi] \times \mathbb{R})).$$

We say that $\overline{f} \in \mathscr{E}^1(\overline{\Sigma})$ is *exact symplectic* if there is a function $H \in C^1(\overline{\Sigma})$, and hence 2π-periodic in θ, such that

$$dH = r_1 d\theta_1 - r d\theta.$$

Using Exercise 2.2 we observe that exact symplectic maps are always symplectic. Since not all closed forms are exact in the cylinder, we can expect that there are symplectic maps which are not exact.

Exercise 2.5. Prove that $\overline{f} \in \mathscr{E}^1(\overline{\Sigma})$ is exact symplectic if and only if it is symplectic and

$$\int_0^{2\pi} G(\theta, r_*) \frac{\partial F}{\partial \theta}(\theta, r_*) d\theta = 2\pi r_*$$

for some $r_* \in]a, b[$.

The notion of exact symplectic map can also be characterized in terms of measure theory. Given an arbitrary Jordan curve $\Gamma \subset \overline{\Sigma}$ which is C^1, regular and non-contractible, the image $\Gamma_1 = \overline{f}(\Gamma) \subset \mathbb{T} \times \mathbb{R}$ is another Jordan curve enjoying the same properties. Let us fix some $r_0 \in \mathbb{R}$ such that $\Gamma \cup \Gamma_1 \subset \{r > r_0\}$ and let A and A_1 denote the bounded components of $\{r > r_0\} \setminus \Gamma$ and $\{r > r_0\} \setminus \Gamma_1$, respectively. Then, if $\overline{f}$ is exact symplectic, $\mu(A) = \mu(A_1)$.

Exercise 2.6. Prove that the previous property is a characterization of exact symplectic maps.

To illustrate the above definitions we consider some simple maps in the cylinder and the corresponding lifts in the plane.

Example 1. $f(\theta, r) = (\theta + \omega, r)$ for fixed $\omega \in]0, 2\pi[$.
   In the plane this map is a translation in the horizontal direction. It can be seen as the lift of a rotation.

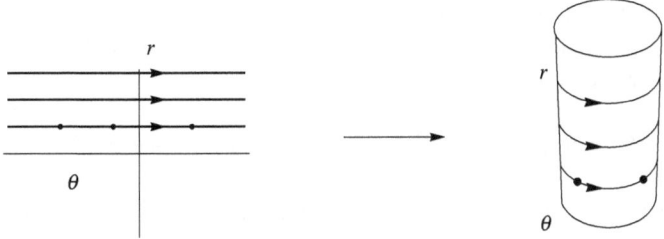

From $r_1 d\theta_1 - r d\theta = r d(\theta + \omega) - r d\theta = 0$ we deduce that the condition (4) holds with $H \equiv 0$. Hence, rotations are exact symplectic maps.

Example 2. $f(\theta, r) = (\theta, r + \lambda)$ for fixed $\lambda \in \mathbb{R} \setminus \{0\}$.
   This map can be interpreted as a vertical translation.

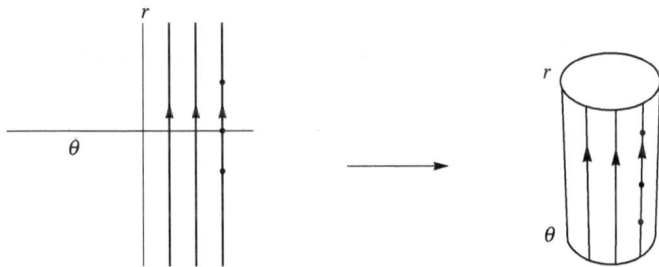

Now, $r_1 d\theta_1 - r d\theta = (r + \lambda)d\theta - r d\theta = \lambda d\theta$. In the plane the condition (4) holds with $H(\theta, r) = \lambda \theta$. The differential form $\lambda d\theta$ is not exact in the cylinder and so $\overline{f}$ is symplectic but not exact.

   We finish this section with another characterization of exact symplectic maps. It is less standard but it is useful to suggest how to introduce a related notion in the plane.

Exercise 2.7. Assume that $\overline{f} \in \mathscr{E}^1(\overline{\Sigma})$ is symplectic and $H \in C^1(\Sigma)$ is such that $dH = r_1 d\theta_1 - r d\theta$. Prove that the three conditions below are equivalent:

   (i) $\overline{f}$ is exact symplectic
   (ii) H is 2π-periodic in θ

(iii) H is bounded on each strip $\mathbb{R} \times [A, B]$ with $a < A < B < b$.

Let us consider now a general map $f \in \mathcal{E}^1(\Sigma)$, possibly not 2π-periodic in θ. We say that f is E-*symplectic* if there exists a function $H \in C^1(\Sigma)$ satisfying

$$dH = r_1 d\theta_1 - r d\theta$$

and

$$\sup\{|H(\theta, r)| : \theta \in \mathbb{R}, \ A \le r \le B\} < \infty$$

for each A, B with $a < A < B < b$.

  In the introduction we already presented some examples of E-symplectic maps. As a further example consider the map

$$\theta_1 = \theta + r, \quad r_1 = r + a + b \sin(\theta + r) + c \sin \sqrt{2}(\theta + r),$$

with $a, b, c \in \mathbb{R}$. In this case $G(\theta, r)$ is not 2π-periodic, if b and c do not vanish, but in the plane it satisfies (4) taking

$$H(\theta, r) = \frac{1}{2} r^2 + a(\theta + r) - b \cos(\theta + r) - \frac{c}{\sqrt{2}} \cos \sqrt{2}(\theta + r).$$

Therefore f is E-symplectic when $a = 0$. Recall that θ is an unbounded variable.

3   The Twist Condition and the Generating Function

A map $f = (F, G) \in \mathcal{E}^1(\Sigma)$ has *twist* if

$$\frac{\partial F}{\partial r}(\theta, r) > 0 \quad \forall (\theta, r) \in \Sigma. \tag{6}$$

Geometrically this means that vertical segments are twisted to the right.

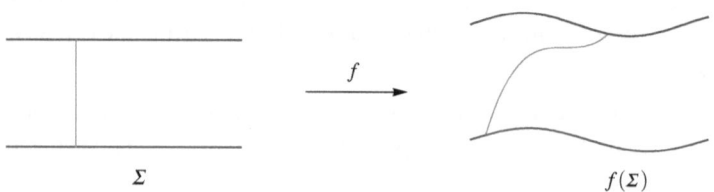

From an analytic point of view the condition (6) is employed to solve the implicit function problem

$$\theta_1 = F(\theta, r). \tag{7}$$

In this way a function $r = R(\theta, \theta_1)$ is obtained. It is defined on the region

$$\Omega = \{(\theta, \theta_1) \in \mathbb{R}^2 : F(\theta, a) < \theta_1 < F(\theta, b)\}$$

where

$$F(\theta, a) = \lim_{r \downarrow a} F(\theta, r), \quad F(\theta, b) = \lim_{r \uparrow b} F(\theta, r).$$

Notice that

$$-\infty \le F(\theta, a) < F(\theta, b) \le +\infty \quad \text{for each } \theta \in \mathbb{R}.$$

Exercise 3.1. Prove that Ω is open and connected. Hint: $\Omega = \bigcup \Omega_\epsilon$, $\Omega_\epsilon = \{(\theta, \theta_1) \in \mathbb{R}^2 : F(\theta, a + \epsilon) < \theta_1 < F(\theta, b - \epsilon)\}$.

The function R is in $C^1(\Omega)$ and, by implicit differentiation,

$$F_\theta \circ \mathscr{R} + (F_r \circ \mathscr{R})R_\theta = 0, \quad (F_r \circ \mathscr{R})R_{\theta_1} = 1, \tag{8}$$

where $\mathscr{R}(\theta, \theta_1) = (\theta, R(\theta, \theta_1))$.

Assuming that there exists a function $H \in C^1(\Sigma)$ such that $dH = r_1 d\theta_1 - r d\theta$, the *generating function* of f is defined as

$$h(\theta, \theta_1) = -H(\theta, R(\theta, \theta_1)), \quad (\theta, \theta_1) \in \Omega.$$

Combining the identities (5), (8), and differentiating $h = H \circ \mathscr{R}$ we obtain

$$\frac{\partial h}{\partial \theta}(\theta, \theta_1) = R(\theta, \theta_1), \quad \frac{\partial h}{\partial \theta_1}(\theta, \theta_1) = -G(\theta, R(\theta, \theta_1)). \tag{9}$$

In a less formal language we can say that the map f given by $\theta_1 = F(\theta, r)$, $r_1 = G(\theta, r)$, is now expressed as

$$\frac{\partial h}{\partial \theta}(\theta, \theta_1) = r, \quad \frac{\partial h}{\partial \theta_1}(\theta, \theta_1) = -r_1.$$

This formula says that the map, originally defined in terms of two functions F and G, can be given in terms of a single function only, the generating function. This is reminiscent of the role played by the Hamiltonian function in the theory of Hamiltonian systems. The above formulas also have a consequence for the regularity of the generating function, because (9) implies that h is in $C^2(\Omega)$. Moreover,

$$\frac{\partial^2 h}{\partial \theta \, \partial \theta_1} > 0 \quad \text{in } \Omega.$$

This follows from the twist condition together with (9) and (8), since

$$h_{\theta\theta_1} = R_{\theta_1} = 1/(F_r \circ \mathscr{R}) > 0.$$

Assume now that $\overline{f} \in \mathscr{E}^1(\overline{\Sigma})$ is a map in the cylinder whose coordinates satisfy

$$F(\theta + 2\pi, r) = F(\theta, r) + 2\pi, \quad G(\theta + 2\pi, r) = G(\theta, r).$$

When the lift is symplectic and has twist, the domain Ω and the function R are invariant under the translation

$$T(\theta, \theta_1) = (\theta + 2\pi, \theta_1 + 2\pi).$$

This means that $T(\Omega) = \Omega$ and $R \circ T = R$. The second identity is a consequence of the uniqueness of solution for the implicit function problem (7) and the generalized periodicity of F. The generating function is not always invariant under T. Indeed the identity $h \circ T = h$ is equivalent to

$$H(\theta + 2\pi, R(\theta, \theta_1)) = H(\theta, R(\theta, \theta_1)),$$

as can be deduced from the definition of h and the periodicity of R. For fixed θ the function $\theta_1 \mapsto R(\theta, \theta_1)$ maps the interval $]F(\theta, a), F(\theta, b)[$ onto $]a, b[$. Hence the above identity, valid for all θ and θ_1, is equivalent to

$$H(\theta + 2\pi, r) = H(\theta, r) \quad \forall (\theta, r) \in \mathbb{R} \times]a, b[.$$

This is just the periodicity of H with respect to θ and, from the definition of h, we deduce that

$$h(\theta + 2\pi, \theta_1 + 2\pi) = h(\theta, \theta_1), \quad (\theta, \theta_1) \in \Omega, \tag{10}$$

holds whenever $\overline{f}$ is exact symplectic.

To illustrate the previous notions let us go back to the example at the end of the last section. We consider the symplectic map in $\Sigma = \mathbb{R}^2$ given by

$$\theta_1 = \theta + r, \quad r_1 = r + a + b\sin(\theta + r) + c\sin\sqrt{2}(\theta + r).$$

Since $\frac{\partial \theta_1}{\partial r} = 1$ the map has twist. Moreover $\Omega = \mathbb{R}^2$ and $R(\theta, \theta_1) = \theta_1 - \theta$. The generating function is

$$h(\theta, \theta_1) = -\frac{1}{2}(\theta_1 - \theta)^2 - a\theta_1 + b\cos\theta_1 + \frac{c}{\sqrt{2}}\cos\sqrt{2}\theta_1.$$

When $c = 0$ this map can be defined on the cylinder, and h satisfies the periodicity condition (10) in the case where $c = 0$ and $a = 0$.

Exercise 3.2. Assume that $h \in C^2(\mathbb{R}^2)$, $h = h(\theta, \theta_1)$, is a function satisfying $\frac{\partial^2 h}{\partial \theta \partial \theta_1} > 0$ and, for some numbers a, b,

$$\inf_{\theta_1 \in \mathbb{R}} \frac{\partial h}{\partial \theta}(\theta, \theta_1) \le a < b \le \sup_{\theta_1 \in \mathbb{R}} \frac{\partial h}{\partial \theta}(\theta, \theta_1)$$

for each $\theta \in \mathbb{R}$. Then there exists a twist symplectic map $f \in \mathscr{E}^1(\Sigma)$, $\Sigma = \mathbb{R} \times]a, b[$, such that h is its generating function. Here it is understood that h is restricted to an appropriate domain.

4   The Variational Principle

We will construct a functional such that its critical points are in correspondence with the orbits generated by symplectic twist maps. First we present a concrete example, arising in solid state physics.

4.1   The Frenkel–Kontorowa Model

Let us imagine an infinite chain of atoms placed on a line, the positions of the atoms being described by bi-infinite sequences $(\theta_n)_{n \in \mathbb{Z}}$. It is assumed that every atom n is attracted by its neighbors $n - 1$ and $n + 1$, according to Hooke's law (with constant C). In addition there is a force derived from a potential $V = V(\theta)$ acting on the real line.

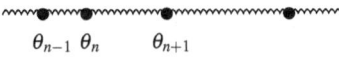

$$\theta_{n-1}\ \theta_n \qquad \theta_{n+1}$$

To find the equilibrium positions of the chain it is enough to impose that the sum of forces acting on each atom vanishes. That is,

$$C(\theta_{n-1} - \theta_n) + C(\theta_{n+1} - \theta_n) - V'(\theta_n) = 0.$$

We arrive at the second order difference equation

$$\theta_{n+1} + \theta_{n-1} - 2\theta_n = \frac{1}{C} V'(\theta_n), \quad n \in \mathbb{Z}, \tag{11}$$

which can be seen as a discrete counterpart of the Newtonian equation $\ddot{\theta} = \frac{1}{C} V'(\theta)$.

Alternatively we can look for critical points of the potential energy

$$\Phi((\theta_n)_{n\in\mathbb{Z}}) = \sum_{n\in\mathbb{Z}} \left[\frac{C}{2}(\theta_{n+1} - \theta_n)^2 + V(\theta_n)\right].$$

It is straightforward to check that the system of conditions $\frac{\partial\Phi}{\partial\theta_n} = 0$ leads to (11). Of course this computation is purely formal since typically the series defining Φ will be divergent. One way to proceed rigorously is to consider finite strings $(\theta_n)_{|n|\leq N}$ and to assume that the end points are fixed and known, say $\theta_{-N} = A_{-N}$ and $\theta_N = A_N$. Then we can consider the truncated potential energy

$$\Phi_N((\theta_n)_{n<|N|}) = \sum_{-N\leq n<N} \left[\frac{C}{2}(\theta_{n+1} - \theta_n)^2 + V(\theta_n)\right].$$

The critical points of Φ_N satisfy (11) for $|n| < N$.

Let us now assume that the potential V is in $C^2(\mathbb{R})$ and let us interpret the function

$$h(\theta, \theta_1) = -\frac{C}{2}(\theta_1 - \theta)^2 - V(\theta)$$

as the generating function of a symplectic twist map. This makes sense since $h_{\theta\theta_1} = C > 0$ and so Exercise 3.2 is applicable. The associated map is defined by

$$r = \frac{\partial h}{\partial\theta}(\theta, \theta_1), \quad r_1 = -\frac{\partial h}{\partial\theta_1}(\theta, \theta_1),$$

or

$$f: \quad \theta_1 = \theta + \frac{1}{C}r + \frac{1}{C}V'(\theta), \quad r_1 = r + V'(\theta).$$

The previous discussion leads to an interesting conclusion: given a "critical point" $(\theta_n^*)_{n\in\mathbb{Z}}$ of Φ, the sequence $(\theta_n^*, r_n^*)_{n\in\mathbb{Z}}$ with $r_n^* = C(\theta_{n+1}^* - \theta_n^*) + V'(\theta_n^*)$ is an f-orbit. The process can be also reversed.

Exercise 4.1. Prove that the map f defined above is E-symplectic in $\mathbb{R}^2$ when $\|V\|_\infty + \|V'\|_\infty < \infty$. Under what conditions is there an induced exact symplectic map $\bar{f}$ in the cylinder?

Exercise 4.2. Prove that f is conjugate to the "standard map" $\theta_1 = \theta + \frac{1}{C}r, r_1 = r + V'(\theta_1)$. Hint: $r \mapsto r + V'(\theta)$.

4.2 A General Framework

Assume now that $f \in \mathscr{E}^1(\Sigma)$ is a twist symplectic map and let $h = h(\theta, \theta_1)$ denote its generating function. Given $N \geq 1$ consider the function

$$\Phi_N((\theta_n)_{|n|\leq N}) = \sum_{-N\leq n<N} h(\theta_n, \theta_{n+1}),$$

where $\theta_{\pm N} = A_{\pm N}$ are fixed numbers. This function is of class C^3 on the domain

$$\Omega_N = \{(\theta_n)_{|n|<N} : (\theta_n, \theta_{n+1}) \in \Omega, \ -N \le n < N\}.$$

Exercise 4.3. Prove that Ω_N is an open and connected subset of $\mathbb{R}^{2N-1}$.

Critical points of Φ_N are solutions of

$$\partial_1 h(\theta_n, \theta_{n+1}) + \partial_2 h(\theta_{n-1}, \theta_n) = 0, \ |n| < N, \ \theta_{\pm N} = A_{\pm N}. \tag{12}$$

The sequence $(\theta_n)_{|n|\le N}$ obtained in this way leads to a segment of f-orbit with the definitions $r_N = \partial_2 h(\theta_{N-1}, \theta_N)$ and $r_n = \partial_1 h(\theta_n, \theta_{n+1})$ if $-N \le n < N$. Actually, from the definition of the function $R(\theta, \theta_1)$ and (9) we deduce that $a < r_n < b$ if $-N \le n < N$. This inequality also holds for $n = N$, as follows from (12). Putting together (9), (12) and the definition of R it is easy to deduce that $f(\theta_n, r_n) = (\theta_{n+1}, r_{n+1})$ for $-N \le n < N$.

The previous process can be reversed in order to obtain critical points of Φ_N from f-orbits. Our goal is to construct complete f-orbits with certain additional properties. To this end we will prove the existence of critical points of Φ_N and let $N \to \infty$. This is clarified by the following result, valid for the general second order difference equation

$$E(\theta_{n-1}, \theta_n, \theta_{n+1}) = 0 \tag{13}$$

where $E : S \to \mathbb{R}$ is a continuous function defined on

$$S = \{(\theta_{-1}, \theta_0, \theta_1) \in \mathbb{R}^3 : \delta \le \theta_0 - \theta_{-1} \le \Delta \text{ and } \delta \le \theta_1 - \theta_0 \le \Delta\}$$

for some $\Delta > \delta \ge 0$.

Lemma 4.4. *Assume that for* $N \ge N^*$ *there exists a finite sequence* $(\theta_n^{[N]})_{|n|\le N}$ *satisfying (13) for* $|n| < N$. *Moreover, assume that*

$$\lim_{N \to +\infty} \theta_{\pm N}^{[N]} = \pm\infty.$$

Then there exists a complete solution of (13). This means that there is a sequence $(\theta_n)_{n\in\mathbb{Z}}$ *satisfying (13) for all* $n \in \mathbb{Z}$.

Proof. Let us consider the space of sequences $\mathbb{R}^\mathbb{Z}$ endowed with the product topology. We recall that this space is metrizable and the associated convergence is just the convergence of each coordinate. Inside $\mathbb{R}^\mathbb{Z}$ we consider the space

$$K_\infty = \{\Theta = (\theta_n)_{n\in\mathbb{Z}} : \delta \le \theta_{n+1} - \theta_n \le \Delta \ \forall n \in \mathbb{Z}, \ |\theta_0| \le \Delta\}.$$

This space is compact because it can be viewed as a closed subset of

$$\hat{K}_\infty = \{\Theta = (\theta_n)_{n\in\mathbb{Z}} : \theta_n \in [n\delta - \Delta, (n+1)\Delta] \ \forall n \in \mathbb{Z}\}$$

and $\hat{K}_\infty$ is compact by Tichonoff's Theorem on the product of compact spaces. We will look for a solution of (13) lying in K_∞.

For large N we can assume $\theta_N^{[N]} > 0 > \theta_{-N}^{[N]}$ and so there exists an integer $\nu = \nu(N), -N \leq \nu < N$, such that $\theta_\nu^{[N]} < 0 \leq \theta_{\nu+1}^{[N]}$. Since $0 < \theta_{\nu+1}^{[N]} - \theta_\nu^{[N]} \leq \Delta$ we conclude that

$$|\theta_\nu^{[N]}| \leq \Delta. \tag{14}$$

Also we notice that

$$\lim_{N \to +\infty} [\pm N - \nu(N)] = \pm\infty. \tag{15}$$

For the sign $+$ this limit is justified using the estimates

$$\theta_N^{[N]} - \Delta \leq \theta_N^{[N]} - \theta_\nu^{[N]} = \sum_{n=\nu}^{N-1} (\theta_{n+1}^{[N]} - \theta_n^{[N]}) \leq (N - \nu)\Delta.$$

The case of the sign $-$ is treated similarly.

Equation (13) is autonomous and so the shifted sequence $\tilde{\theta}_n^{[N]} = \theta_{n-\nu}^{[N]}$, satisfies (13) for $|n - \nu| < N$. Next we complete the finite sequence $(\tilde{\theta}_n^{[N]})$ so that it becomes a point $\tilde{\Theta}^{[N]}$ of K_∞. A simple way to achieve this is to define

$$\tilde{\theta}_n^{[N]} = \omega(n - N - \nu) + \tilde{\theta}_N^{[N]}, \quad \text{if } n > N + \nu$$

and

$$\tilde{\theta}_n^{[N]} = \omega(n + N - \nu) + \tilde{\theta}_{-N}^{[N]}, \quad \text{if } n < -N + \nu,$$

where $\omega = \frac{1}{2}(\delta + \Delta)$. Using (14) it is easy to check that $\tilde{\Theta}^{[N]}$ is contained in K_∞. By compactness we can extract a convergent subsequence $(\tilde{\Theta}^{[\sigma(N)]})$, where $\sigma : \mathbb{N} \to \mathbb{N}$ is increasing. We claim that the limit $\Theta = (\theta_n)_{n \in \mathbb{Z}}$ is a complete solution (13). To check this assertion we observe that for fixed n and large N, $\tilde{\theta}_k^{[N]} = \theta_{k-\nu}^{[N]}$ if $|n - k| \leq 1$. This is a consequence of (15). Then

$$E(\tilde{\theta}_{n-1}^{[\sigma(N)]}, \tilde{\theta}_n^{[\sigma(N)]}, \tilde{\theta}_{n+1}^{[\sigma(N)]}) = 0, \quad N \text{ large},$$

and we can pass to the limit ($N \to +\infty$) using the continuity of E. $\qquad\square$

We will apply the previous lemma to the difference equation

$$\partial_1 h(\theta_n, \theta_{n+1}) + \partial_2 h(\theta_{n-1}, \theta_n) = 0, \quad \delta \leq \theta_{n+1} - \theta_n \leq \Delta,$$

where δ and Δ are positive numbers such that

$$F(\theta, a) < \delta < \Delta < F(\theta, b) \quad \forall \theta \in \mathbb{R}.$$

After finding solutions for $|n| < N$ as critical points of Φ_N, we will pass to the limit.

This section will be finished with a discussion of the nature of the critical points of Φ_N in the simplest instance. Consider the map $\theta_1 = \theta + r$, $r_1 = r$. The generating function is

$$h(\theta, \theta_1) = -\frac{1}{2}(\theta_1 - \theta)^2.$$

After fixing $A_{\pm N} \in \mathbb{R}$ we observe that the function $-\Phi_N$ is coercive on $\mathbb{R}^{2N-1}$. To establish this claim, if $\Theta_N = (\theta_n)_{|n|<N}$ is a generic point in $\mathbb{R}^{2N-1}$ and n_0 is an integer such that $|n_0| < N$ and $|\theta_{n_0}| = \max_{|n|<N} |\theta_n| = \|\Theta_N\|_\infty$, then

$$-4\Phi_N(\Theta_N) \geq 2\sum_{n_0 \leq n < N}(\theta_{n+1} - \theta_n)^2 \geq \left(\sum_{n_0 \leq n < N} |\theta_{n+1} - \theta_n|\right)^2.$$

The last term dominates $|\theta_{n_0} - A_N|^2$ and, since $|\theta_{n_0} - A_N| \geq \|\Theta_N\|_\infty - |A_N|$, we deduce that $\Phi_N(\Theta_N) \to -\infty$ as $\|\Theta_N\|_\infty \to \infty$. The conditions $\frac{\partial \Phi_N}{\partial \theta_n} = 0$ lead to the discrete Dirichlet problem

$$\begin{cases} \theta_{n+1} + \theta_{n-1} - 2\theta_n = 0, \\ \theta_{-N} = A_{-N}, \ \theta_N = A_N. \end{cases}$$

This problem has the unique solution $\theta_n^{[N]} = n\omega + c$ for $\omega = \frac{A_N - A_{-N}}{2N}$ and $c = \frac{1}{2}(A_N + A_{-N})$. As a consequence $\Theta_N^* = (\theta_n^{[N]})_{|n|<N}$ is the unique critical point of Φ_N and

$$\max \Phi_N = \Phi_N(\Theta_N^*).$$

Let us fix $\delta < \Delta$ and assume that $\delta < \frac{A_N - A_{-N}}{2N} < \Delta$. Then Θ_N^* is in the interior of the compact set

$$S_N = \{\Theta_N = (\theta_n)_{|n|<N} : \delta \leq \theta_{n+1} - \theta_n \leq \Delta, \text{ if } -N \leq n < N\}.$$

This observation will be relevant later.

5   Existence of Complete Orbits

In this section we fix two positive numbers $\Delta > \delta > 0$ and consider the strip

$$S = \{(\theta, \theta_1) \in \mathbb{R}^2 : \delta \leq \theta_1 - \theta \leq \Delta\}$$

and a given function $h = h(\theta, \theta_1)$ in $C^1(S)$. Our goal is to prove the existence of a complete orbit of the difference equation

$$\partial_1 h(\theta_n, \theta_{n+1}) + \partial_2 h(\theta_{n-1}, \theta_n) = 0, \quad n \in \mathbb{Z}. \tag{16}$$

Notice that this setting implicitly implies that the complete solution satisfies $(\theta_n, \theta_{n+1}) \in S$ for each n. The prototype of function h will be $h_*(\theta, \theta_1) = -\alpha(\theta - \theta_1)^2$ with α a positive constant. We will impose two conditions that roughly say that h is close to h_* and the strip S is sufficiently wide, "width" being measured by the quotient Δ/δ.

Theorem 5.1. *Assume that $h \in C^1(S)$ and there are two positive numbers $\overline{\alpha}$, $\underline{\alpha}$ with $\overline{\alpha} < 2\underline{\alpha}$ and*

$$-\overline{\alpha}(\theta_1 - \theta)^2 \leq h(\theta_1, \theta) \leq -\underline{\alpha}(\theta_1 - \theta)^2 \quad \forall (\theta, \theta_1) \in S. \tag{17}$$

Then there exists a number $\sigma \geq 1$, depending only on the quotient $\overline{\alpha}/\underline{\alpha}$, such that if $\sigma^2 \delta \leq \Delta$ then (16) has a complete solution.

As will be seen from the proof, the number σ can be computed explicitly. To obtain results of qualitative nature it is enough to interpret $\sigma = \sigma(q)$ as an increasing function depending on $q = \overline{\alpha}/\underline{\alpha} \in [1, 2[$. This is illustrated by the following consequence on the existence of equilibria for the Frenkel–Kontorowa model.

Corollary 5.2. *Assume that the potential V is bounded and of class C^1. Then the equation*

$$\theta_{n+1} + \theta_{n-1} - 2\theta_n = V'(\theta_n), \quad n \in \mathbb{Z},$$

has infinitely many complete solutions $(\theta_{n,N})_{n\in\mathbb{Z}}$ with $N = 1, 2, \ldots$. Moreover, the upper and lower rotation numbers

$$\underline{\omega}_N := \liminf_{|n|\to\infty} \frac{\theta_{n,N}}{n} \leq \limsup_{|n|\to\infty} \frac{\theta_{n,N}}{n} =: \overline{\omega}_N$$

satisfy $\overline{\omega}_N < \infty$ for each N and $\underline{\omega}_N \to +\infty$ as $N \to +\infty$.

Proof. To prove the corollary we select the number $\sigma_0 = \sigma(3/2)$ corresponding to $\overline{\alpha}/\underline{\alpha} = 3/2$ and work on the region $S : \delta \leq \theta_1 - \theta \leq \Delta$, where $\delta > 0$ is a parameter to be adjusted and $\Delta = \sigma_0^2 \delta$. Our equation is just (16) for

$$h(\theta, \theta_1) = -\frac{1}{2}(\theta - \theta_1)^2 - V(\theta).$$

Moreover, if $(\theta, \theta_1) \in S$,

$$h(\theta, \theta_1) \leq -\frac{1}{2}(\theta - \theta_1)^2 + \|V\|_\infty \leq -\frac{1}{2}(\theta - \theta_1)^2 + \frac{\|V\|_\infty}{\delta^2}(\theta - \theta_1)^2.$$

A similar lower estimate can be obtained to see that the condition (17) holds for

$$\underline{\alpha} = \frac{1}{2} - \frac{\|V\|_\infty}{\delta^2}, \quad \overline{\alpha} = \frac{1}{2} + \frac{\|V\|_\infty}{\sigma_0^4 \delta^2}.$$

For large δ the inequality $\bar{\alpha}/\alpha \leq 3/2$ holds and so the constant $\sigma = \sigma(\bar{\alpha}/\alpha)$ given by the theorem satisfies $\sigma \leq \sigma_0$. Then $\sigma^2\delta \leq \sigma_0^2\delta = \Delta$ and the theorem is applicable. This shows the existence of an equilibrium for the Frenkel–Kontorowa model $(\theta_n^\delta)_{n\in\mathbb{Z}}$ with $\delta \leq \theta_{n+1} - \theta_n \leq \sigma^2\delta$, $n \in \mathbb{Z}$. Letting $\delta = N$ we obtain infinitely many equilibria, and the assertions on the rotation numbers are easily seen to be verified. $\qquad\qquad\qquad\qquad\qquad\qquad\qquad\qquad\qquad\qquad\qquad\qquad\quad$ $\square$

Later we will present other applications of the theorem or of some variant of it. In all cases h will be the generating function of a twist symplectic map f. Indeed the condition (17) automatically implies that f is E-symplectic.

Proof of Theorem 5.1. For each $N \geq 3$ we select two numbers $A_{\pm N}$ satisfying

$$A_{-N} = -A_N, \quad N\delta \leq A_N \leq N\Delta, \tag{18}$$

and consider the following subset of $\mathbb{R}^{2N-1}$:

$$S_N = \{\Theta_N = (\theta_n)_{|n|\leq N} : \delta \leq \theta_{n+1} - \theta_n \leq \Delta \text{ for each } n = -N, \ldots, N - 1\},$$

with the convention $\theta_{\pm N} = A_{\pm N}$. This set is non-empty since it contains at least the point $(\frac{n}{N}A_N)$. It is easily proved that S_N is closed and contained in the ball $\|\Theta_N\|_\infty \leq A_N$ and, since we are in finite dimension, we can deduce that this set is compact. The continuous function

$$\Phi_N : S_N \to \mathbb{R}, \quad \Phi_N(\Theta_N) = \sum_{-N\leq n<N} h(\theta_n, \theta_{n+1}),$$

attains its maximum at some point $\Theta_N^* \in S_N$,

$$\Phi_N(\Theta_N^*) = \max_{S_N} \Phi_N.$$

We will prove that, for an appropriate choice of the sequence A_N, the point Θ_N^* is in the interior of S_N. Hence this is a critical point of Φ_N that can be also interpreted as a solution of (12). Finally we can apply Lemma 4.4 to complete the proof. From now on we will concentrate on the claim

$$\Theta_N^* \in \text{int}(S_N). \tag{19}$$

To this end we make a couple of observations on the configuration of the atoms of Θ_N^*.

(i) *There exists $L > 1$ such that*

$$\frac{1}{L}(\theta_{n+1}^* - \theta_n^*) \leq \theta_{n+2}^* - \theta_{n+1}^* \leq L(\theta_{n+1}^* - \theta_n^*)$$

for each $n = -N, \ldots, N - 2$. Moreover L only depends on the quotient $\bar{\alpha}/\underline{\alpha}$.

To prove this assertion we modify Θ_N^* by replacing θ_{n+1}^* with the mid-point between θ_n^* and θ_{n+2}^*; that is,

$$\hat{\Theta}_N = (\hat{\theta}_n)_{|n|\le N}, \quad \hat{\theta}_m = \theta_m^* \text{ if } m \ne n+1, \quad \hat{\theta}_{n+1} = \frac{1}{2}(\theta_n^* + \theta_{n+2}^*).$$

The new point $\hat{\Theta}_N$ also belongs to S_N. Indeed

$$\hat{\theta}_{n+2} - \hat{\theta}_{n+1} = \hat{\theta}_{n+1} - \hat{\theta}_n = \frac{1}{2}(\theta_{n+2}^* - \theta_n^*) = \frac{1}{2}(\theta_{n+2}^* - \theta_{n+1}^*) + \frac{1}{2}(\theta_{n+1}^* - \theta_n^*)$$

and these differences remain between δ and Δ. The maximizing property of Θ_N^* implies that $\Phi_N(\Theta_N^*) \ge \Phi_N(\hat{\Theta}_N)$, leading to

$$h(\theta_n^*, \theta_{n+1}^*) + h(\theta_{n+1}^*, \theta_{n+2}^*) \ge h(\theta_n^*, \hat{\theta}_{n+1}) + h(\hat{\theta}_{n+1}, \theta_{n+2}^*).$$

As a consequence

$$-\underline{\alpha}[(\theta_{n+1}^* - \theta_n^*)^2 + (\theta_{n+2}^* - \theta_{n+1}^*)^2] \ge -\bar{\alpha}\frac{(\theta_{n+2}^* - \theta_n^*)^2}{2}.$$

Assume that $\ell := \frac{\theta_{n+2}^* - \theta_{n+1}^*}{\theta_{n+1}^* - \theta_n^*} \ge 1$, otherwise we would define ℓ as the inverse fraction. Using that $\theta_{n+2}^* - \theta_n^* = (1+\ell)(\theta_{n+1}^* - \theta_n^*)$ we are led to the inequality

$$\varphi(\ell) := \frac{2(1 + \ell^2)}{(1 + \ell)^2} \le \bar{\alpha}/\underline{\alpha}.$$

The function $\varphi : [1, +\infty[\to [1, 2[$ is an increasing homeomorphism and so $\ell \le \varphi^{-1}(\bar{\alpha}/\underline{\alpha})$. This implies that (i) holds with $L = \varphi^{-1}(\bar{\alpha}/\underline{\alpha})$. Notice that at this point we are using $\bar{\alpha} < 2\underline{\alpha}$.

(ii) *There exists $\sigma > 1$ such that*

$$\frac{\Delta^*}{\delta^*} \le \sigma,$$

where $\Delta^* = \max_{-N \le n < N}(\theta_{n+1}^* - \theta_n^*)$ *and* $\delta^* = \min_{-N \le n < N}(\theta_{n+1}^* - \theta_n^*)$. *Moreover σ only depends on the quotient $\bar{\alpha}/\underline{\alpha}$.*

Let us assume that $\Delta^* = \theta_{M+1}^* - \theta_M^*$ and $\delta^* = \theta_{m+1}^* - \theta_m^*$ with $m, M \in \{-N, \ldots, N - 1\}$. If $|m - M| \le 1$ we can apply the previous step and deduce that $\Delta^*/\delta^* \le L$. From now on we assume that $|m - M| \ge 2$, say $M \ge m + 2$.

We modify Θ_N^* in a new way: After eliminating θ_{m+1}^* a new atom is inserted between θ_M^* and θ_{M+1}^*. Let $\tilde{\Theta}_N = (\tilde{\theta}_n)_{|n|<N}$ be defined as $\tilde{\theta}_n = \theta_n^*$ if $n \leq m$ or $n > M$, $\tilde{\theta}_n = \theta_{n+1}^*$ if $m < n < M$ and $\tilde{\theta}_M = \frac{1}{2}(\theta_M^* + \theta_{M+1}^*)$. We prove that $\tilde{\Theta}_N \in S_N$ as soon as $\Delta^*/\delta^* \geq L + 1$, where L is given by step (i). Actually,

$$\tilde{\theta}_{m+1} - \tilde{\theta}_m = \theta_{m+2}^* - \theta_m^* \leq (L+1)\delta^* \leq \Delta^* \leq \Delta$$

and

$$\tilde{\theta}_{M+1} - \tilde{\theta}_M = \tilde{\theta}_M - \tilde{\theta}_{M-1} = \frac{\Delta^*}{2} \geq \frac{L+1}{2}\delta^* \geq \delta^* \geq \delta,$$

so that $\tilde{\Theta}_N \in S_N$. Then from $\Phi_N(\Theta_N^*) \geq \Phi_N(\tilde{\Theta}_N)$ we deduce that

$$h(\theta_m^*, \theta_{m+1}^*) + h(\theta_{m+1}^*, \theta_{m+2}^*) + h(\theta_M^*, \theta_{M+1}^*)$$

$$\geq h(\theta_m^*, \theta_{m+2}^*) + h(\theta_M^*, \tilde{\theta}_M) + h(\tilde{\theta}_M, \theta_{M+1}^*).$$

Hence

$$-\underline{\alpha}[(\theta_m^* - \theta_{m+1}^*)^2 + (\theta_{m+1}^* - \theta_{m+2}^*)^2 + (\theta_M^* - \theta_{M+1}^*)^2]$$

$$\geq -\overline{\alpha}[(\theta_m^* - \theta_{m+2}^*)^2 + 2(\tilde{\theta}_M - \theta_M^*)^2]$$

and, using again (i), we are led to $\psi_L(\Delta^*/\delta^*) \leq \overline{\alpha}/\underline{\alpha}$, where the function ψ_L is defined as

$$\psi_L(q) = \frac{1 + L^{-2} + q^2}{(1+L)^2 + \frac{1}{2}q^2}.$$

This function is strictly increasing on the interval $[1 + L, \infty[$ and satisfies $\psi_L(1+L) < 1$ as well as $\psi_L(\infty) = 2$. Hence the inequality $\psi_L(\Delta^*/\delta^*) \leq \overline{\alpha}/\underline{\alpha}$ is equivalent to $\Delta^*/\delta^* \leq \psi_L^{-1}(\overline{\alpha}/\underline{\alpha})$, i.e., we have proved (ii) taking

$$\sigma = \max\{1 + L, \psi_L^{-1}(\overline{\alpha}/\underline{\alpha})\}.$$

Now that we have shown (ii) we can complete the proof of the theorem. Define

$$A_N = \frac{1}{2}(\sigma^{-1}\Delta + \sigma\delta)N.$$

From the assumption $\sigma^2\delta < \Delta$ we observe that $\delta < \sigma\delta < \sigma^{-1}\Delta < \Delta$, and A_N is the mid point of the interval $[\sigma\delta, \sigma^{-1}\Delta]$. This implies that (18) holds. We are going to prove that for this choice of the sequence $\{A_N\}$ the claim (19) holds.

By contradiction assume that $\varDelta^* = \varDelta$ or $\delta^* = \delta$. Then either $\delta^* \geq \frac{1}{\sigma}\varDelta$ or $\varDelta^* \leq \sigma\delta$. To fix ideas let us consider the first case $\varDelta^* = \varDelta$, $\delta^* \geq \frac{1}{\sigma}\varDelta$. Then

$$2A_N = \sum_{n=-N}^{N-1} (\theta_{n+1}^* - \theta_n^*) \geq \frac{2N\varDelta}{\sigma},$$

and this contradicts the definition of A_N. The case $\delta^* = \delta$ is treated similarly.

$\square$

Exercise 5.3. Show that the previous proof allows to compute $\sigma = \sigma(\overline{\alpha}/\underline{\alpha})$ explicitly. Hint: $\sigma = \frac{4}{3}\sqrt{\frac{85}{3}}$ for $\overline{\alpha}/\underline{\alpha} = 5/4$.

Exercise 5.4. Compute two numbers δ and $\varDelta$ such that the equation

$$\theta_{n+1} - 2\theta_n + \theta_{n-1} = \sin\theta_n + \cos(\sqrt{2}\theta_n), \quad n \in \mathbb{Z}$$

has a solution lying in $\delta \leq \theta_{n+1} - \theta_n \leq \varDelta$.

Exercise 5.5. Prove that the conclusion of Theorem 5.1 still holds when the condition (17) is replaced by

$$-\overline{\alpha}(\theta_1 - \theta)^k \leq h(\theta_1, \theta) \leq -\underline{\alpha}(\theta_1 - \theta)^k \quad \forall (\theta, \theta_1) \in S \qquad (20)$$

with $k > 1$ and $\overline{\alpha} < 2^{k-1}\underline{\alpha}$. Hint: $\varphi(\ell) = \frac{2^{k-1}(1+\ell^k)}{(1+\ell)^k}$, $\psi_L(q) = \frac{1+L^{-k}+q^k}{(1+L)^k+2^{1-k}q^k}$.

Exercise 5.6. Prove that the conclusion of Theorem 5.1 also holds when the condition (17) is replaced by

$$-\overline{\alpha}(\theta_1 - \theta)^{-k} \leq h(\theta_1, \theta) \leq -\underline{\alpha}(\theta_1 - \theta)^{-k} \quad \forall (\theta, \theta_1) \in S \qquad (21)$$

with $k > 0$ and $\overline{\alpha} < 2^k\underline{\alpha}$. Hint: $\varphi(\ell) = \frac{(1+\ell^{-k})(1+\ell)^k}{2^{k+1}}$, $\psi_L(q) = \frac{1+L^{-k}+q^{-k}}{2^{-k}+2^{1+k}q^{-k}}$, $\theta_{m+2}^* - \theta_m^* \geq 2\delta^*$.

In the applications of Theorem 5.1 or the variants given by the previous exercises we must know how to compute h or at least how to estimate it in order to verify (17), (20) or (21). The next two sections are devoted to the computation of generating functions in two interesting mechanical situations.

6   The Action Functional of a Newtonian Equation

Consider the differential equation

$$\ddot{x} = -V_x(t, x), \quad t \in [0, 1], \quad x \in \mathbb{R}, \qquad (22)$$

where the potential $V : [0, 1] \times \mathbb{R} \to \mathbb{R}$ is continuous and has two partial derivatives with respect to x, V_x and V_{xx}, which are also continuous functions of (t, x). It will be assumed that the Cauchy problem is globally well posed. This can be guaranteed if V_x has linear growth, that is

$$|V_x(t, x)| \leq A|x| + B, \quad (t, x) \in [0, 1] \times \mathbb{R},$$

for some $A, B > 0$. Given $x_0, v_0 \in \mathbb{R}$, the solution satisfying $x(0) = x_0$, $\dot{x}(0) = v_0$, will be denoted by $x(t; x_0, v_0)$. If we interpret these initial conditions as coordinates, say $\theta = x_0$ and $r = v_0$, then we can define the Poincaré map

$$f : \mathbb{R}^2 \to \mathbb{R}^2, \quad \theta_1 = x(1; \theta, r), \quad r_1 = \dot{x}(1; \theta, r).$$

The classical theorems on the Cauchy problem[1] imply that f is a C^1-diffeomorphism. Moreover f is symplectic. This can be justified using Liouville's theorem on Hamiltonian flows but we will prove it in a different way. Consider the Lagrangian function

$$L(t, \theta, r) = \frac{1}{2} \dot{x}(t; \theta, r)^2 - V(t, x(t; \theta, r))$$

and its time average

$$H(\theta, r) = \int_0^1 L(t, \theta, r) \, dt.$$

This function is of class C^1 with partial derivatives

$$H_\theta = \int_0^1 L_\theta \, dt = \int_0^1 \{\dot{x} \frac{\partial \dot{x}}{\partial \theta} - V_x \frac{\partial x}{\partial \theta}\} \, dt,$$

$$H_r = \int_0^1 L_r \, dt = \int_0^1 \{\dot{x} \frac{\partial \dot{x}}{\partial r} - V_x \frac{\partial x}{\partial r}\} \, dt.$$

Commuting ∂_t with ∂_θ and ∂_r and integrating by parts,

$$\int_0^1 \dot{x} \frac{\partial \dot{x}}{\partial \theta} dt = [\dot{x} \frac{\partial x}{\partial \theta}]_{t=0}^{t=1} - \int_0^1 \ddot{x} \frac{\partial x}{\partial \theta} dt, \quad \int_0^1 \dot{x} \frac{\partial \dot{x}}{\partial r} dt = [\dot{x} \frac{\partial x}{\partial r}]_{t=0}^{t=1} - \int_0^1 \ddot{x} \frac{\partial x}{\partial r} dt.$$

From (22) we conclude that $dH = r_1 d\theta_1 - r d\theta$ and so f is symplectic.

The map f will induce a map $\bar{f}$ on the cylinder if the potential satisfies

$$V(t, x + 2\pi) = V(t, x) + p(t), \quad (t, x) \in [0, 1] \times \mathbb{R}, \tag{23}$$

---

[1]Notice that no smoothness in t has been assumed.

for some function $p : [0, 1] \to \mathbb{R}$. This condition of generalized periodicity implies that

$$x(t; \theta + 2\pi, r) = x(t; \theta, r) + 2\pi, \quad \dot{x}(t; \theta + 2\pi, r) = \dot{x}(t; \theta, r),$$

and letting $t = 1$, $f(\theta + 2\pi, r) = f(\theta, r) + (2\pi, 0)$. Hence $\overline{f}$ is symplectic.

Exercise 6.1. Prove that the Poincaré map $\overline{f}$ associated to

$$\ddot{x} + a \sin x = p(t)$$

is exact symplectic if and only if $\int_0^1 p(t)dt = 0$. Here $a > 0$ is a parameter and $p : [0, 1] \to \mathbb{R}$ is a given continuous function.

Exercise 6.2. Assume that, instead of (23), the potential satisfies

$$V(t, x) = B(t, x) + p(t)x$$

where $p : [0, 1] \to \mathbb{R}$ is continuous and B, B_x are bounded. Prove that the Poincaré map is E-symplectic if $\int_0^1 p(t)dt = 0$. Hint: The kinetic energy $T(t) = \frac{1}{2}\dot{x}(t)^2$ satisfies $|\dot{T}| \le CT^{1/2}$ and $\int_0^1 p(t)x(t)dt = -\int_0^1 P(t)\dot{x}(t)dt$ for $P(t) = \int_0^t p(s)ds$.

Next we are going to discuss under what conditions the Poincaré map will satisfy the twist condition. The partial derivative $\frac{\partial F}{\partial r} = \frac{\partial \theta_1}{\partial r}$ can be expressed as

$$\frac{\partial \theta_1}{\partial r} = y(1)$$

where $y(t)$ is the solution of the variational equation

$$\ddot{y} + V_{xx}(t, x(t; \theta, r))y = 0$$

such that $y(0) = 0$, $\dot{y}(0) = 1$. The twist condition becomes $y(1) > 0$ and can be proved using Sturm comparison theory. Actually it holds when the potential satisfies

$$V_{xx}(t, x) < \pi^2, \quad (t, x) \in [0, 1] \times \mathbb{R}. \tag{24}$$

In this case our solution $y(t)$ must oscillate less than the solution of the comparison equation $\ddot{z} + \pi^2 z = 0$, which is $z(t) = \sin \pi t$. This implies that $y(t) > 0$ if $t \in]0, 1]$ and therefore the twist holds.

Another hypothesis implying the twist condition is

$$(2n\pi)^2 < V_{xx}(t, x) < ((2n + 1)\pi)^2, \quad (t, x) \in [0, 1] \times \mathbb{R}, \tag{25}$$

for some $n = 1, 2, \ldots$. Now the oscillations of $y(t)$ are between those of $z_-(t) = \sin(2n\pi t)$ and $z_+(t) = \sin((2n + 1)\pi t)$. Hence $y(t)$ has exactly $2n$ zeros on $]0, 1[$ and $y(1)$ is positive.

Exercise 6.3. Find conditions on the parameters a and ω to guarantee that the Poincaré map associated to $\ddot{x} + \omega^2 x + a \sin x = p(t)$ is a twist symplectic map.

Once we know that the twist condition holds, to compute the generating function we solve the equation

$$x(1; \theta, r) = \theta_1$$

and find r for given θ and θ_1. This is equivalent to finding $r = \dot{x}(0)$, where $x(t)$ is the solution of the Dirichlet problem

$$\ddot{x} = -V_x(t, x), \quad x(0) = 0, \ x(1) = \theta_1. \tag{26}$$

The conditions (24) or (25) are sufficient to guarantee that this problem has at most one solution. This is obviously a consequence of the twist condition. However these conditions are not sufficient for the existence of solution.

Exercise 6.4. Find θ and θ_1 such that

$$\ddot{x} + \pi^2 x - \arctan x = 0, \quad x(0) = 0, \ x(1) = \theta_1,$$

has no solution. Hint: Multiply the equation by $\sin \pi t$ and integrate between $t = 0$ and $t = 1$.

A classical result in the theory of nonlinear boundary value problems says that (26) has a unique solution, if (24) or (25) are replaced by the corresponding stronger conditions

$$V_{xx}(t, x) \leq \Gamma < \pi^2, \quad (t, x) \in [0, 1] \times \mathbb{R}, \tag{27}$$

$$(2n\pi)^2 < \gamma \leq V_{xx}(t, x) \leq \Gamma < ((2n + 1)\pi)^2, \quad (t, x) \in [0, 1] \times \mathbb{R}, \tag{28}$$

where $n = 1, 2, \ldots$ and γ and Γ are given constants. From now on we assume that (27) or (28) are satisfied and so the set Ω defined in Sect. 3 is $\mathbb{R}^2$. The solution of (26) will be denoted by $\xi(t; \theta, \theta_1)$ and the discussions of Sect. 3 on the regularity of the function $R = R(\theta, \theta_1)$ together with the standard theorems on differentiability with respect to initial conditions imply that ξ and $\dot{\xi}$ are of class C^1 in $[0, 1] \times \mathbb{R}^2$. Notice that

$$\xi(t; \theta, \theta_1) = x(t; \theta, R(\theta, \theta_1)) \quad \text{and} \quad R(\theta, \theta_1) = \dot{\xi}(0; \theta, \theta_1).$$

The generating function $h(\theta, \theta_1) = -H(\theta, R(\theta, \theta_1))$ is well defined on the whole plane by

$$h(\theta, \theta_1) = -\int_0^1 \left[\frac{1}{2} \dot{\xi}(t; \theta, \theta_1)^2 - V(t, \xi(t; \theta, \theta_1)) \right] dt.$$

The reader who is familiar with the classical theory of the Calculus of Variations will recognize this expression. Up to a sign, the generating function is the restriction of the action functional to fields of extremals defined by $\xi = \xi(t; \theta, \theta_1)$. More precisely, if we consider the Sobolev space $H^1(0, 1)$ and the functional

$$\mathscr{A} : H^1(0, 1) \to \mathbb{R}, \quad \mathscr{A}[x] = \int_0^1 \left[\frac{1}{2} \dot{x}(t)^2 - V(t, x(t)) \right] dt,$$

then

$$h(\theta, \theta_1) = -\mathscr{A}[\xi(\cdot; \theta, \theta_1)].$$

Exercise 6.5. Compute the generating function associated to $\ddot{x} + \omega^2 x = 0$ with $2n\pi < \omega < 2(n+1)\pi$ for some $n = 1, 2, \ldots$.

Finally we propose a more difficult exercise dealing with an application of Theorem 5.1 to the framework of this section.

Exercise 6.6. Prove that the equation $\ddot{x} + a \sin x = p(t)$ with $0 < a < \pi^2$ and $p(t+1) = p(t)$, $\int_0^1 p(t)dt = 0$, has a solution satisfying $\delta \leq x(t+1) - x(t) \leq \Delta$ for some $\Delta > \delta > 0$.

7 Impact Problems and Generating Functions

Let us consider a particle moving on the half-line $x = x(t) \geq 0$. It satisfies a Newtonian law for $x > 0$ but at the end point $x = 0$ there is an obstacle and the particle bounces elastically.

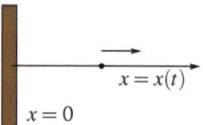

The function $x(t)$ is a solution of the impact problem

$$\begin{cases} \ddot{x} = -V_x(t, x), \quad t \in \mathbb{R}, \\ x(t) \geq 0, \\ x(\tau) = 0 \Rightarrow \dot{x}(\tau^+) = -\dot{x}(\tau^-), \end{cases} \tag{29}$$

where $V : \mathbb{R} \times \mathbb{R} \to \mathbb{R}$ is continuous and has two partial derivatives in x, V_x and V_{xx}. Moreover it is assumed that both derivatives are continuous with respect to both variables (t, x). In short, $V \in C^{0,2}(\mathbb{R} \times \mathbb{R})$.

The exact meaning of the above impact problem is clarified by the following definition. A *bouncing solution* of (29) is a continuous function $x : \mathbb{R} \to [0, \infty[$ and a sequence of times $(t_n)_{n \in \mathbb{Z}}$ satisfying

(i) $\inf_{n\in\mathbb{Z}}(t_{n+1} - t_n) > 0$,

(ii) $x(t_n) = 0$ and $x(t) > 0$ for $t \in]t_n, t_{n+1}[$ and $n \in \mathbb{Z}$,

(iii) The restriction of $x(t)$ to $[t_n, t_{n+1}]$ is of class C^2 and satisfies the differential equation,

(iv) $\dot{x}(t_n^+) = -\dot{x}(t_n^-)$ for $n \in \mathbb{Z}$.

This solution will be *bounded* if furthermore

(v) $\sup_{t\in\mathbb{R}} |x(t)| + \text{ess sup}_{t\in\mathbb{R}} |\dot{x}(t)| < \infty$,

(vi) $\sup_{n\in\mathbb{Z}}(t_{n+1} - t_n) < \infty$.

Notice that $\dot{x}(t)$ is well defined for $t \neq t_n$ and so the essential supremum makes sense.

Exercise 7.1. Compute the bouncing solutions for a linear spring with obstacle, $V(t, x) = \frac{1}{2}x^2$. Prove that all of them are bounded.

We will present a method for the construction of bouncing solutions. The first step will be the study of a boundary value problem.

7.1   The Dirichlet Problem

Let us consider the problem

$$\ddot{x} = -V_x(t, x), \quad x(t_0) = x(t_1) = 0. \tag{30}$$

From now on we will assume that the potential satisfies two additional conditions:

(C1) $V_{xx}(t, x) \leq 0$ for each $(t, x) \in \mathbb{R}^2$.

(C2)   There exist two numbers $c_1, c_2 \in \mathbb{R}$ and two functions $\psi, \phi \in C^2(\mathbb{R})$ such that

$$\ddot{\psi}(t) + c_1 \leq V_x(t, x) \leq \ddot{\phi}(t) + c_2 \quad \text{for each } (t, x) \in \mathbb{R}^2.$$

Moreover, $\sup_{t\in\mathbb{R}} |\dot{\psi}(t)| < \infty$.

These assumptions have strong consequences for the problem (30). First of all we present a result showing that there is a unique solution.

Lemma 7.2. *Assume that (C1) and (C2) hold. Then problem (30) has a unique solution on the interval $[t_0, t_1]$ for each $t_1 - t_0 > 0$.*

Proof. The existence will be obtained via the method of upper and lower solutions. Let $\alpha(t)$ and $\beta(t)$ be the solutions of the linear problems

$$\ddot{\alpha} = -\ddot{\phi}(t) - c_2, \ \alpha(t_0) = \alpha(t_1) = 0 \ \text{ and } \ \ddot{\beta} = -c_1 - \ddot{\psi}(t), \ \beta(t_0) = \beta(t_1) = 0.$$

From **(C2)** we deduce that $-\ddot{\beta} \leq -\ddot{\alpha}$ and so, by the Maximum Principle, $\alpha(t) \geq \beta(t)$ everywhere. Moreover, using **(C2)**,

$$-\ddot{\alpha}(t) = c_2 + \ddot{\phi}(t) \geq V_x(t, \alpha(t)), \quad -\ddot{\beta}(t) = c_1 + \ddot{\psi}(t) \leq V_x(t, \beta(t)).$$

This shows that $\alpha(t)$ and $\beta(t)$ is a couple of ordered upper and lower solutions. Therefore the problem (30) has a solution lying in $\alpha \geq x \geq \beta$.

For the uniqueness we assume that $x_1(t)$ and $x_2(t)$ are two solutions of (30). We notice that the difference $y(t) = x_1(t) - x_2(t)$ satisfies the linear problem

$$\ddot{y} + a(t)y = 0, \quad y(t_0) = y(t_1) = 0, \tag{31}$$

where $a(t) = \int_0^1 V_{xx}(t, \lambda x_1(t) + (1-\lambda)x_2(t)) \, d\lambda$. The condition **(C1)** implies that $a \leq 0$ everywhere. By Sturm comparison theory we deduce that (31) is disconjugate and so we arrive at a contradiction unless y vanishes identically and $x_1 = x_2$. Instead of using comparison techniques we can also prove the uniqueness just by multiplying the equation with y and using integration by parts. □

Exercise 7.3. Prove the Maximum Principle used above: Let $y(t)$ be the solution of $-\ddot{y} = p(t)$, $y(t_0) = y(t_1) = 0$ with $p \in C[t_0, t_1]$. If $p(t) \geq 0$ and $\int_{t_0}^{t_1} p(t)dt > 0$ then $y(t) > 0$ for $t \in]t_0, t_1[$.

To guarantee the positivity of the solution of (30) it is enough to know that the lower solution $\beta(t)$ is positive. This will be the case provided that $t_1 - t_0$ is large enough. To check this fact it is convenient to employ the explicit formula for β given by

$$\beta(t) = \frac{c_1}{2}(t_1 - t)(t - t_0) + \frac{\psi(t_1) - \psi(t_0)}{t_1 - t_0}(t - t_0) + \psi(t_0) - \psi(t).$$

Exercise 7.4. Prove that $\beta(t) > 0$ for each $t \in]t_0, t_1[$ if $t_1 - t_0 > \frac{8}{c}\|\dot{\psi}\|_\infty$. Hint: First study the interval $]t_0, \frac{t_0+t_1}{2}[$.

To complete our study of the Dirichlet problem we present a result on differentiability with respect to the end points. The unique solution of (30) will be denoted by $x_D(t; t_0, t_1)$.

Lemma 7.5. The map $(t; t_0, t_1) \in D \mapsto (x_D(t; t_0, t_1), \dot{x}_D(t; t_0, t_1)) \in \mathbb{R}^2$ is of class C^1, where $D = \{(t; t_0, t_1) \in \mathbb{R}^3 : t_1 - t_0 > 0, t_0 \leq t \leq t_1\}$.

Proof. Let $x(t; t_0, x_0, v_0)$ be the solution of

$$\ddot{x} = -V_x(t, x), \quad x(t_0) = x_0, \quad \dot{x}(t_0) = v_0.$$

Since $\ddot{\psi}(t) + c_1 \leq V_x(t, x) \leq \ddot{\phi}(t) + c_2$, this solution is well defined and smooth for $t \in]-\infty, +\infty[$ and $(t_0, x_0, v_0) \in \mathbb{R}^3$. Let us consider the equation in v_0,

$$x(t_1; t_0, 0, v_0) = 0.$$

It is equivalent to solving (30) and so we know that it has a unique solution $v_0 = v_0(t_0, t_1)$. The Implicit Function Theorem will imply that $v_0(t_0, t_1)$ is of class C^1 if we prove that

$$\frac{\partial x}{\partial v_0}(t_1; t_0, 0, v_0) > 0 \quad \text{for } t_1 > t_0 \text{ and } v_0 \in \mathbb{R}.$$

The function $y(t) = \frac{\partial x}{\partial v_0}(t; t_0, 0, v_0)$ is a solution of the initial value problem

$$\ddot{y} + V_{xx}(t, x(t; t_0, 0, v_0))y = 0, \quad y(t_0) = 0, \quad \dot{y}(t_0) = 1.$$

From **(C1)** we deduce that this linear equation is disconjugate and so $y(t_1)$ has to be positive. □

7.2 The Condition of Elastic Bouncing

A naive approach for the construction of bouncing solutions could consist in juxtaposing solutions of Dirichlet problems for prescribed sequences of impact times. Given a sequence $(t_n)_{n \in \mathbb{Z}}$, the function

$$x(t) := x_D(t; t_n, t_{n+1}) \quad \text{for } t \in [t_n, t_{n+1}], \ n \in \mathbb{Z}, \tag{32}$$

would be the candidate for a bouncing solution. Indeed, if we assume that the sequence satisfies

$$t_{n+1} - t_n > \frac{8}{c_1} \|\dot{\psi}\|_\infty, \quad n \in \mathbb{Z}, \tag{33}$$

then the conditions (i), (ii), and (iii) of the definition are satisfied. Here we are using the previous discussions, in particular Exercise 7.4. In most cases this procedure does not lead to a bouncing solution because the elasticity condition given by (iv) does not necessarily hold. Next we present a method for the construction of a judicious sequence of impacts.

Consider the function

$$h(t_0, t_1) = \int_{t_0}^{t_1} L(t, x_D(t; t_0, t_1), \dot{x}_D(t; t_0, t_1)) \, dt, \tag{34}$$

where L is the Lagrangian function associated to $\ddot{x} = -V_x$. More precisely, here

$$L(t, x, \dot{x}) = \frac{1}{2}\dot{x}^2 - V(t, x) + V(t, 0).$$

We recall that the Newtonian equation can be expressed in the Lagrangian framework as

$$\partial_x L - \frac{d}{dt}(\partial_{\dot{x}} L) = 0. \tag{35}$$

The function h is of class C^1 in the region $\{(t_0, t_1) \in \mathbb{R}^2 : t_1 - t_0 > 0\}$. This is a consequence of Lemma 7.5. An integration by parts leads to

$$\partial_{t_0} h(t_0, t_1)$$

$$= -L(t_0, x_D(t_0; t_0, t_1), \dot{x}_D(t_0; t_0, t_1)) + \int_{t_0}^{t_1} \{(\partial_x L)\frac{\partial x_D}{\partial t_0} + (\partial_{\dot{x}} L)\frac{\partial \dot{x}_D}{\partial t_0}\} dt$$

$$= -\frac{1}{2}\dot{x}_D(t_0; t_0, t_1)^2 + [(\partial_{\dot{x}} L)\frac{\partial x_D}{\partial t_0}]_{t=t_0}^{t=t_1} + \int_{t_0}^{t_1} [(\partial_x L) - \frac{d}{dt}(\partial_{\dot{x}} L)]\frac{\partial x_D}{\partial t_0} dt.$$

From $x_D(t_0; t_0, t_1) = x_D(t_1; t_0, t_1) = 0$ we deduce that

$$\dot{x}_D(t_0; t_0, t_1) + \frac{\partial x_D}{\partial t_0}(t_0; t_0, t_1) = \frac{\partial x_D}{\partial t_0}(t_1; t_0, t_1) = 0.$$

These identities together with (35) imply that

$$\partial_{t_0} h(t_0, t_1) = \frac{1}{2}\dot{x}_D(t_0; t_0, t_1)^2.$$

After differentiating with respect to t_1 we arrive at

$$\partial_{t_1} h(t_0, t_1) = -\frac{1}{2}\dot{x}_D(t_1; t_0, t_1)^2.$$

Assume now that (t_n) is a sequence solving

$$\partial_{t_0} h(t_n, t_{n+1}) + \partial_{t_1} h(t_{n-1}, t_n) = 0, \quad n \in \mathbb{Z}. \tag{36}$$

If the condition (33) holds, then the function defined by (32) is non-negative and it satisfies $\dot{x}(t_n^+)^2 = \dot{x}(t_n^-)^2$. Then $\dot{x}(t_n^-) \leq 0 \leq \dot{x}(t_n^+)$ and so the condition (iv) holds and $x(t)$ becomes a bouncing solution.

Exercise 7.6. Assume that $(t_n)_{n\in\mathbb{Z}}$ satisfies (33), (36) and $\sup(t_{n+1} - t_n) < \infty$. Moreover, let $\sup_{t\in\mathbb{R}} |\dot{\phi}(t)| < \infty$. Prove that $x(t)$ is bounded.

7.3 A Bouncing Ball

Let us apply the previous discussions to a concrete model. Assume that a horizontal plate (the racket) is moving according to some prescribed protocol and a particle (the ball) is in free fall until hitting the plate, when it bounces elastically. In more analytic terms assume that the unknown $z = z(t)$ is the vertical position of the particle and the given function $w(t)$ is the position of the plate. For $z > w(t)$ the free fall is modelled by $\ddot{z} = -g$, where $g > 0$ is the gravitational constant. The elastic impact is easily modelled through the relative position $x(t) = z(t) - w(t)$,

$$x(\tau) = 0 \Rightarrow \dot{x}(\tau^+) = -\dot{x}(\tau^-).$$

Assuming that $w(t)$ is of class C^2 we find that $x(t)$ is a solution of the impact problem (29) with

$$V(t, x) = (g + \ddot{w}(t))x. \tag{37}$$

From now on we assume that the position and velocity of the plate are bounded; that is,

$$w \in C^2(\mathbb{R}) \quad \text{and} \quad \|w\|_\infty + \|\dot{w}\|_\infty < \infty. \tag{38}$$

This is sufficient to guarantee that **(C1)** and **(C2)** are satisfied for $c_1 = c_2 = g$ and $\phi = \psi = w$.

In this case the simplicity of the potential allows for explicit computations.

Exercise 7.7. Determine $x_D(t; t_0, t_1)$ in terms of $w(t)$. Hint: $\beta(t)$.

Exercise 7.8. Use the previous exercise together with (34) to prove that the generating function is

$$h(t_0, t_1) = -\frac{g^2}{24}(t_1 - t_0)^3 - \frac{g}{2}(w(t_1) + w(t_0))(t_1 - t_0)$$
$$+ \frac{(w(t_1) - w(t_0))^2}{2(t_1 - t_0)} + g \int_{t_0}^{t_1} w(t)\, dt - \frac{1}{2} \int_{t_0}^{t_1} \dot{w}(t)^2\, dt.$$

Hint: $\int_{t_0}^{t_1} \dot{x}_D^2(t)\, dt = -\int_{t_0}^{t_1} x_D(t)\ddot{x}_D(t)\, dt$

We do not need this exact formula for h, since for our purposes it is sufficient to determine the dominant term as $t_1 - t_0 \to \infty$. From the above exercise,

$$h(t_0, t_1) = -\frac{g^2}{24}(t_1 - t_0)^3 + R(t_0, t_1)$$

where

$$|R(t_0, t_1)| \leq C(t_1 - t_0) \quad \text{for } t_1 > t_0.$$

Here C is a constant depending only on $\|w\|_\infty + \|\dot{w}\|_\infty$.

We are going to apply Exercise 5.5 with $k = 3$ and fixed numbers $\underline{\alpha}, \overline{\alpha}$ satisfying $\underline{\alpha} < \frac{g^2}{24} < \overline{\alpha}$ and $\overline{\alpha} < 4\underline{\alpha}$. Then there exists $d > 0$ such that

$$-\overline{\alpha}(t_1 - t_0)^3 \leq h(t_1 - t_0) \leq -\underline{\alpha}(t_1 - t_0)^3 \quad \text{for } t_1 - t_0 \geq d.$$

The number σ associated to $\underline{\alpha}$ and $\overline{\alpha}$ can be computed in order to find complete orbits of the difference equation (36) lying in $\delta \leq t_{n+1} - t_n \leq \sigma^2 \delta$ if $\delta \geq d$. These sequences of impact times lead to bouncing solutions. Actually, the conditions (v) and (vi) are also satisfied and so these solutions are bounded. The condition (vi) is automatic from the construction of the sequence (t_n). To verify (v) we notice that $x(t)$ is a solution of the Dirichlet problem

$$\ddot{x} = -(g + \ddot{w}(t)), \quad x(t_n) = x(t_{n+1}) = 0$$

and, going back to Exercise 7.7, we obtain a bound for $\|x\|_\infty + \|\dot{x}\|_\infty$ in terms of g, σ, δ and $\|w\|_\infty + \|\dot{w}\|_\infty$. Alternatively we could apply Exercise 7.6.

We sum up the previous discussions.

Theorem 7.9. *Assume that $w(t)$ satisfies (38) and consider the impact problem (29) with potential given by (37). Then there exist positive constants $\sigma > 1$ and d such that for each $\delta \geq d$ there exists a bounded solution with impact times $(t_n^\delta)_{n \in \mathbb{Z}}$ satisfying*

$$\delta \leq t_{n+1}^\delta - t_n^\delta \leq \sigma^2 \delta, \quad n \in \mathbb{Z}.$$

8   Comments and Bibliographical Remarks

1. Introduction. In [38] Zharnitsky replaced standard angles on $\mathbb{S}^1$ by angles on a torus $\mathbb{S}^1 \times \ldots \times \mathbb{S}^1$. These generalized angles were then employed to reformulate KAM theory for quasi-periodic maps. The paper [38] motivated us to consider non-periodic angles.

Standard versions of the Poincaré–Birkhoff and Aubry–Mather theorems concern diffeomorphisms mapping the cylinder onto itself. In our setting the image of the map T_ϵ is not necessarily contained in the region $a < r < b$. This is not a serious problem since one can first apply KAM theory in order to find invariant curves. Then the region between two of these invariant curves is mapped onto itself. This region can be symplectically deformed into a compact cylinder of the type $A \leq r \leq B$ where the standard theory applies. An alternative is the use of more sophisticated versions of P-B and A-M theorems. See for instance [10, 19, 31].

The map $T_{\epsilon,3}$ was presented as an example in [15]. In that paper there also some results on the existence of orbits with rotation number for certain maps with non-periodic angle.

The mechanical model described in the introduction is sometimes called Fermi-Ulam ping-pong. The problem of deciding whether the velocity can become unbounded is of physical significance in connection with the so-called Fermi acceleration. This is explained in a paper by Dolgopyat [9]. The relativistic version of the model is probably more significant in physics and has been considered, in the periodic case, by Pustyl'nikov, see [32].

2. Symplectic maps in the plane and in the cylinder. Given a homeomorphism $f : \mathbb{T} \times \mathbb{R} \to \mathbb{T} \times \mathbb{R}$, the theory of covering spaces allows to construct a lifting $\tilde{f} : \mathbb{T} \times \mathbb{R} \to \mathbb{R}^2$. Then $p \circ \tilde{f}$ is a homeomorphism of the plane and this is what we mean by a lift from the cylinder to the plane. More details on this point can be found in [4]. The condition (3) can be formulated in a slightly more abstract language as $f^*\omega = \omega$, where $f^*\omega$ is the pull-back of the two form $\omega = d\theta \wedge dr$. Symplectic manifolds are the natural setting to define symplectic maps, the plane and the cylinder endowed with the form ω are just two simple examples of this class of manifolds. We refer to [2, 13, 25, 29] for the general theory. The notion of E-symplectic map was introduced in [14].

3. The twist condition and the generating function. Twist maps are also studied in more degrees of freedom. See the papers by Herman [12] and by Bialy–MacKay [6]. The presentation of the notion of generating function on the cylinder is inspired by the frameworks defined by Mather in [23] and by Moser in [28]. Generating functions can be defined on general symplectic manifolds. We refer to the textbooks [2, 26, 29] on Hamiltonian Dynamics. Usually generating functions are associated to symplectic maps by local constructions based on the implicit function theorem.

4. The variational principle. The presentation of the Frenkel–Kontorowa model follows Aubry's paper [5]. This paper contains interesting results on the dynamics of the standard map when the potential V is a trigonometric function. The variational principle is just an adaptation of classical ideas in Aubry–Mather theory. The only difference is that A–M theory is concerned with generating functions satisfying the periodicity condition

$$h(\theta + 2\pi, \theta_1 + 2\pi) = h(\theta, \theta_1).$$

Mather arrived at a different variational principle in [22]. It was concerned with functions of the continuous variable θ instead of discrete sequences (θ_n). Mather's original motivation was to understand what remains of invariant curves after perturbations when KAM theory does no longer apply. The book by Moser [28] presents a general view of the different variational principles connected with Aubry–Mather theory. The use of results about compactness in spaces of sequences in Lemma 4.4 is influenced by the paper [1]. There Angenent extended Aubry–Mather theory to more general classes of difference equations using a method of upper and lower solutions. Finally we mention the paper by MacKay et al. [21], where the variational principle is employed for a non-standard application.

5. Existence of complete orbits. Theorem 5.1 first appeared in [14]. In the original version the condition $\overline{\alpha} < 2\underline{\alpha}$ was replaced by the more restrictive condition $\overline{\alpha} < (3/2)\underline{\alpha}$. The two variants of the theorem presented in Exercises 5.5 and 5.6 are taken from [15] and [16]. Again there is some improvement in the conditions on the quotient $\overline{\alpha}/\underline{\alpha}$. The steps (i) and (ii) of the proof of Theorem 5.1 are inspired by the techniques employed by Terracini and Verzini in [35]. See also [30].

6. The action functional of a Newtonian equation. The standard versions of the theorem on differentiability with respect to initial conditions assume the differentiability of the vector field with respect to both variables t and x. However, the differentiability in t is not required to differentiate the solution at a fixed time $t = t_1$. See for instance the book by Lefschetz [20].

Alternative proofs of the result in Exercise 6.1 can be found in the papers by Franks [10] and You [36].

If (25) is replaced by

$$((2n + 1)\pi)^2 < V_{xx}(t, x) < (2n\pi)^2,$$

then there is backward twist, which means that the derivative $\frac{\partial \theta_1}{\partial r}$ is negative. A classical result in the theory of boundary value problems, see [18], says that the problem

$$\ddot{z} = -\tilde{V}_z(t, z), \quad z(0) = z(1) = 0,$$

has a unique solution if $\tilde{V}_{zz} \leq \Gamma < \pi^2$ or $(n\pi)^2 < \gamma \leq \tilde{V}_{zz} \leq \Gamma < ((n+1)\pi)^2$. This problem has homogeneous boundary conditions but the previous result is applicable to (26) after the change of variables $x(t) = z(t) + (\theta_1 - \theta)t + \theta$.

Very clear discussions on the connection between the generating function and the action functional can be found in Moser's course [28]. They apply to general Lagrangian systems but we have preferred to restrict ourselves to Newtonian equations to make the discussion simpler and also to present a non-local formulation. In [33] Radmazé considered the problem of minimization of the action functional in a space of periodic functions, say $H^1(\mathbb{R}/\mathbb{Z})$. He found that there is a minimizer if and only if the function $\mathscr{R}(\theta) := h(\theta, \theta)$ reaches a minimum at some θ^*. Moreover, the minimizer is precisely $\xi(t; \theta^*, \theta^*)$. This is also described in Caratheodory's book [8].

Exercise 6.6 is a particular case of some results in [15] for equations of the type

$$\ddot{x} + V'(x) = p(t), \quad p(t + 1) = p(t), \quad \int_0^1 p(t)dt = 0,$$

with $V(x)$ bounded. An alternative way to solve it is to find a generalized periodic solution of the type $x(t + 1) = x(t) + 2\pi$ by minimization of the action functional. More information on this approach can be found in the paper by Mawhin on the pendulum equation [24].

7. Impact problems and generating functions. The connections between the generating function and the action functional for impact problems is discussed in [15]. The key is the formula (34), which also explains the connection between the so-called Nehari method and our method of construction of bouncing solutions via the Eq. (36). The basic idea of Nehari's method (see [35]) is to consider the function

$$h(t_0, t_1) = \inf \left\{ \int_{t_0}^{t_1} L(t, x(t), \dot{x}(t)) \, dt : x \in H_0^1(t_0, t_1), \ x > 0 \text{ on }]t_0, t_1[\right\},$$

and then to obtain a solution satisfying the boundary conditions $x(t_0) = x(t_{n+1}) = 0$ and having exactly n zeros in $]t_0, t_{n+1}[$ by finding a maximum of the function

$$\Phi(t_1, t_2, \ldots, t_n) = \sum_{k=0}^{n} h(t_k, t_{k+1}).$$

From formula (34) and the uniqueness of solution to the Dirichlet problem we observe that h is precisely the generating function and the critical points of Φ are the solutions of (36) on a finite interval of indexes.

The problem of the bouncing ball with gravity was considered in [32] by Pustyl'nikov in the case where the motion of the racket $w(t)$ is periodic. In particular he obtained an interesting result on the existence of motions with unbounded velocity for certain functions $w(t)$ which are periodic, smooth and have large norm.

References

1. S. Angenent, Monotone recurrence relations, their Birkhoff orbits and topological entropy. Ergodic Theor. Dyn. Syst. **10**, 15–41 (1990)
2. V.I. Arnold, *Mathematical Methods of Classical Mechanics*. Graduate Texts in Mathematics, vol. 60 (Springer, Berlin, 1978)
3. V.I. Arnold, A. Avez, *Ergodic Problems of Classical Mechanics* (Benjamin, New York, 1968)
4. D. Arrowsmith, C. Place, *An Introduction to Dynamical Systems* (Cambridge University Press, London, 1990)
5. S. Aubry, The concept of anti-integrability: definition, theorems and applications to the standard map, in *Twist Mappings and Their Applications*. IMA Vol. Math. Appl. 44 (Springer, Berlin, 1992), pp. 7–54
6. M. Bialy, R. MacKay, Symplectic twist maps without conjugate points. Israel J. Math. **141**, 235–247 (2004)
7. G. Birkhoff, *Dynamical Systems* (American Mathematical Society, Providence, 1927)
8. C. Carathéodory, *Calculus of Variations* (Chelsea Publ., New York, 1982)
9. D. Dolgopyat, Fermi acceleration, in *Geometric and Probabilistic Structures in Dynamics*, ed. by K. Burns, D. Dolgopyat, Y. Pesin (American Mathematical Society, Providence, 2008), pp. 149–166
10. J. Franks, Generalizations of the Poincaré-Birkhoff theorem. Ann. Math. **128**, 139–151 (1988)
11. Ch. Golé, *Symplectic Twist Maps* (World Scientific, Singapore, 2001)
12. M. Herman, Dynamics connected with indefinite normal torsion, in *Twist Mappings and Their Applications*. IMA Vol. Math. Appl. 44 (Springer, Berlin, 1992), pp. 153–182
13. H. Hofer, E. Zehnder, *Symplectic Invariants and Hamiltonian Dynamics* (Birkhäuser, Boston, 2011)
14. M. Kunze, R. Ortega, Complete orbits for twist maps on the plane. Ergodic Theor. Dyn. Syst. **28**, 1197–1213 (2008)
15. M. Kunze, R. Ortega, Complete orbits for twist maps on the plane: extensions and applications. J. Dyn. Differ. Equat. **23**, 405–423 (2011)
16. M. Kunze, R. Ortega, Complete orbits for twist maps on the plane: the case of small twist. Ergodic Theor. Dyn. Syst. **31**, 1471–1498 (2011)
17. S. Laederich, M. Levi, Invariant curves and time-dependent potentials. Ergodic Theor. Dyn. Syst. **11**, 365–378 (1991)
18. A. Lazer, D. Leach, On a nonlinear two-point boundary value problem. J. Math. Anal. Appl. **26**, 20–27 (1969)
19. P. Le Calvez, J. Wang, Some remarks on the Poincaré-Birkhoff theorem. Proc. Am. Math. Soc. **138**, 703–715 (2010)

20. S. Lefschetz, *Differential Equations: Geometric Theory* (Dover, New York, 1977)
21. R. MacKay, S. Slijepčević, J. Stark, Optimal scheduling in a periodic environment. Nonlinearity **13**, 257–297 (2000)
22. J. Mather, Existence of quasi-periodic orbits for twist homeomorphisms on the annulus. Topology **21**, 457–467 (1982)
23. J. Mather, Variational construction of orbits of twist diffeomorphisms. J. Am. Math. Soc. **4**, 207–263 (1991)
24. J. Mawhin, Global results for the forced pendulum equation, in *Handbook of Differential Equations*, vol. 1, chap. 6 (Elsevier, Amsterdam, 2004), pp. 533–589
25. D. McDuff, D. Salamon, *Introduction to Symplectic Topology*, 2nd edn. (Oxford University Press, London, 1998)
26. K.R. Meyer, G.R Hall, D. Offin, *Introduction to Hamiltonian Dynamical Systems and the N-Body Problem*, 2nd edn. (Springer, Berlin, 2009)
27. J. Moser, Stability and nonlinear character of ordinary differential equations, in *Nonlinear Problems* (Proc. Sympos., Madison, WI, 1962) (University of Wisconsin Press, Madison, 1963), pp. 139–150
28. J. Moser, in *Selected Chapters in the Calculus of Variations*. Lecture Notes by O. Knill, Lectures in Mathematics, ETH Zürich (Birkhäuser, Boston, 2003)
29. J. Moser, E. Zehnder, *Notes on Dynamical Systems* (American Mathematical Society, Prvidence, 2005)
30. R. Ortega, G. Verzini, A variational method for the existence of bounded solutions of a sublinear forced oscillator. Proc. Lond. Math. Soc. **88**, 775–795 (2004)
31. M. Pei, Aubry-Mather sets for finite-twist maps of a cylinder and semilinear Duffing equations. J. Differ. Equat. **113**, 106–127 (1994)
32. L. Pustyl'nikov, Poincaré models, rigorous justification of the second law of thermodynamics from mechanics, and the Fermi acceleration mechanism. Russ. Math. Surv. **50**, 145–189 (1995)
33. A. Radmazé, Sur les solutions périodiques et les extremales fermées du calcul des variations. Math. Ann. **100**, 63–96 (1934)
34. C. Siegel, J. Moser, *Lectures on Celestial Mechanics* (Springer, Berlin, 1971)
35. S. Terracini, G. Verzini, Oscillating solutions to second order ODEs with indefinite superlinear nonlinearities. Nonlinearity **13**, 1501–1514 (2000)
36. J. You, Invariant tori and Lagrange stability of pendulum-type equations. J. Differ. Equat. **85**, 54–65 (1990)
37. V. Zharnitsky, Instability in Fermi-Ulam ping-pong problem. Nonlinearity **11**, 1481–1487 (1998)
38. V. Zharnitsky, Invariant curve theorem for quasiperiodic twist mappings and stability of motion in the Fermi Ulam problem. Nonlinearity **13**, 1123–1136 (2000)

List of Participants

1. Anichini Giuseppe
   giuseppe.anichini@unifi.it
2. Barutello Vivina Laura
   vivina.barutello@unito.it
3. Benedetti Irene
   irene.benedetti@dmi.unipg.it
4. Benevieri Pierluigi
   pierluigi.benevieri@unifi.it
5. Bisconti Luca
   luca.bisconti@unifi.it
6. Bonanno Gabriele
   bonanno@unime.it
7. Boscaggin Alberto
   boscaggi@sissa.it
8. Calamai Alessandro
   calamai@dipmat.univpm.it
9. Candito Pasquale
   pasquale.candito@unirc.it
10. Capietto Anna
    anna.capietto@unito.it (**lecturer**)
11. Corsato Chiara
    corsato.chiara@gmail.com
12. Fabbri Roberta
    roberta.fabbri@unifi.it
13. Garrione Maurizio
    garrione@sissa.it
14. Giacobbe Andrea
    giacobbe@math.unipd.it
15. Infante Gennaro
    g.infante@unical.it

A. Capietto et al., *Stability and Bifurcation Theory for Non-Autonomous Differential Equations*, Lecture Notes in Mathematics 2065, DOI 10.1007/978-3-642-32906-7, © Springer-Verlag Berlin Heidelberg 2013

16. Johnson Russell
    johnson@dsi.unifi.it (**editor**)
17. Kanigowski Adam
    hannibal@mat.umk.pl
18. Kloeden Peter
    kloeden@math.uni-frnkfurt.de (**lecturer**)
19. Kuna Mariel
    marielku@yahoo.com.ar
20. Livrea Roberto
    roberto.livrea@unirc.it
21. Lorenz Thomas
    lorenz@math.uni-frankfurt.de
22. Lukasiak Renata
    renatka@mat.umk.pl
23. Marò Stefano
    stefano.maro@unito.it
24. Maurette Manuel
    maurette@dm.uba.ar
25. Matranga Mattia
    mattia@libero.it
26. Matucci Serena
    serena.matucci@unifi.it
27. Maurette Manuel
    maurette@dm.uba.ar
28. Mawhin Jean
    jean.mawhin@uclouvain.be (**lecturer**)
29. Molica Bisci Giovanni
    gmolica@unirc.it
30. Montoro Luigi
    montoro@mat.unical.it
31. Mugelli Francesco
    francesco.mugelli@unifi.it
32. Novo Sylvia
    sylnovo@matem.eis.uva.es (**lecturer**)
33. Ortega Rafael
    rortega@ugr.es (**lecturer**)
34. Panasenko Elena
    panlena_t@mail.ru
35. Pera Maria Patrizia
    mpatrizia.pera@unifi.it (**editor**)
36. Pizzimenti Pasquale Francesco
    ppizzimenti@unime.it
37. Poggiolini Laura
    laura.poggiolini@unifi.it

38. Ruiz-Herrera Alfonso
    alfonsoruiz@ugr.es
39. Sciammetta Angela
    asciammetta@unime.it
40. Sfecci Andrea
    sfecci@sissa.it
41. Soave Nicola
    nicola.soave@gmail.com
42. Spadini Marco
    marco.spadini@math.unifi.it
43. Taddei Valentina
    valentina.taddei@unimore.it
44. Tavernise Marianna
    tavernise@mat.unical.it
45. Verzini Gianmaria
    gianmaria.verzini@polimi.it
46. Vilasi Luca
    lvilasi@unime.it
47. Zampogni Luca
    zampoglu@dmi.unipg.it

LECTURE NOTES IN MATHEMATICS  Springer

Edited by J.-M. Morel, B. Teissier; P.K. Maini

Editorial Policy (for Multi-Author Publications: Summer Schools / Intensive Courses)

1. Lecture Notes aim to report new developments in all areas of mathematics and their applications - quickly, informally and at a high level. Mathematical texts analysing new developments in modelling and numerical simulation are welcome. Manuscripts should be reasonably selfcontained and rounded off. Thus they may, and often will, present not only results of the author but also related work by other people. They should provide sufficient motivation, examples and applications. There should also be an introduction making the text comprehensible to a wider audience. This clearly distinguishes Lecture Notes from journal articles or technical reports which normally are very concise. Articles intended for a journal but too long to be accepted by most journals, usually do not have this "lecture notes" character.

2. In general SUMMER SCHOOLS and other similar INTENSIVE COURSES are held to present mathematical topics that are close to the frontiers of recent research to an audience at the beginning or intermediate graduate level, who may want to continue with this area of work, for a thesis or later. This makes demands on the didactic aspects of the presentation. Because the subjects of such schools are advanced, there often exists no textbook, and so ideally, the publication resulting from such a school could be a first approximation to such a textbook. Usually several authors are involved in the writing, so it is not always simple to obtain a unified approach to the presentation.

   For prospective publication in LNM, the resulting manuscript should not be just a collection of course notes, each of which has been developed by an individual author with little or no coordination with the others, and with little or no common concept. The subject matter should dictate the structure of the book, and the authorship of each part or chapter should take secondary importance. Of course the choice of authors is crucial to the quality of the material at the school and in the book, and the intention here is not to belittle their impact, but simply to say that the book should be planned to be written by these authors jointly, and not just assembled as a result of what these authors happen to submit.

   This represents considerable preparatory work (as it is imperative to ensure that the authors know these criteria before they invest work on a manuscript), and also considerable editing work afterwards, to get the book into final shape. Still it is the form that holds the most promise of a successful book that will be used by its intended audience, rather than yet another volume of proceedings for the library shelf.

3. Manuscripts should be submitted either online at www.editorialmanager.com/lnm/ to Springer's mathematics editorial, or to one of the series editors. Volume editors are expected to arrange for the refereeing, to the usual scientific standards, of the individual contributions. If the resulting reports can be forwarded to us (series editors or Springer) this is very helpful. If no reports are forwarded or if other questions remain unclear in respect of homogeneity etc, the series editors may wish to consult external referees for an overall evaluation of the volume. A final decision to publish can be made only on the basis of the complete manuscript; however a preliminary decision can be based on a pre-final or incomplete manuscript. The strict minimum amount of material that will be considered should include a detailed outline describing the planned contents of each chapter.

   Volume editors and authors should be aware that incomplete or insufficiently close to final manuscripts almost always result in longer evaluation times. They should also be aware that parallel submission of their manuscript to another publisher while under consideration for LNM will in general lead to immediate rejection.

4. Manuscripts should in general be submitted in English. Final manuscripts should contain at least 100 pages of mathematical text and should always include

   – a general table of contents;
   – an informative introduction, with adequate motivation and perhaps some historical remarks: it should be accessible to a reader not intimately familiar with the topic treated;
   – a global subject index: as a rule this is genuinely helpful for the reader.

   Lecture Notes volumes are, as a rule, printed digitally from the authors' files. We strongly recommend that all contributions in a volume be written in the same LaTeX version, preferably LaTeX2e. To ensure best results, authors are asked to use the LaTeX2e style files available from Springer's web-server at
   ftp://ftp.springer.de/pub/tex/latex/svmonot1/ (for monographs) and
   ftp://ftp.springer.de/pub/tex/latex/svmultt1/ (for summer schools/tutorials).
   Additional technical instructions, if necessary, are available on request from:
   lnm@springer.com.

5. Careful preparation of the manuscripts will help keep production time short besides ensuring satisfactory appearance of the finished book in print and online. After acceptance of the manuscript authors will be asked to prepare the final LaTeX source files and also the corresponding dvi-, pdf- or zipped ps-file. The LaTeX source files are essential for producing the full-text online version of the book. For the existing online volumes of LNM see:
   http://www.springerlink.com/openurl.asp?genre=journal&issn=0075-8434.
   The actual production of a Lecture Notes volume takes approximately 12 weeks.

6. Volume editors receive a total of 50 free copies of their volume to be shared with the authors, but no royalties. They and the authors are entitled to a discount of 33.3 % on the price of Springer books purchased for their personal use, if ordering directly from Springer.

7. Commitment to publish is made by letter of intent rather than by signing a formal contract. Springer-Verlag secures the copyright for each volume. Authors are free to reuse material contained in their LNM volumes in later publications: a brief written (or e-mail) request for formal permission is sufficient.

Addresses:
Professor J.-M. Morel, CMLA,
École Normale Supérieure de Cachan,
61 Avenue du Président Wilson, 94235 Cachan Cedex, France
E-mail: morel@cmla.ens-cachan.fr

Professor B. Teissier, Institut Mathématique de Jussieu,
UMR 7586 du CNRS, Équipe "Géométrie et Dynamique",
175 rue du Chevaleret, 75013 Paris, France
E-mail: teissier@math.jussieu.fr

For the "Mathematical Biosciences Subseries" of LNM:

Professor P. K. Maini, Center for Mathematical Biology,
Mathematical Institute, 24-29 St Giles,
Oxford OX1 3LP, UK
E-mail : maini@maths.ox.ac.uk

Springer, Mathematics Editorial I,
Tiergartenstr. 17,
69121 Heidelberg, Germany,
Tel.: +49 (6221) 4876-8259
Fax: +49 (6221) 4876-8259
E-mail: lnm@springer.com